The Steamboat Era

ALSO BY S.L. KOTAR AND J.E. GESSLER
AND FROM MCFARLAND

Yellow Fever: A Worldwide History (2017)
Cholera: A Worldwide History (2014)
Smallpox: A History (2013)
The Rise of the American Circus, 1716–1899 (2011)
Ballooning: A History, 1782–1900 (2011)

The Steamboat Era

A History of Fulton's Folly on American Rivers, 1807–1860

S.L. KOTAR and J.E. GESSLER

McFarland & Company, Inc., Publishers
Jefferson, North Carolina

The present work is a reprint of the illustrated case bound edition of The Steamboat Era: A History of Fulton's Folly on American Rivers, 1807–1860, *first published in 2009 by McFarland.*

All illustrations are from the collection of the Public Library of Cincinnati and Hamilton County.

LIBRARY OF CONGRESS CATALOGUING-IN-PUBLICATION DATA

Kotar, S.L.
The steamboat era : a history of Fulton's folly on American rivers, 1807–1860 / S.L. Kotar and J.E. Gessler.
p.   cm.
Includes bibliographical references and index.

ISBN 978-1-4766-8368-3
softcover : acid free paper ∞

1. Steamboats — United States — History — 19th century.
2. River steamers — United States — History — 19th century.
I. Gessler, J.E. II. Title.
VM461.K645 2020   386'.22436 — dc22   2009027458

British Library cataloguing data are available

© 2009 S.L. Kotar and J.E. Gessler. All rights reserved

No part of this book may be reproduced or transmitted in any form or by any means, electronic or mechanical, including photocopying or recording, or by any information storage and retrieval system, without permission in writing from the publisher.

On the cover: Mississippi river steamboat
(Heliotype Printing Co., Boston);
rope border © 2020 Shutterstock

Printed in the United States of America

*McFarland & Company, Inc., Publishers
Box 611, Jefferson, North Carolina 28640
www.mcfarlandpub.com*

This book is lovingly dedicated to Mary E. Kotar, whose faith in us throughout the years has never wavered.

Acknowledgments

The authors are indebted to Diane Mallstrom, reference librarian, Cincinnati Room, at the Public Library of Cincinnati and Hamilton County, for her assistance in obtaining the numerous illustrations from this source used in the text.

We would also like to express our appreciation to Betsy Bennett for her assistance in our final preparations.

Table of Contents

Acknowledgments	vi
Preface	1
1. The "Invention of the Devil"	5
2. Keelboats and Barges: The "Arks" of the New World	16
3. Onward to the "Father of Waters": The Fulton-Livingston Patent	22
4. The Meschasipi — Or Is That the "Mississipi"?	29
5. "St. Looy" and the Upper Mississippi	41
6. A World Unto Their Own: The Mississippi and Ohio Rivers	58
7. Navigating the Inland Western Waterways	72
8. The Trades and the Trade-Offs	77
9. Economic Conditions During the Steamboat Era	88
10. Making Money on the Rivers	100
11. Man Overboard! Steamboat Disasters on Western Waterways	116
12. The Steamboat Race Is On!	137
13. The Development of Steamboat Crews	145
14. Deckhands: A Distinctive Class of Casual Workers	167
15. Steamboat Gothic: The Florid and the Ornate	177
16. "My satisfaction was complete": First-Class Passengers	186
17. Steamboat Diversions	197
18. "Just the bar' necessities, ma'am"	206
19. The Nameless Masses: Deck Passengers	217

20. An International Incident	223
21. Potions, Purging and Practitioners	230
22. (In)Famous Steamboat Cities	245
23. The Roaring '50s: The Railroads Come Calling	259
Appendix A: Glossary	265
Appendix B: Original Accounts of Steamboat Disasters	277
Chapter Notes	289
Bibliography	293
Index	297

Preface

The Steamboat Era is one of the most fascinating and perhaps most romanticized times in American history. It encapsulates the spirit of adventure, an entirely new mode of transportation, westward migration, the development of commerce (and incidentally, the shifting of power from east to west), a period of wild land speculation, financial panics and depressions, and epidemics. It oversaw the influx of infamous riverboat gamblers, the creation of legendary pirates, a subculture of deckers and *nymphes du fleuve*, and the introduction of such oddities as the flush toilet. Finally, in the "Roaring '50s," a time when Floating Palaces were at their height, in sneaked the evil railroad Robber Barons, who succeeded in sinking the Steamboat Gothic boats faster than boiler explosions or snags.

Having figuratively lived and worked in the mid–19th century for decades, we learned one lesson above all others: no one can understand the events of a bygone age without thoroughly comprehending the times in which they occurred. This, as much as anything, compelled us to write this book. There are numerous texts on steamboating (interestingly, of the thousands of contemporary accounts we read, only two instances, both from 1839, actually used the words "river boat"), but none of our acquaintance which make a serious attempt to intertwine the times and the culture with the nuts and bolts, facts and figures, of the early to middle 1800s.

No man, woman or child is independent of the times in which he or she lives. People are influenced by their home life, religion, family and friends, as well as the opinion of others. All have egos, and self-worth is based, in part, on the work they perform, their health, monetary concerns, politics and the development of technology. It is all very well to encapsulate an era by statistics of disasters and charts of commerce, but to truly comprehend who the individuals were and why they acted—and reacted—as they did, a backdrop of the times is essential.

We have attempted to present a unique oversight, while weaving a picture of the technical development of the steamboat as a way of explaining the glamorous and sometimes convoluted events that comprised the era of 1807–1860. To do this, we have selected articles from newspapers of the time, feeling it is important to present situations in the actual words and expressions of the day. Researching extensively in periodicals of this bygone age, we drew from thousands of original articles and included the entire text whenever practical. By this method, we hoped to give the flavor, the excitement, and frequently the heartbreak of those directly or peripherally involved with steamboating.

Historical misspellings, contractions and punctuation have been retained. Occasionally, charts and tables of figures do not add up: they were left as stated. Names of cities, rivers and newspapers evolved through the years: Pittsburg was typically spelled without an "h" at the end of the word; Nachitoches spelled without a "t" (Natchitoches); Galvezton with a "z" instead of the now-familiar "s"; "Arkansaw" was spelled phonetically. Words such as "steamboat" were printed variously as "steam boat" or "steam-boat." One, both or neither of the beginning letters may be capitalized. At times usage varied within the same article. The same holds true for "pilothouse," "wheelhouse," and "paddle wheels." Even the spelling of newspapers' names varied, going from "Centinal" to "Centinel" to "Sentinel." Many newspapers compiled articles from other papers and noted the original source at the end of the article. These notations have been retained.

The word "Negro" was seldom capitalized, and the names of Native American tribes went through a score of different versions (Native Americans themselves were invariably either termed "Indians" or "Red Men"). In rare instances, where we have been unable to decipher a word from an old newspaper, a question mark and the short notation that the text was unreadable has been inserted. Where applicable, we italicized pertinent passages; this is always noted.

Armed with natural curiosity and a desire to present a lucid and readable account with an eye toward continuity — always being aware that subjects frequently crisscrossed boundaries — we set about to answer myriad questions.

How did a portrait painter become the father of steam-powered vessels? What were contemporary opinions of the Great Mississippi, and where did that peculiar name originate? Why were "river tramps" vital to commerce? How did our ancestors deal with newfangled "crappers" when outhouses were the order of the day? Why have films and television series pictured captains as the ultimate object of veneration when pilots and even clerks were more lauded in the local press of the day?

Researching this book has been like a treasure hunt: one question led to another and often led us away on an entirely different tangent. Investigating Mark Twain's reference to Captain Basil Hall led to Frances Trollope; she took us to an anonymous book written by "an American," that in turn lent itself to the chapter called "An International Incident."

The Fulton-Livingston monopoly brought us to piracy and the U.S. Supreme Court; an investigation of piracy underscored the differences between ocean, river and land bandits and the creation of popular fictionalized heroes. Delving into law sent us back to maritime versus civil cases, the tradition of the seas and, ultimately, the status and relative importance of steamboat officers.

Researching accidents (see the "Man Overboard! Steamboat Disasters on Western Waterways" chapter 11) gave us casualty figures, explained why a "raft" had to be cleared rather than floated, and introduced us to "steam guards," which had less to do with "guards" than "gauges." Casualty lists pointed directly to immigration and the differences between the moneyed and the poor. Turning another page, we plunged into profits and losses: the accumulation of wealth depended on seasons, tides, trades and luck. Cutthroat competition directed the path to the rise of the "Steamboat Gothic" style with its gingerbread and gilt. Gilt is gold and gold is specie. Specie took us to the Specie Circular, Andrew Jackson and the mad land grab. Unbridled speculation in the opening lands of the West circled around economic panics and depressions. They, in turn, adversely affected steamboat construction and western commerce.

The use of steamboats for military transport was directly linked to the Steamboat Scandal and the treatment of Native Americans. Oppression took us through the uncomfortable issues of slavery and the treatment of African Americans as stokers and passengers. Investigating the major cities of the era lured us into the history of New Orleans, which was full of gambling parlors, dens of sin and Voodoo. Natchez-Under-the-Hill and Natchez-Over-the-Hill contrast as two sides of the same 19th-century coin.

How much did it cost to travel from New Orleans to St. Louis? What did a first-class cabin look like? How was a steamboat navigated through the twists and turns of the "Big Muddy"? Why did boilers explode and why would anyone in their right mind want to be an engineer? What did a contemporary bar stock and what did the patrons drink? Not the film favorite "red eye," but a series of sweetened concoctions garnished with fresh fruit. What were steamboat races really like and why were they considered by some authorities to be safer than regular voyages?

Why did towns hang out yellow flags and what were the "cures" for the horrendous epidemics sweeping the lower Southern states? Who traveled the Fashionable Tour? Why, at the pinnacle of success in the 1850s, were steamboats already being superseded by the railroads of the Robber Barons?

In the modern day of research, the Internet proved a valuable tool, but nothing can surpass the many beloved, well-fingered books we used to prepare this text. Primary among them are *Steamboats on the Western Rivers* by Louis C. Hunter, *Steamboating on the Upper Mississippi* by William J. Petersen, and *Old Times on the Upper Mississippi* by George Byron Merrick. To these scholars, we owe a debt of gratitude, as their compelling and comprehensive studies are a vital source for anyone desiring an in-depth understanding of the steamboat era.

As a supplement to the text we have compiled a glossary of steamboat terms and expressions. From "A1" (pilot rating) to "wrecked and floundered," terms are concisely defined (see Appendix A).

Oliver Evans was an early steamboat advocate who designed a boat in Philadelphia in 1802 and assembled it in New Orleans. Before he could make a trial run, however, a flood carried the vessel away. The disaster did not dissuade him from his goal and, although ultimately unsuccessful in making a significant mark (his sole boat, the *Oliver Evans*, exploded), he did pen an eerily accurate future. As printed in the *Adams Sentinel,* June 19, 1843, the text reads as follows:

Predictions Fulfilled.—In 1813, Oliver Evans predicted:

1st. That the time would come when people would travel in stages moved by steam engines from one city to another, almost as fast as birds fly, 15 or 20 miles an hour.

2nd. A carriage will set out for Washington in the morning, the passengers will breakfast at Baltimore, and sup in New York, the same day.

To accomplish this two sets of railway will be laid, traveled by night as well as day, and the passengers will sleep as comfortably as they now do on steamboats.

3rd. A steam engine, consuming from a quarter to a half a cord of wood, will drive a carriage 180 miles in twelve hours, with 20 or 30 passengers, and will consume six gallons of water.—[Technically incorrect but fascinating insight]

Steam engines will drive boats 10 or 12 miles an hour, and there will be as many as 100 steamboats on the Mississippi and other Western waters, as prophesied 30 years ago.

An article such as this stimulates the imagination and brings the times to life. We hope our text has done the same in addressing the routes, twists and turns of the 19th century in a factual and readable manner that encapsulates the excitement, the high times, the miseries and uncertainties of a turbulent period. The steamboat era can fairly be said to have begun with Robert Fulton in 1807, come to fruition in the 1830s and 1840s, achieved glory in the 1850s, and all but disappeared with the Civil War. What happened within those decades is a fascinating look at America, Americana and the men and women of all nationalities who went along for the ride.

SLK • JEG • Fall 2009

1

The "Invention of the Devil"

They called it "Fulton's Folly," and it changed the world.

Perhaps the two most significant inventions in the late 18th and early 19th Century were Eli Whitney's cotton gin and Robert Fulton's steamboat. One made the harvesting of cotton practical and profitable (and inadvertently ensured the continuance of human bondage). The other opened up inland waterways, providing for the rapid transportation of passengers, produce and manufactured goods. Both prompted the United States to expand its borders, at once deepening the conflict between slave and free states and providing Americans and emigrants with not only rich new lands but the unprecedented opportunity for travel, advancement and, not inconsequentially, unbridled speculation.

Prior to the era of steam-powered boats, western merchants were compelled to send raw material down the Ohio and Mississippi rivers to New Orleans on hundreds of small rafts and flatboats. The process was time-consuming and labor-intensive. Upon arrival at New Orleans, cargo was loaded on sailing vessels and sent on a roundabout, 3,000-mile waterway journey across the Gulf of Mexico and up the Atlantic coast to the industrial centers of New York and Boston.

August 17, 1807 — the day Fulton's steamboat traveled from New York City to Albany — became a day to remember. But some 20 years earlier, there were precursors to Fulton's historic steamboat. On February 9, 1785, Joseph Bramah of Piccadilly, England, patented the application of a paddle-wheel to the stern of a vessel driven by a steam engine. Another early pioneer was James Rumsey, who employed a system of hydraulic jet propulsion (also proposed by others, including John Allen as early as 1730) to draw water in at the bow and force it out the stern, thus creating a steam-propelled forward movement. In 1785, 1787 and 1788, Rumsey demonstrated the technique, achieving a speed of four miles per hour while running the Potomac.[1]

John Fitch of Connecticut received exclusive privileges from New Jersey to operate steam-powered boats in that state, and in 1786, he created a stock association called the Steamboat Company, for that purpose. On August 22, 1787, he presented a 45-foot boat moved by cranks, that turned six paddles per side. In May 1790, Fitch steamed his boat from Burlington, Pennsylvania, to Philadelphia in three and a quarter hours. During the summer, this boat actually ran a passenger and freight service on the Delaware River between Philadelphia and Bordentown, New Jersey. Since she ran only one summer, it is presumed the weight of the machinery left little room for passengers and cargo, and was not commercially viable.[2]

A particularly touching and informative article on this early pioneer was printed in the *Ariel,* and later published in the Norwalk *Reporter and Huron Advertiser,* March 22, 1828:

THE FIRST STEAM BOAT

In the year 1788, the bosom of the Delaware was first ruffled by a Steam Boat. The projector at that early day was *John Fitch*. He first conceived the design in 1785; and being poor and illiterate, a multitude of difficulties which he did not foresee, occurred to render abortive every effort of a most persevering mind, to construct and float a steam boat. Applying to Congress for assistance, he was refused, and then, without success, offering his invention to the Spanish Government for the purpose of navigating the Mississippi. He at last succeeded in forming a company, by the aid of whose funds he launched his first rude effort a steamboat in the year '98. The idea of wheels had not occurred to Mr. Fitch, but oars, working in a frame, were used in place of them. The crude ideas which he entertained, and the want of experience, subjected this unfortunate man to difficulties of the most humbling character. Regarded by many as a mere visionary, his project was discouraged by those whose want of all motive for such a course rendered their opposition more barbarous: while those whose station in life placed it in their power to assist him, looked coldly on; barely listening to his elucidations, and receiving him with an indifference that chilled him to the heart. By a perseverance as unwearied as it was unrewarded, his darling project was at length sufficiently matured, and a steamboat was seen floating at the wharves of Philadelphia near forty years ago. So far his success amid the most mortifying discouragements had been sufficient to prove the merit of the scheme.—But a reverse awaited him as discouraging as it was unexpected.—The boat performed a trip to Burlington, a distance of twenty miles, when as she was rounding at the bend, a boiler burst. The next [words missing] her back to the city; and after great difficulty, a new boat was procured. In October, 1788, she again performed her trip to Burlington. But it was discovered that she traveled only three miles an hour, although assisted by the tide, and as it was important that a greater degree of expedition should be obtained, and as the experience of getting better machinery, was a difficulty absolutely insurmountable, the important enterprise was suddenly abandoned! The boat was laid up as useless, rotting silently and unnoticed in the docks of Kensington. Fitch became more embarrassed by his creditors, and after procuring three manuscript volumes, which he deposited in the Philadelphia Library to be opened thirty years after his death, he was carried off by the yellow fever in 1793. Such was the unfortunate termination of this early conceived project of the steamboat. Fitch was no doubt an original inventor of the steam boat.—He was certainly the first that ever applied steam to the propulsion of steam boats in America. Though it was reserved to Fulton to advance its application to a degree of perfection which has made his name immortal: yet to the unfortunate Fitch belongs the honor of completing and navigating the first American steam boat.

His three manuscript volumes were opened about three years ago. Although they exhibit him an illiterate man, yet they indicate the possession of a strong mind of much mechanical ingenuity; he describes his many difficulties and disappointments with a degree of feeling which cannot fail to win the sympathy of every reader causing him to wonder and regret that so much time and talent should have been so unprofitably devoted. Though the project failed—and it failed for the want of funds—yet he never for a moment doubted its practicability. He tells us that in less than a century we shall see our western rivers swarming with steam boats, and that his darling wish is to be buried on the margin of the romantic Ohio, where the song of the boatsmen may sometimes penetrate into the stillness of his everlasting resting place and and [sic] the music of the steam engine echo over the sod that shelters him for ever.

At the dawn of the 19th century, experiments with steam-powered boats were also progressing across the Atlantic. These would have a profound effect on the ultimate development of the American steamboat.

The *Edinburgh Advertiser* of June 26, 1801, reported on a "*Steam Boat,*" "which was, with care and dispatch," navigated from Carron to Grangemouth, a distance of two or three miles.

> It is intended, by the power of *Steam*, to drag vessels up the Canal between *Clyde* and *Grangemouth*, which it appears well fitted to do.... The nice and effectual manner in which the machinery is applied, is an additional proof of the merit of Mr. Symington, the engineer, and the whole plan is highly honourable to Lord Dundas, the patron and promoter.

William Symington had already engaged in successful trials with pleasure steamboats as far back as 1788–89, and by 1801, had received financial assistance from Thomas, Lord Dundas of Kerse. Robert Fulton actually visited Symington some time after 1803 to inspect the steam tug *Charlotte Dundas*. (Symington claimed Fulton came in July 1801, but a record of Fulton's movements indicates he could not have been in Scotland before 1804.)[3]

Unfortunately for Symington, a majority of co-proprietors in the Forth and Clyde Canal declined use of the tug on the grounds that the wash of the paddles would be destructive to the canal banks. The *Charlotte Dundas* was eventually laid up at Bainsford, near Carron Ironworks, where she remained until 1861, when she was broken up.

Critics of Robert Fulton claim he stole all of Symington's ideas, but there were significant differences between the two men's work. Symington had a horizontal engine with a cylinder 22 inches in diameter, by 4-foot stroke, acting directly on a paddlewheel in a recess in the stern, while Fulton utilized a vertical beam engine 24 inches in diameter by a 4-foot stroke, connected by gearing through a flywheel to drive the side paddle-wheels. Symington's design was actually the better of the two, and after nearly fifty years, steamboat engineers went back to Symington's direct-acting plan.[4]

Other men of genius experimented with steam-powered boats (including William Murdock, of the Soho Foundry, Birmingham, who has often been credited with designing the bell-crank engine design used by Fulton),[5] but it was Robert Fulton who demonstrated—and delivered—the first steamboat capable of durability and economic success.

Robert Fulton: From Miniatures to Patents

Robert Fulton was born in Lancaster County, Pennsylvania, on November 14, 1765. His father (Robert, Sr.) held various public positions, including secretary of the Union Fore Office, before buying a 364-acre farm on Conowingo Creek in Little Britain. The elder Fulton did not succeed as a farmer and mortgaged the property in 1766.

Educated by his mother, Mary, until the age of eight, Robert, Jr., was then sent to the school of Caleb Johnson, a Quaker with Tory sentiments. The boy evinced a talent for painting, but as this was not considered a reputable occupation for a young man, he had difficulty procuring proper materials.[6]

In 1782, Robert left for Philadelphia, a city renowned for the arts. By 1785, he had established his reputation as a miniatures painter and met with considerable financial success. Meeting many of the prominent men of the day, he not only painted Benjamin Franklin's portrait but took with him a letter of introduction from that worthy when he moved to England at the end of 1786 or the beginning of 1787.

In the years that followed, Fulton gave up painting to concentrate on engineering. He devoted much time and energy to canals, eventually taking out a patent in 1794. It was titled, "A machine or Engine for conveying Boats and Vessels and their Cargoes to and from the different levels in and upon Canals, without the Assistance of Locks or the other Means now known and used for that purpose." That same year, he made his first known query to the firm of Boulton and Watt, which would later design the engine for the *Clermont*—the steamboat that would make a historic voyage on August 17, 1807.

Years without success followed, and although Fulton styled himself a "Civil Engineer," he never actually earned that title by completing the requisite work.[7]

Taking advantage of a brief respite in the ongoing war between England and France, Fulton crossed the channel in the summer of 1797. Staying with friends John Barlow and his wife, Ruth Baldwin, for the next seven years, Fulton pursued work on canal development, earning several patents. The first came in 1798, for "*Des nouveaux de construire des canaux navigables*" [A new manner for the construction of small canal navigation]. While never completely abandoning his system of "small canals," Fulton then directed his energies toward the creation of a type of gunpowder that would explode underwater. In conjunction with that project, he expanded its usefulness by developing an underwater submarine, designed specifically to destroy enemy (in this case, British) navies.

Believing that his *Nautilus* would end war by rendering it impractical (the invention being so terrible and so effective no navy could defeat it), Fulton wrote to J. N. F. Barras, a member of the French directory on October 27, 1798, in defense of his idea:

> If, at first glance, the means that I propose seem revolting, it is only because they are extraordinary; they are anything but inhuman. It is certainly the gentlest and the least bloody method that the philosopher can imagine to overturn this system of brigandage and of perpetual war which has always vexed maritime nations: To give at last peace to the earth, and to restore men to their natural industries, and to a happiness now unknown.[8]

Fulton made repeated attempts to sell his theories and his ideas to the ministers of France. With their tacit approval, he succeeded in creating a prototype vessel, and ultimately attempted to destroy the English fleet near Isigny. Fortunately for England, his plans were made known to the British Admiralty, and the warships escaped before the *Nautilus* was put to a practical application. Napoleon I ultimately considered the project impracticable, but that did not stop Fulton from peddling the idea back to the English.

An article in the *London Times,* on May 15, 1813, sums up Fulton's dealings with the British:

> As to Mr. Fulton, his agreement with Mr. Pitt and Lord Melville came under his (Earl Grey's) review, when at the Admiralty, and he took credit to himself for a comparison by which he saved the public money, though if he had thought the object desirable, he should have had no thought of sparing. Every examination and opportunity was given to Mr. Fulton. A ship was blown up indeed; but many precautionary measures were necessary such as would not be afforded by an enemy ship. Compressed gunpowder placed under a ships' bottom might certainly blow it up; but it was like the catching of a bird by

putting salt on its tail. In the centre of the experiment, the night was sometimes not dark enough; at other times the sea was too rough or too calm; for placing the explosives. As to Boonaparte [*sic*], was it to be supposed he did not examine before he rejected? He himself felt some uneasy nights at the idea of a plan involving such dreadful waste of human life.

Fulton and Livingston's Voyage Begins

The groundwork for Robert "Toot" Fulton's work (ironically, an affectionate nickname bestowed by his friends the Barlows long before steamboat whistles were heard on the Mississippi), had already been laid by Robert Livingston, chancellor to France under President Thomas Jefferson. Long interested in steamboats, Livingston had used his considerable influence to obtain for himself, in March 1798, the "exclusive right and privilege of navigating all kinds of boats which might be propelled by the force of steam or fire on all the waters within the territory or jurisdiction of the State of New York for the term of twenty years from the passing of the Act; upon condition that he should, within a twelvemonth build such a boat, the mean of whose progress should not be less than four miles per hour.[9] These were prophetic words for a prophetic tandem that would bring the two men fame and fortune under virtually the same provisions.

Livingston met Fulton through his brother, and the two formed a partnership on October 10, 1802, to embark on a steamboat enterprise, with Livingston backing the project. Their ultimate aim was to make a boat "120 feet long, 8 feet wide and 15 inches draught," constructed "to carry sixty passengers, and to run between New York and Albany at a rate of eight miles per hour in still water," operating under Livingston's New York grant. They agreed Fulton would take out a United States patent in his own name and divide its worth into 100 shares, with 50 going to each partner. All profits were to be shared equally.[10]

After lengthy correspondence with the firm of Boulton and Watt, Fulton finally decided upon the specifications of a steam-powered engine and arranged to have it transported across the Atlantic to New York City.

The "Long Island Skiff"

The yet-unchristened boat designed to hold the Boulton and Watt engine was constructed by Charles Browne at the boatyard at Corlear's Hook on the East River. After launching, the vessel was taken to Pauler's Hook Ferry, where Fulton had a workshop. Extensive fittings for the engine were put in place, but cost overruns forced Fulton to seek additional monies. When all necessary specifications were complete, Fulton made a test run on August 9, 1807. "Much," he wrote the following day, in a letter to Livingston, "has been proved by this experiment."[11]

As constructed, the boat was described as looking like a "Long Island skiff." Designated the *Clermont,* after Livingston's New York mansion, the craft was little more than a barge, housing a large engine in the hold, open to view. A rude cabin covered the boiler. Side wheels were equipped with twelve huge paddles. The rudder was similar to those used on sailing ships and operated with a tiller. A thirty-foot smokestack rose nearly as high as the two masts, to be used for auxiliary power, should the engine fail. The prototype was heavy and cumbersome.

Starting from Greenwich Village at one o'clock on the afternoon of August 17, 1807, the *Clermont* carried about forty guests, most relatives or close friends of the developers. Rousing cheers and not a little derision were heard from shore. With expectations at fever pitch, the boat steamed up the Hudson River at an unprecedented 5 miles per hour and reached Albany, a distance of 150 miles, in 32 hours. The subsequent return voyage downstream was accomplished in 30 hours, but not after a brief controversy. The engineer, a Scotsman, imbibed too heavily in celebratory cheer, prompting his immediate dismissal. Fortunately, the assistant engineer knew the workings of the engine sufficiently well, and was able to navigate the craft back to her home port.

The Clermont

Built: 1807 in New York
Hull: wood, constructed by Charles Brown
Length: 133 feet
Breadth of beam: 18 feet
Depth of hold: 7.1 feet

The original caption to this reads: "The above engraving represents the first Fulton Boat built in America." The *Clermont* had both a paddle-wheel and sails, to used in case of emergency. It is likely this was made some time after the original voyage as there is considerable question about the actual appearance of the *Clermont*.

Engine: Built in England by Boulton and Watt
   Diameter of cylinder: 24 inches
   Length of piston stroke: 4 feet
Boiler: Copper, low pressure
   Length: 20 feet
   Height: 7 feet
   Width: 8 feet
Wheels: 15 inches in diameter, 8 buckets to each wheel
   Length: 4 feet
   Dip: 2 feet

After her successful voyage, the *Clermont* was enlarged, her name changed to *North River* and she went into public service. The following notice appeared in the Albany, New York, *Gazette* of September 4, 1807:

> The North River Steam Boat
> Will leave Pauler's Hook Ferry on Friday, the 4th of September at 6 in the morning and Arrive at Albany at 6 in the afternoon. Provisions, good berths and accommodations are provided.

And thus, the germ of future "Floating Palaces" was planted. The *North River* continued to ply the Hudson trade for the Fulton-Livingston Line as a passenger boat before finally being broken up.[12]

Interestingly, a report of the first voyage of the *Clermont* ran in the Milwaukie *Sentinel,* June 9, 1845. Note the last sentence, which comes from a faulty remembrance:

> THE FIRST STEAMBOAT VOYAGE ON THE HUDSON — The editor of the Evening Journal gives the following account of the first voyage by FULTON on the Hudson.
> It was an event not to be forgotten, any more than the mode resorted to for seeing the wonder by some truant boys. The present Steamboat Landing at Catskill was then an Island, the long wharf which now connects it to the main land not having ever been thought of. Hearing that a vessel which could make headway against wind and tide, without sails, was to run from New York to Albany, some boys of us determined to await its approach upon the Island, and as no boat could be procured to take us there, we put our clothes in our hats, lashed the hats to a board and swam off to the Island pushing the board before us. But we watched in vain during the first day. On the following day, having reached the Island in the same manner, after long watching, we saw what seemed to be a thing from the infernal regions, vomiting smoke and fire, and moving in a mysterious manner through the water. She was a rude, unfinished, black, ill looking craft, and, compared, in magnitude and beauty, with the steamers now upon the Hudson, as the ugly mudturtle compares with the graceful salmon. This was in the summer of 1807. Fulton, with his fast, but anxious friends, were on board of this boat. She was called the "North River," and thus made "the first steamboat voyage ever performed."

Robert Fulton's steamboat garnered him $1,000 for three months' work, and he anticipated earning $10,000 the following season. He immediately put into operation plans for a second boat, with the object being to have one "depart from Albany and one from New York every other day and carry all the passengers." There were problems, of course. Not only did his operation have to deal with those who would purloin his patent and attempt their own lines, but local men, jealous of his success, frequently attempted "malicious injury" to the steamboat, finally prompting the New York state legislature to pass an act providing remedies against those guilty parties.

Interestingly, despite the notoriety of the *Clermont*'s first voyage, New York's twenty newspapers, nearly half of them dailies, paid scant attention to this new mode of travel. Aside from letters Fulton and some passengers wrote, the most an inquisitive citizen could find was located in the advertising pages. There he would discover a list of fares from the city, upriver to:

> West Point — $2.00
> Poughkeepsie — $3.00
> Kinderhook — $5.75
> Albany — $7.00

All other passengers to pay at a rate of $1 for every twenty miles. No one can be taken on board and put on shore however short the distance for less than 1 dollar. Young persons from two to ten years of age to pay half price; Children under two years one fourth price. Servants who use a berth, two thirds price; half price if none.[13]

Fulton's Hudson Steamboat Company eventually designed and put into service the *Car of Neptune*, the *Paragon*, *Firefly*, *Richmond* and the *Chancellor Livingston*.

The Fulton *Onward to New Orleans, Charleston, New Haven and Beyond!*

The success of the *Clermont* prompted Fulton to proclaim:

> Whatever may be the fate of steamboats for the Hudson, everything is completely proved for the Mississippi, and the object is immense.[14]

Indeed, while Fulton and Livingston had eyes for running steamboats on New York waterways, their goal had always been the Mississippi River, for therein lay the greatest profit. Fulton also had a keen interest in obtaining exclusive steamboat rights in England, Russia and India, and made plans to secure monopolies in those countries, as well.

Robert Fulton wrote to John Quincy Adams (then ambassador to Russia), asking his help in obtaining a twenty-year contract granting him exclusive rights to operate a steamboat line between St. Petersburg and Cronstadt. In 1812, Fulton did obtain Imperial permission to introduce steamboats onto Russian waters, and, in fact, the *Emperor of Russia* was being constructed at the time of his death. While it is felt that the name was more honorary than actually representative of a boat being built for Russia (there being no means of getting it there)[15] the following extracts from contemporary newspapers paint an interesting picture. The first was carried in the *Adams Sentinel* on Wednesday, August 14, 1816, taken from the *Columbian* of an earlier date.

New York, August 3, 1816

A NOBLE ENTERPRISE

It is reported and believed that a distinguished barrister of this city, together with Capt. Bunker of the steam packet Fulton, have resolved to cross the Atlantic to England, and proceed thence to Russia in the new steam-boat. This grand undertaking we understand, is in fulfillment, or acceptance of a contract offered to Mr. Fulton by the Emperor of Russia, allowing him the exclusive navigation of Steam-Boats in the Russian empire for 25 years. As the vessel is built as substantial and strong as a sloop of war little or no doubt is entertained by naval men of the practicality of the attempt. We are delighted with the prospect of a Steam Boat propelled across the Atlantic ocean, by Americans "the first." There is no doubt of the *expedition*: it is determined; and, since rumor is busy on the subject, we make free to mention that Mr. Colden* is the gentleman alluded to.

On September 17, 1816, the Edinburgh *Advertiser* mentioned the story, remarking:

> The *Boston* Paper of the 7th August has the following paragraph — The *Steam-boat*, FULTON, is under contract to sail for *Russia* from *New York*, and to arrive in *Russia* by the 1st of December.

The *Fulton* never went to Russia. Actually, the first effort to navigate the Neva was made in November 1815, by Charles Baird, with a barge that had been redesigned for the purpose.[16]

Nor did Robert Fulton extend his steamboat empire to India, although he entered into an agreement with Captain Law to introduce steamboats on the Ganges. On April 16, 1812, Fulton wrote to Law:

> I agree to make the Ganges enterprise a joint concern. You will please to send me a plan how you mean to proceed to secure a grant for twenty years and find funds to establish the first boat.

He ends the letter, "Keep the Ganges Secret."[17]

The project apparently never got beyond the planning stages, and the first steamboat to ply the Indian waters came from England.

All of Fulton's early steamboats were constructed like the *Clermont*: flat-bottomed and wall-sided. The *Fulton*, meant for navigating Long Island Sound, was the first constructed "ship-shaped." The design proved successful, and all subsequent boats adopted that style. Even if the *Fulton* never made it to foreign locales, it did have its own unique history.

NEW YORK, APRIL 17, 1820

> *Steam Ship Robert Fulton*—This beautiful ship in hauling down town yesterday to take in coal for her intended voyage, gave a specimen of her celerity. She gratified the citizens assembled on the wharves from one end of the city to the other. She proceeded from East into the North River, from thence to Gravesend Bay, and returned in two hours and thirty minutes, a distance of 22 to 24 miles without a sail. Judging from this small trial, she is likely to prove herself one of the fastest vessels ever propelled by the use of steam. She leaves this port in two or three days on her intended voyage for N. Orleans, touching at Charleston and Havana, merely to land and receive on board passengers. She is truly one of the wonders of the present age.[18]

The saga continued:

> The experiment of navigating the ocean by steam, has succeeded to admiration. The elegant steam ship *Robert Fulton*, which went from New York some time since for Havana, is on her way back; via New Orleans and Charleston.[19]

And on June 27, 1821, the *Republican Compiler* reported:

> STEAM SHIP— The clouds of prejudice and fear appear to recede at the test of experiment and reason. The steam-ship *Robert Fulton* was full of passengers from New Orleans and Charleston, and each trip which this splendid vessel makes, not only gives ample proof of safety and dispatch, but adds to the number of converts in favour of steam navigation on the ocean. What a gigantic improvement in science! Those who are in the habit

---

Robert Fulton's friend and later biographer.

of crossing the Atlantic, who are compelled to pass days and weeks in a dead calm, when the unruffled surface of the ocean reflects like a mirror, and when the sun pours down its fierce and intolerable rays, and the sails flap to and fro, can well imagine what their feelings must be at seeing the steam-ship pass them rapidly; the wheels in quick motion, and the smoke rolling in curled volumes from the furnace; to see myriads of well dressed and contented passengers walking the decks. They must wish that steam navigation was universally preferred, and long to be on board.[20]

Back in New York, the *Republican Compiler,* on July 26, 1820 (irrespective of its June 27th comment on safety), reported on an "Extraordinary Escape" of July 5, when the steamboat *Fulton,* under Captain Law, from New London was going up the harbor under full sail and a powerful steam. The captain came upon a pleasure craft going from New Haven to the lighthouse.

> From some miscalculation on the part of the person who was steering the sail boat, in attempting to clear the steam boat, the latter struck her in the middle, and passed instantly over her. Thirty of the persons in the boat caught hold of the steam boat's bowsprit and bows and saved themselves.... No blame was imputable to capt. L. for the accident, but every exertion was made by him and his men for the preservation of the persons in the boat.

Robert Fulton and His Legacy

Numerous court battles began over exclusive rights and state boundaries, particularly between New York, New Jersey and Connecticut (see chapter 3). After testifying at one of those trials, Fulton was exposed to wet weather and subsequently contracted inflammation of the lungs (presumably pneumonia) and died early in the morning of February 23, 1815, at the age of 50.

Two years after his death, the struggle over patents continued, as reprinted in the *Adams Sentinel,* July 23, 1817, from the New London *Gazette:*

> Another Steam-Boat Accident "Explained"
>
> The account given of the accident which befell the "Norwich" steamboat on the 2nd instant, and running through the public papers, is calculated to mislead those who are unacquainted with steamboats. The facts are as follows:
>
> The boat in question is a small vessel, lately built and owned by a few individuals at Norwich to ply between Norwich and New-London, the proprietors of which, to save the expense of Fulton and Livingston's patent right, and an expensive engine, have put into her a simple engine, upon a new construction, and entirely experimental, with high pressure cylinders, and extraordinary as it may appear, *wooden boilers!* without condensers, safety-valves or balance wheels. As was predicted by many, her *wooden* boilers burst, and 3 persons hurt, but not dangerously. It is a fact worthy of notice, that the steamboats upon the North River and Long Island sound constructed on the Fulton and Livingston plan, have been running ten years, without a single person ever being injured, and it is impossible any serious injury should happen to them, their safety valves being calculated to relieve an excess of steam spontaneously.

While others attempted to navigate rivers by steam-powered engines, it took a man of genius as well as fortitude, persistence, and finally, knowing the right people, to prove the concept feasible. A painter (Fulton patented the concept of the Panorama), canal designer, investigator and designer of the "plunging boat" (or submarine) with underwater-discharged torpedoes, and the man who constructed the world's first steam-propelled war vessel, was buried with honors in the Livingston family tomb.

Shortly after Fulton's death, a note was written in the Sandusky *Centinel,* March 19, 1821:

> The benefit of Fulton's genius is felt in every part of the world. Navigation by steam is now in operation in almost every state in Europe. Such are the inestimable advantages, which a single great man may confer upon the world. The name of Fulton will hereafter be associated with the greatest benefactors of the human family, and will be a noble part of that heritage of honor, which belongs to our countrymen.

Robert Fulton ultimately achieved the wealth and fame he had sought all his life, and in so doing, opened the way west for thousands of adventurous pioneers. His invention provided an inexpensive and relatively safe journey along the world's rivers and tributaries. The opening of America's western commerce also had the unintended effect of shifting the population and thus the balance of power from the East, which had, until the steamboat era, dominated national politics. How that story developed would have surprised even the great inventor, the man to whose name popular history has affixed the word "folly."

2

Keelboats and Barges: The "Arks" of the New World

The creation of the Northwest Territory in 1787 prompted Americans to flood into the Ohio Valley. On April 30, 1803, Napoleon of France sold the Louisiana Territory to the United States. Thomas Jefferson was president, and the signature of Robert Fulton's future partner Robert Livingston was on the document. These vast acquisitions set in motion a land grab unparallel in the history of the world.

In the late 18th century, the easiest and most practical way to travel across large areas was via waterways. Americans were fortunate, for there existed a vast array of rivers, streams and creeks traversable by boat. Down the Missouri, the Ohio and Mississippi rivers they came, following the twisted channels, running the challenging falls, forging around and over river obstructions, at low and high water levels. Regardless of weather, earthquakes, disease and myriad dangers from unfriendly inhabitants, these early pioneers persisted, carrying what few possessions they owned on the decks of flat-bottomed craft.

From the 1500s on, rafting offered explorers the opportunity of inexpensive transportation. The type of vessel employed depended on need. Log canoes were copied from the Native Americans and could be purchased from dealers for three dollars or less. They were unwieldy and limited the number of persons and cargo that could be transported. Larger canoes were called pirogues. These were often forty to fifty feet in length and from six to eight feet wide. These could handle a family and several tons of household goods. Neither found favor, however, because they were difficult to navigate and they offered no protection from Indian attack.

The flat-bottomed skiff fit two or three people and, like the bateau, a larger version of the basic skiff, were propelled by several pairs of long oars, or sweeps. From Natchez to New Orleans, passage by skiff took nine days. The flatboat, a creation of the Ohio Valley, was the favored craft of migrating families because it was easier to handle and constructed more substantially, with bedrooms, a dining area, occasionally separate quarters for the crew and a fireplace for cooking. The outside walls also fended off arrows and bullets, a necessity for early immigrants.

Small flatboats were called "Kentucky boats," after their destination, while larger ones were named "New Orleans boats" for like reason. Flatboats cost from one dollar to one dollar and fifty cents for each foot of length and generally averaged twenty to sixty feet long, and ten to twenty feet wide. Flatboats rose three to six feet above the water and drew from one to two and a half feet of water depending on weight carried. Arks were shallow of hull and either square or V-shaped at the ends, usually lacking housing, but reaching lengths of one hundred feet. Canoes, skiffs, flatboats and arks were used for downriver travel only, being incapable of fighting the upstream current.

Businesses sprang up along river towns, offering boats for sale, and their proprietors made a lively living. In order to save their hard-earned money, however, many families and crews opted to construct their own craft. As late as 1831, for example, when Abraham Lincoln and his associates were commissioned to take a load of live hogs, corn and barreled pork down the Mississippi, they first had to build their own boat, using wood cut down along the Sangamon River in central Illinois.[1] Typically, once the family or the cargo reached its destination, the boats were broken up and sold as wood.

The keelboat, so called because of its heavy, four-inch-square timbers extending from bow to stern (designed to absorb the shock of underwater collisions), was sharp at front and rear. They were usually forty-five to seventy-five feet long and from seven to nine feet wide, capable of carrying twenty to forty tons of freight. Sometimes equipped with masts and sails, the keelboat required a crew of three: the captain, acting as steersman, and two men at the sweeps. On larger rafts, the captain stood at the bow and studied the river, shouting orders to the steersman at the tiller, who manipulated the rudder with brute force. Barges ranged from thirty to seventy feet long and seven to twelve feet wide and cost five dollars for every foot of length. Four long oars, two to a side, were employed for added speed. Headed downstream, a barge could travel at four to five miles per hour.

Prior to the introduction of steamboats, keelboats and barges were the only vessels capable of the upstream return. Although sails were occasionally utilized, they were of minimal value. Heavy growth of trees and shrubs along the Mississippi banks blocked off whatever breeze might have formed, forcing crews to rely on physical exertion to work against the stiff current. Progress of two miles an hour was considered good time. In 1818, keelboats could run from Louisville to New Orleans in six weeks, but the return trip took from three to four and a half months. Louisville to Pittsburg required at least four weeks. Barges required three months of backbreaking effort to shove and pull the vessel from Louisville to Pittsburg.*

Hard tasks called for hardened men, and the boatman usually began his day with a generous measure of whiskey. Fighting the current often meant pushing with poles and at other times, pulling the boat along by means of brush growing along the edge of the river bank in a method called "bushwhacking." When that proved ineffectual, the crew was compelled to attach a rope to a distant tree or rock. The men would then take turns grasping the rope and moving, hand-over-hand toward the bow, literally pulling the boat forward in a technique called "warping." Progress was excruciatingly slow and by dusk it was not uncommon to be within sight of the previous day's rest. This method took thirty-one days for a keelboat to go from St. Louis to Galena, a 400-mile trip, making an average of 13 miles per day, or less than one mile per hour.

---

*The town of Pittsburg was not then spelled with the "h" at the end of the word. The h-less spelling is used throughout this text.

The Half Wild Horse and Half Cockeyed Alligator Pirates

In 1809, the only points of importance between Pittsburg (population 4,000+) and New Orleans (population 17,242 by 1810) were Cincinnati (population 2,500), Louisville (population 1,250) and Natchez, which was then described as "a land of fevers, alligators, niggers and cotton bales, slave auctions, heavy drinking and constant fighting."[2]

The reputation of Natchez was based on the keelboat and bargemen, among the earliest navigators. These tough frontiersmen earned their living bringing goods and produce down the waterways of the Ohio and Mississippi on their way to New Orleans. The often dangerous journey across rapids, around obstructions and through extremes of temperature bred a class of rivermen known for short tempers, heavy drinking, wild carousing and fierce loyalties.

As a primary port of call, Natchez-Under-the-Hill (the unsavory area below the bluff where boats loaded and unloaded cargo) assumed importance. An entire subculture developed, with bitter rivalries between keelboat and barge gangs often resulting in violent clashes. Each boat typically had its champion and when challenged by the least provocation, they would meet in vicious combat. After blood was spilled, honor served and wages spent, these boatmen would then hire themselves out for the unenviable task of working a keelboat or barge back upstream.

Among the larger-than-life keelboatmen was Mike Fink, who boasted, "I'm a Salt River roarer! I'm a ring-tailed squealer! I'm a reg'lar screamer from the ol' Massassip! Whoop! I'm the very infant that refused his milk before its eyes were open, and called out for a bottle of old Rye!" Self-described as half wild horse and half cockeyed alligator, the rest of him being "crooked snags an' red-hot snappin' turtle,"[3] Fink and men like him ruled the inland waterways in the 1700s and into the early part of the 19th century.

Not to be outdone, the Upper Mississippi had its own class of keelboat and bargemen. In Prescott, Wisconsin, where raftsmen continued to work the lumber trade from St. Croix, they gathered to take in supplies for the trip downriver.

> The hundreds of rough men who handled the great steering oars on these rafts spent their money in the saloons which lined the river front and adjacent streets, filling themselves with noxious liquors, and often ending their "sprees" with a free fight between rival crews. A hundred men would join in the fray, the city marshal sitting on a "snubbing post," revolver in hand, watching the affair with the enlightened eye of an expert and the enjoyment of a connoisseur.[4]

After steam-powered vessels gained dominance with their speed, large freight capacity and ability to navigate upstream, keelboatmen lost considerable business. Those who opted to throw their lot in with the steamboat trades formed the backbone of early crews. The more highly skilled became captains or mates, while the bulk comprised those employed as deckhands or rousters. Clearly, they were well adapted to work the "invention of the devil."

Other rivermen opted for a life of crime. Gathering in such places at Natchez-Under-the-Hill, they found it easier to prey on unwary travelers, emigrants and operators of tramp steamers. Feared as much for their daring as their bloodthirsty ways, early captains often posted watches to keep an eye out for these river thieves. Many of these pirates hid on the innumerable small islands of the rivers, waiting their chance to snatch a prize. Stolen goods were easily sold locally or passed down to New Orleans, where a clever man could earn a fine reward. Or not. A report from the Ohio *Repository,* June 1, 1820, stands as a stark reminder of the consequences of such a lifestyle:

> On Thursday last *Eighteen* men, convicted of Piracy, were to have been hung on board the U.S. schooners Louisiana and Alabama, in the river opposite New-Orleans.

The threat of death did not deter the river, sea and land pirates, however. The *Torchlight & Public Advertiser,* September 4, 1821, gives a "blow by blow" description of what Americans on all borders of the new country feared:

Piracy

> The brig *Frances,* who arrived at New Castle, Delaware, was robbed by a boat's crew of Buccaneers, of several nations and of the most ferocious character, near Cape Antonia, on the 31st July, 1821. She was chased at dark, and when the shot began to fall thick around her, hove to at 9:00 P.M. Her crew was ordered on board the pirate; the men sent retained; and the boat returned with armed people, who jumped on board, and drove all below, cutting all in their way. They threatened to run the brig on shore, unless all money, watches and apparel, was delivered to them; and enforced their demand with strokes of a sabre, & a threat of general massacre. There was surrendered up to them wearing apparel worth $1,000, Cash in specie, 800 dollars, four valuable gold and silver watches and seals, chains, &c. valued at 500 dollars, writing desk, spy-glass, gun, pistols, swords, all the sea clothing, beds and bedding, compass, studding-sails, colors, stock wines, &c. leaving nothing but the clothes on the backs of the *Frances'* crew. After the depredation, the marauders went a shore, intoxicated, and promised to return and massacre the crew. The latter, however, took possession of the brig again, tacked and stood off and saw them no more.

Piracy continued to be an issue throughout the decades leading up to the Civil War. On January 4, 1845, a report of such lawless activity was printed in the *Milwaukee Daily Sentinel.* (Note the spelling of the city, which had not yet become standardized.)

> The Cincinnati Gazette says that there is a body of men increasing along the settled parts of the Mississippi, who when accidents happen to steamers near them, take *salvage* in the largest sense of that word. The Warrior, sank lately at island 95, is said to have suffered largely from their depredations. They relieved her of whiskey, oil, &c. and then sold them to the lower towns as their own. The Gazette calls for prompt Government interference.

An even more distasteful article on piratical salvage came from the New York *Daily Times,* April 7, 1852:

> The Wreck of the General Warren. The search for the bodies of such as perished by the wreck of the steamer *General Warren,* at Oregon has been vigorously prosecuted. The Oregon *Times* pillories F. GERHEART, Justice of the Peace at Clatsop County, S. CONDIT, James Cook and Joel Welch as claimants for salvage for money found upon the bodies of the dead. GERHEART and CONDIT are indicted as the parties who barely avoided a fight about the money found upon the corpse of D.H. LUTHER, but who finally comprised the matter by taking a hundred and five dollars salvage, and thirty dollars for burying the body in a rough box.

Robert Fulton, Pirate

While the Mike Finks of the day were easily associated with the concept of "pirate," early rivermen were not the only ones who earned the distinction of the title. Before his days of celebrity as designer of the *Clermont,* Robert Fulton himself might be considered the first pirate of the steamboat era. As early as 1797, he conceived the idea of a submarine boat. Then living in Paris, he offered his idea to General Bonaparte and the Executive Directory, in the "6th Year of the Republic," at which time France and England were at war.

> Considering the great importance of diminishing the Power of the British Fleets, I have Contemplated the Construction of a Mechanical Nautilus. A Machine which flatters me with the hope of being Able to Annihilate their Navy; hence feeling confident that practice will Bring the apparatus to perfection.

For this purpose, Fulton established a company willing to bear the expense of the project. He enclosed six terms that were to be met, the last of which expressly states his concern:

> And whereas fire Ships or other unusual means of destroying Navies are Considered Contrary to the Laws of war, And persons taken in Such enterprise are Liable to Suffer death, it will be an object of Safety if the Directory give the Nautilus Company Commissions Specifying that all persons taken in the Nautilus or Submarine expedition Shall be treated as Prisoners of war, and in Case of Violence being offered; the Government will Retaliate on the British Prisoners in a four fold degree.[5]

The proposal was delivered to the minister of marine, who accepted them with provisions: that the requested prize money for destroying enemy ships be halved; that the reimbursement to the company, in case of peace, be refused; and finally, an outright denial of Fulton's request for a commission in the French navy, in that the operation was outside the customs of civilization, *and the crew no better than pirates.*[6*]

Robert Fulton was not yet done with pirating and pirates. During the ongoing battle over his eastern monopoly, the state of New York passed a resolution permitting the Fulton-Livingston concern to enforce seizure of any boats violating their rights. They might:

> ... take and hold possession of the forfeited property without any preceding process of law, if they can accomplish that object without a breach of the Peace.

On January 25, 1811, the state of New Jersey countered by enacting legislation permitting its citizens to use vessels of all kinds on the waters between the states. The resolution continued by avowing that if any New Jersey citizens had their boats confiscated, they had the right to retaliate upon any New York steamboats entering New Jersey waters, thus tacitly granting permission for outright piracy.

The situation continued to deteriorate and lawsuits abounded. On June 28, 1811, Fulton commented on one case involving a steamboat company working in opposition to his own:

> My time is now occupied in building North River and Steam ferry boats, and in an interesting lawsuit to crush 22 Pirates who have clubbed their purses and copied my boats and have actually started my own Inventions in opposition to me by running one trip to Albany.[7]

The monopoly was eventually lost, but not before the Father of Steamboats had dabbled on both sides of the piracy issue.

---

* Italics added

2. Keelboats and Barges

An Inexpensive Means of Transportation

Although flatboats ultimately lost the passenger trade to the developing steamboat, they did not altogether disappear from the rivers. To the contrary, the use of flatboats to transport heavy and less expensive products, such as sand, stone, coal, salt, bricks, lumber, whisky and smoked meats, actually reached its height in 1846–47, with the arrival of 2,792 flatboats during that period at New Orleans. Cincinnati became the center of flatboat operations on the Ohio River, where keelboats brought in large quantities of produce for consumption and processing.

Keelboats were also used in conjunction with steamboats, especially in the early years, as reported by the *Enquirer:*

> St. Louis, July 9 [1819] The postscript to the last paper, which states the return of the military expedition is incorrect. No part of it has returned. The Steam Boats, however, get on slowly, and with many stoppages. The keel boats go on rapidly, and passed St. Charles (18 miles from Belle Fontaine) two days ahead of the former. The contents of the steamboats will probably be transferred to keels and barges. The river is in fine order: the July flood is coming down, and swelling the stream with the last of the tributes. The failure of the steamboats is attributable to their construction or management, and not to the rivers.

By the 1850s, the flatboat trade finally fell off, although they were used to haul fresh produce between New Orleans and Bayou Sara (known as "the coast"), an area that included most of the sugar and rice country.

Rafts also continued to provide an inexpensive means of transportation. They were particularly effective on the smaller tributaries that were shallow and/or difficult for steamboats to navigate. They carried lumber, shingles and wood products down from the Allegheny, Kanawha, Big Sandy and the Upper Tennessee on the Ohio, from St. Croix and Wisconsin on the Upper Mississippi, and the Ouachita and Black rivers on the Lower Mississippi. Their principal destination was Pittsburg, but occasionally some went all the way to New Orleans.

With their demise, a way of life and a peculiarly American character passed into the pages of history. But they left behind themselves a legacy and a begrudged respect, for the roughhousing keelboatmen pioneered the inland waterways. Their experience, sweat and strength charted the rivers, created and sustained infant towns, gave rise to western commerce, and eventually filled the decks of the noisy, smoky mechanical boats that would literally change the landscape of the emerging nation.

3

Onward to the "Father of Waters": The Fulton-Livingston Patent

Fulton wrote that

> Steamboats will give a cheap and quick conveyance to the merchandise on the Mississippi, Missouri and other great rivers, which are now laying open their treasures to the enterprise of our countrymen; and although the prospect of personal emolument has been some inducement to me, yet I feel infinitely more pleasure in reflecting on the immense advantage my country will derive from the invention.[1]

After the Louisiana Purchase, William C. C. Claiborne was appointed governor of the Orleans Territory (present-day Louisiana). The remaining wilderness between the Mississippi River and the Rocky Mountains, all the way to the Canadian border, was designated the Louisiana Territory. It was Governor Claiborne whom the team of Fulton-Livingston approached with their ideas about steam-powered boats on the Lower Mississippi. They brought to the table their patent and one very powerful word: monopoly.

Arguing that he wished to extend steamboat navigation into the western territories, Robert Livingston stated the benefits in a letter dated August 20, 1810: A "capital approaching to two hundred thousand dollars will be required, which capital must be raised by subscription." With that warning came the caveat: "...subscribers cannot be obtained until an effectual law presents a fair prospect of securing to them such exclusive rights as will return emolument equal to the risk and trouble."[2]

Chancellor Livingston had already obtained exclusive rights to New York State waterways thirteen years earlier, and experience proved key in his present negotiations. Pressing the case with consummate skill, he succeeded, almost to the letter, in getting what he wanted. On April 19, 1811, the legislature of the Orleans Territory passed an act granting the Fulton-Livingston partnership "the sole and exclusive right and privilege to build, construct, make use, employ and navigate boats, vessels and water crafts, urged or propelled through the water by fire or steam, in all the creeks, rivers, bays and waters whatsoever, within the jurisdiction of the territory, during eighteen years from the first of January, 1812."[3]

Although this became the only official sanction received from their pleas, Fulton and Livingston had their patents, their monopoly on the Lower Mississippi and their capital.

"The British Are Coming!"

Nicholas J. Roosevelt was hired by the firm of Fulton-Livingston to travel down the Mississippi from Pittsburg to New Orleans, appraising conditions on the river for steamboat navigation. He constructed a flatboat suitable to his needs and those of his bride, Lydia Latrobe of Baltimore, and they left in the summer of 1809. It rapidly became apparent that although the western waters posed unique challenges, the prospect of running steam-powered craft seemed feasible.

Returning to New York with this eagerly anticipated news, Fulton and Livingston hurriedly incorporated the Ohio Steamboat Navigation Company under their exclusive patents. Roosevelt went back to Pittsburg and set about establishing a boatyard on the Monongahela River. Soon, construction of the first western steamboat was underway.

The boat was christened the *New Orleans*. It was built of native white pine. Its copper boiler was made in New York and transported across the Alleghenies, with eastern engineers brought to install it. The engine and smokestack stood exposed in the center with a cabin in the rear. The living quarters were divided, one aft containing four berths for the ladies and a larger one forward for gentlemen. The vessel was a side-wheeler, designed to carry freight stowed in the bow. Because she carried two masts, a gentleman of the times described her as "a large and heavy boat combining incongruous features of marine and river craft."[4] In fact, all steamboats until 1815 were merely reworked keelboats with the addition of engines and boilers in the middle of the boat.

The New Orleans

Length: 116 feet
Beam: 20 feet
Tonnage: 371 tons
Cylinder: 34" diameter
Cost: $38,000[5]

The *New Orleans* left Pittsburg on Sunday, October 20, 1811, under the command of Captain Nicholas Roosevelt. The engineer was Nicholas Baker and the pilot was Andrew Jack. Also aboard were "six hands, two female servants, a man waiter, a cook, and an immense Newfoundland dog, named Tiger."[6] (Clearly, "Floating Palaces" could not be far behind.) Traveling with the flow of the current, the *New Orleans* achieved a speed of eight to ten miles per hour, reaching Cincinnati in two days and arriving at Louisville sixty-four hours after departure from Pittsburg.

Families along the waterways and backwoodsmen, living by what would soon become thriving wooding stops, witnessed the new mode of transportation with less wonder than alarm. Describing it as everything from a floating saw mill (ironic, in that one of Fulton's earliest inventions was a "mill for sawing marble or other stones"[7]), to a "contraption," many feared the boat to be an invasion by the British (the War of 1812 was just around the corner), or the devil, whichever most aroused their particular fears.

Surviving the New Madrid earthquake, encounters with the Chickasaw and a nearly disastrous fire, the *New Orleans* reached her namesake's destination on the evening of January 10, 1812, eighty-two days out of Pittsburg. (Of that time, the captain estimated the craft spent only ten days and nineteen hours actually running the river.)

The *New Orleans* worked the 270-mile New Orleans–Natchez trade (a relatively safe route) until July 14, 1814, when she hit a stump two miles above Baton Rouge. The boiler and part of her machinery were salvaged and used aboard the second *New Orleans*. Built in 1815 and weighing 325 tons, this vessel was even more successful, achieving a net profit of $4,000 from one trip along the trade in 1817.[8]

In addition to the second *New Orleans,* the Ohio Steamboat Navigation Company quickly followed with the *Vesuvius,* 340 tons (not a particularly inspiring name for a type of vessel already establishing a reputation for explosions), and the *Etna,* 360 tons.

The fate of the *Etna* was given in the *Republican Compiler,* August 23, 1820:

> A letter received from New York, from a respectable house in New Orleans, under the date of the 18th ult., mentions the total loss of the steam boat Etna, with a very valuable cargo, on her passage up the river, by running on a snag, near Fort Adams, 260 miles above New Orleans. The Etna was one of the oldest boats on the river, an excellent vessel, and about 380 tons burthen.

Henry Shreve and the War of 1812

Westerners did not take kindly to the Fulton-Livingston patent, and no sooner had the *New Orleans* made her maiden voyage than opposition arose. There was too much opportunity, too much money to be made to allow an "eastern" group to dominate the new "western" mode of transportation. Steamboats represented adventure, challenge and, not incidentally, profit. "Open the waterways to all," they cried. "America is not about patents and exclusive rights!"

Henry Miller Shreve was born on October 21, 1785, to Israel and Mary (née Cokely) in New Jersey. After the British ransacked and looted their homestead in 1788, Israel moved his family to western Pennsylvania. The patriarch died in 1799[9] or 1802[10]), Henry hired on as a flatboat crewman and began his career as a riverman. By 1807, he purchased his own keelboat and entered into the beaver pelt business. That same year, Shreve took his 35-ton barge to St. Louis, where he made the acquaintance of Auguste Chouteau, one of the city's leading citizens, who made a strong impression on him. The youth returned to Pittsburg with one of the first cargoes of furs destined for Philadelphia via that water route, and he continued in that trade for three more years.[11] In 1810, he navigated his barge to the Fever River and returned with a cargo of lead, marking the beginning of lead traffic on the Upper Mississippi. Investing the $11,000 gained from that venture, Shreve started a barge business between Pittsburg and New Orleans.[12]

An incident destined to alter the course of his life occurred in 1811 when he witnessed Robert Fulton's steamboat *New Orleans* on her maiden voyage.[13] It did not take him long to realize the future of river transportation belonged to the steam-powered vessel. Working with Daniel French, whom he also financed, they built the *Comet,* a stern-wheeled, steam-powered boat. This act was against the law, and some historical records speculate that the adaptation of the stern-wheel design was done specifically to avoid infringing on the Fulton-Livingston patent.

The Comet

Built: 1813 in Brownsville, Pennsylvania by Daniel French and Henry Shreve
Tonnage: 25 tons

Engine: French oscillating cylinder (vibrating) engine
Design: Sternwheeler

Before the *Comet* departed from her Ohio River home, the first of many lawsuits was filed against her owners. They were ignored, as was a federal act of 1812 that decreed that as a condition of statehood, all the waters on the Mississippi should forever be free.

After reaching Louisville in the summer of 1813, the *Comet* continued to New Orleans the following summer.

The *Comet* made two trips to Natchez before being sold. Her engine was ultimately installed in a cotton factory. The boat had not been a profitable one.

Despite renewed efforts to preserve the Fulton-Livingston monopoly, Daniel French ignored legal threats and launched the *Enterprise* from Brownsville on the Monongahela in 1814. She made two trips to Louisville in the summer of 1814. In December, with Shreve now in command, the boat steamed for New Orleans, reaching the city in fourteen days. The following day, authorities seized the boat. With the services of A. L. Duncan, the only lawyer in New Orleans not on the payroll of the Fulton-Livingston Corporation, bail was procured. A second seizure occurred several months later, with similar outcome.

The Enterprise

Built: 1814 at Brownsville by Daniel French and Henry Shreve
Tonnage: 75 tons

The *Enterprise* on her fast trip to Louisville in 1815. She was launched from Brownsville in 1814. Under the command of Henry Shreve, she challenged the Fulton-Livingston monopoly and participated in the Battle of New Orleans.

Engine: French oscillating cylinder (vibrating) engine
Ratio of tonnage to cargo: Approximately 2:1

While legal wrangling took place on the Mississippi, a concern of far greater national importance had occurred. America and Great Britain were at war, and Andrew Jackson commanded the strategic city of New Orleans. Much taken with the prospect of utilizing the power of a steamboat (and more than a little impressed with Henry Shreve's reputation as a man of daring), Jackson enlisted the captain's services to navigate upstream and meet a contingency of keelboats floating down the Mississippi with much-needed supplies. Shreve took the *Enterprise* and met the boats, returning with their cargo in six and one half days, covering a combined distance of 654 miles.

General Jackson then directed Captain Shreve to take his boat and deliver supplies to Fort St. Philip. This he did with aplomb, but on the return trip his presence was discovered by the British. They fired on the *Enterprise*, but her deck had been fortified with cotton bales as protection, so what few cannonballs that did strike inflicted little damage. Shreve and the *Enterprise* also participated in the Battle of New Orleans on January 8, 1815, not only earning accolades for her captain, but proving the worth of steam-powered vessels during times of armed conflict.

On May 6, still in conflict with the patent law, the *Enterprise* departed New Orleans for Brownsville, reaching Louisville in twenty-five days. From there, she completed her journey in fifty-four days, becoming the first steamboat to travel up the Mississippi and Ohio rivers.

In 1816, when Shreve finally came up for trial, a state court (Louisiana had entered the Union in 1812) held that the Territorial Legislature had exceeded its authority in granting exclusive rights to the Fulton-Livingston Corporation. Thus ended the monopoly on western waters, and the court's decision may fairly be said to have opened the floodgates of the steamboat era.

A fascinating account of the entire affair from the *New Orleans* to the *Enterprise* was stated in the Burlington (Iowa) *Hawk-Eye* on August 7, 1851:

> The first steamer on the Ohio River, was built at the Pipetown ship yard in Pittsburg, in the fall of 1811. She was intended for the Pittsburg and New Orleans trade, and was called the "Orleans." She was built after the fashion of a ship with port holes in the side — long bow-sprit — painted a sky blue. Her cabin was in the hold. She left in November 1811, for New Orleans, and made the trip down in safety, but was never able to get back over the falls her power being insufficient to propel her against the strong current. Messrs. Winton and McGranaghan of New Port, Ky, in communicating their information on this subject to the *Cincinnati Commercial* say:
>
> "Many persons are of the opinion that the 'Enterprise' was the first boat built for the above trade. Such is not the fact. The 'Enterprise' was the fourth or fifth boat built. The names of the others were 'Aetna' and 'Vesuvius,' built by a company who had a charter for eleven years, renewable for the sole navigation, by steam, of the Ohio and Mississippi rivers. The 'Enterprise' was built at Brownsville by a private company, and on her arrival at New Orleans was attached for an infringement on the chartered rights of the company. A legal investigation followed, and the owners of the 'Enterprise' gained the suit by proving that the plaintiffs had violated their charter. Thus ended the steamboat monopoly on the Ohio and Mississippi rivers."

The "War" between the Northern Atlantic States: No Middle Ground

The foundation of the Fulton-Livingston monopoly in New York rested on the state's supposition that the boundary of the Empire State extended over the waters of the Hudson River up to the low-water mark on the New Jersey shore. This stance was in opposition to common law, which held that when water served as a territorial line, the midpoint of the river became the dividing point. Matters quickly deteriorated, with New York maintaining an aggressive position to maintain what it saw as its rights pertaining to steamboats. The Trenton *True American* (reported in the *Centinal* of March 6, 1811) stated:

> The legislature of New York having by law subjected to seizure and confiscation any Steam Boat owned by citizens of this state [New Jersey] which may be found navigating the waters of that state, the legislature of this state, as called upon by every principle of honor and justice, have passed *unanimously,* an act by which any such seizure and confiscation by citizens of New York of any Steam Boat owned by citizens of this state, the Steam Boats owned by citizens of New York are made liable to seizure and confiscation by citizens of New Jersey if found within our rivers or harbours. A New York paragraphist deems this act a form of "Letters of Marque and Reprisal." Whatever it may be, it is not unprovoked, unjust and arrogant as the law of that state. If their law had not been passed, ours would never have been thought of. If theirs is not carried into effect, ours is a dead letter. But our neighbors must not think that because Jersey is a little state, she will tamely submit to imposition and robbery. While she will be the last to invade the rights of others, she will be the first to maintain her own.

New Jersey was not alone. On June 19, 1822, the *Clarion* reported:

> The legislature of Connecticut retaliatory law, prohibiting the navigation of the waters of that state by the New York steam-boats. It is much to be regretted that the intercourse between the different states should be interrupted by such restrictions. We are of the opinion, that states possessing equal rights, ought to enjoy equal privileges; and until it is fixed upon terms of reciprocity, it will be a fruitful source of contention between citizens of the neighboring states. And we hope New York will repeal or so modify her steam-boat laws, as to put an end to these controversies.

Ultimately, state boundaries did not come into play when deciding the fate of the monopoly. This matter was taken to the Supreme Court in a separate, unrelated case in 1829. The justices appointed a commission to determine the actual lines, fixing the boundary in the middle of the Hudson River. New York and New Jersey ratified the decision in 1834, but that was not the end of the issue. Since the term "middle" was not defined to anyone's satisfaction, the controversy arose again in 1870, when the Central Railroad filled in land along the Hudson at Communipaw, changing the dynamics of the river. Not until 1888 did commissioners from both states exactly pinpoint the boundary lines.

The actual case that ended Fulton and Livingston's exclusive rights began with Colonel Aaron Ogden. He desired to run a steam-powered ferry service between New York and Elizabethtown, New Jersey. Ogden and his partner, Daniel Dod, built a steamboat named the *Sea Horse* and put it into service. Fulton won an injunction, but Ogden, who by this time had become governor-elect of New Jersey, pushed through a resolution granting him exclusive rights to run steamboats on New Jersey waters. Fulton again brought the case to court (Chancellor Livingston having died on February 26, 1813), and the New Jersey law was repealed on February 4, 1815.

Ogden eventually purchased the rights from Robert Fulton and the Livingston estate to operate his line, but soon ran into his own difficulties. Thomas Gibbons, a Southerner known to have a penchant for duels and a hot temper, decided to establish a line with his own steamboats, the *Bellona* and the *Stoudinger*. Employing a young Cornelius Vanderbilt as one of his captains, Gibbons set himself in opposition to the legally authorized steamboat service. Ogden and Gibbons proceeded to sue each another.

Ogden hired Daniel Webster to defend him, and the lawsuit worked its way to the Supreme Court. In 1824, Chief Justice John Marshall spoke for the majority in the case of *Gibbons v. Ogden* by stating that only Congress had the right to regulate interstate commerce. The New York State Legislature responded the following year by revoking all monopoly rights on the state's internal waterways, thus terminating the exclusive privileges that had been granted twenty-seven years earlier.[14]

4

The Meschasipi — Or Is That the "Mississipi"?

Whether spelled phonetically (1697), or without the last "p" (until 1750), by its Indian names (Chucgua, Tamaliseu, Taputa, Mico), Spanish names (the Rio Grande, Palisado, Escondido), or French names (St. Louis, Conception, Baude, Colbert), the vast, sinewy waterway on the border of the American frontier played a dominant role in westward expansion. Known as "Father of Waters" and meaning "Great River" in Ojibway, the term "Mississippi" originally referred solely to the headwaters. French missionaries and fur traders carried the name downstream, until the word and its present-day spelling became the accepted form.

The American waterways comprised 16,000 miles, making them the greatest inland system in the world.[1] A huge part of this belonged to the Mississippi, which traveled southward 660 miles from the Falls of St. Anthony to the mouth of the Missouri River (contributing 19 percent of its volume), which becomes its main tributary. The river continues a 12,000-mile journey from St. Louis, cutting through what would become the heart of the country. At Cairo, the Mississippi is joined by the Ohio River (contributing 31 percent of its water) at a point halfway to the sea. Meandering in crooked-string fashion past such historic towns as Memphis, Vicksburg and Natchez, the waters reach fabled New Orleans. The final journey of 105 miles takes it to the Gulf of Mexico, where the river empties.

Early reactions to the Mississippi were as varied and individualistic as the reporters, themselves. Frances Trollope (who plays a significant role in "An International Incident," chapter 20) had little good to say in 1827, as she approached the Gulf of Mexico from the Atlantic after a voyage of seven weeks and two days from London:

> The first indication of our approach to land was the appearance of this mighty river pouring forth its muddy mass of waters, and mingling with the deep blue of the Mexican Gulf. I never beheld such a scene so utterly desolate as this entrance of the Mississippi. Had Dante seen it, he might have drawn images of another Bolgia from its horrors.[2]

Captain Frederick Marryat, another early traveler, wrote in 1837 that the Mississippi

> is a river of desolation; and instead of reminding you, like other beautiful rivers, of an angel which has descended for the benefit of man, you imagine it a devil, whose energies have been only overcome by the wonderful power of steam.[44]

In 1842, a well-known Englishman, Charles Dickens, made a historic visit to the United States. After traveling through Boston, New York, Philadelphia, Washington, Pittsburg and Cincinnati, he came to Cairo, where the Ohio flows into the Mississippi. Dickens' reaction was also less than flattering:

> But what words shall describe the Mississippi, great father of rivers, who (praise be to heaven) has no young children like him! An enormous ditch, sometimes two or three miles wide, running liquid mud, six miles an hour: its strong and frothy current choked and obstructed everywhere by huge logs and whole forest trees: now twining themselves together in great rafts, from the interstices of which a sedgy, lazy foam works up, to float upon the water's top; now rolling past like monstrous bodies, their tangled roots showing like matted hair; now glancing singly by like giant leeches; and now writhing round and round in the vortex of some small whirlpool, like wounded snakes. The banks low, the trees dwarfish, the marshes swarming with frogs, the wretched cabins few and far apart, their inmates hollow-cheeked and pale, the weather very hot, mosquitoes penetrating into every crack and crevice of the boat, mud and slime on everything: nothing pleasant in its aspect, but the harmless lightning which flickers every night upon the dark horizon.[4]

It took an American, perhaps, to paint in words the majesty experienced by those who cherished the river. The Huron *Reflector,* on July 12, 1836, reprinted the following for its readers:

> The letters from Mr. Brooks, which have been so extensively copied, abound in striking passages. One of these, from a letter written it Italy, is annexed:
> "One of the grandest views, if not the sublimest in the United States, it strikes me is the junction of the Mississippi and the Ohio, not that there is aught in itself in the scenery so very astonishing, but the idea of extent and power inflame the imagination there, when you see a stream of about 3000 miles long mingling with another of about 1200 miles, and the greater absorbing the less, which of itself is a mighty river, without even swelling its stream, or widening its banks, as it seems to the eye, and both then to course on together a thousand miles more to meet the ocean! The almost boundless extent, and the awful but silent power of the seemingly lazy current, affected me more than the noise of Niagara, because extent was associated with power."

From Mark Twain's inspired pen came a more poignant description:

> I had myself called with the four o'clock watch, mornings, for one cannot see too many sunrises on the Mississippi. They are enchanting. First, there is the eloquence of silence; for a deep hush broods everywhere. Next, there is the haunting sense of loneliness, isolation, remoteness from the worry and bustle of the world. The dawn creeps in stealthily; the solid walls of black forest soften to gray, and vast stretches of the river open up and reveal themselves; the water is glass-smooth, gives off spectral little wreaths of white mist, there is not the faintest breath of wind, nor stir of leaf; the tranquility is profound.[5]

In equally moving words, Twain describes another scene.

> I still keep in mind a wonderful sunset which I witnessed when Steamboating was new to me. A broad expanse of the river was turned to blood; in the middle distance the red hue brightened into gold, through which a solitary log came floating, black and conspicuous; in one place a long, slanting mark lay sparkling upon the water; in another the surface was broken by boiling, tumbling rings, that were as many-tinted as an opal.... There were graceful curves, reflected images, woody heights, soft distances; and over the whole scene, far and near, the dissolving lights drifted steadily, enriching it, every passing moment, with new marvels of coloring.[6]

Reflecting on his youthful position as cub apprentice, Twain regretfully wrote that science and technical competence demanded his attention, driving "All the grace, the beauty, the poetry ... out of the majestic river!"[7] Anyone who has moved from the awe of mystery to the intricacies of erudition can fully sympathize. Yet it is to be hoped, after reading the above passages, that the spark of inspiration and love occasionally found its way back into Mark Twain's heart.

Obstructions: Every Voyage an Adventure and Every Pilot a Gambler

Unlike European or inland waterways of the eastern United States where old rivers flow placidly down wide, calm channels, making steamboat navigation relatively easy, the western rivers were another matter entirely. These young, wild passages were filled with dangers undreamed of by rivermen working the Hudson and Great Lakes regions.

On the Missouri and Lower Mississippi, wide, flat lands extend beyond the banks, sometimes for fifty to one hundred miles. Composed of loose alluvial material, these constitute a natural floodplain, where soil has been deposited from floodwaters. The lack of established riverbanks or high rock bluffs hemming in the waters creates an ever-changing landscape, which in turn enhances the speed of the current. Without warning, the rivers may add to their length by cutting huge oxbows through the soft loam and sand of the plains; at other times, the raging waters shorten many miles off their course, as rising floods break over the narrow necks of wide loops, forming cut-offs. These changes bring with them numerous types of obstructions, creating an incredibly unstable environment.

The Huron *Reflector,* July 8, 1831 reported:

> It is stated, that the length of the passage up the Mississippi, from New Orleans, has been shortened 42 miles, by cutting off two large bends in the river — one opposite Red River last winter, and the other 200 miles above Natchez, this spring.

River obstructions came in many shapes, degrees of permanency and potential dangers. The most obvious were the creation of islands, appearing with great regularity. These Channel Islands (so called for their appearance in the body of the rivers), were considered the least formidable of above-water obstructions. Easily spotted by the steamboat navigator, islands affected the river by dividing a section of stream into two separate passageways. Of equal importance, sand bars formed at the heads and feet of the newly created islands. These were not so easily gauged, and it became the pilot's duty to avoid the bars while navigating through the least dangerous and generally deeper of the two channels.

It was estimated that between Pittsburg and the mouth of the Ohio, there was an average of 98 islands. Between the mouth of the Ohio to New Orleans, there might be as many as 126 at any given time. On the Upper Mississippi, it was estimated that from St. Paul to the Illinois River (a distance of 700 miles), there existed 526 islands. Over the course of years, these islands came and went, some disappearing through erosion and others becoming more or less permanent. During the growing season, first shrubs, then trees would sprout. The longer the island remained constant, the more overgrowth occurred, until entire woods grew in the middle of the river. Many were actually occupied by wooders or farm families, who found the soil conducive to crop cultivation.

These Bars Are Not for Drinking

Lesser elevations (whether above or below the surface) were given the catch-all name of "bars." Those hidden from view, or not easily detected by an experienced eye, were the bane of steamboat men. Bars were constantly shifting according to current and precipitation and extended at an angle to, or across, the general direction of the river's course. Higher bars presented difficulty in navigation lower ones were only visible at the lowest stages of water.

Naturally occurring obstructions were obvious deterrents to navigation. Of far greater number than any found on the eastern waterways, they are categorized into three specific types:

Sandbars: These were the most common, being the largest of river obstructions. They formed in the lower portions of both tributaries and trunk lines. Sandbars formed at the head and foot of each island, at the mouth of every significant stream, and at places where the current slowed down enough to permit the sand and silt to drift out of solution and settle on the bottom.

"Grounding," or getting stuck on a sandbar, required immediate action. The delay of even an hour caused shifting sand to form around the hull, making it much more difficult to extricate.

Gravel bars: These were either hard and permanent or loose and shifting. Bars composed of gravel were denser than sandbars and, when struck by a steamboat, caused more damage to the hull. Gravel bars were primarily found on the Upper Mississippi and Ohio rivers.

Rock bars: These were caused by strata tumbling into the river from the high, overhanging bluffs on the Upper Mississippi or Ohio rivers. (Most of the Lower Mississippi was bordered by flat floodplains, so rock bars were not usually encountered in that region.) If a steamboat struck one of these obstructions, the jagged rock could well tear through the hull and sink the boat.

The three main types of bars could also form:

Chains: A series of rock bars.

Reefs: Long sandbars, that rose abruptly on one side. Low water is at the head of the reef, with easy water to the sides. A boat did not like to mount a reef; boats "hated" shoal water, and it required fine handling to navigate past or over a reef. A "little reef" was indicated by fine lines that spread out like ribs on a fan.

Bluff Reefs: These were straight up-and-down reefs, surrounded by plenty of water but with very little over them.

Wind Reef: These were not actual reefs — they were caused by the wind, yet looked exactly like a bluff reef. They could be recognized by instinct and experience.

More prominent bars on the upper rivers were fairly stable and were encountered season after season. Pilots quickly learned the quirks of navigation through these areas and were prepared to face the problems they represented. The more prominent bars achieved reputations and many of the permanent bars were given names.

At high water, bars were fully submerged and presented no serious obstacles to steamboats. During periods of low water, however, bars became, in effect, dams, serving to good effect by creating a series of pools two to three miles in length, and, at times, much longer. While the rest of the river was too low for the deeper draught boats to safely traverse, a skilled pilot could steer a smaller vessel from one pool to the next, benefiting from the greater depth and slow-moving current. During dry seasons, water over the bars was measured at no more than several inches, and the current passed over them at a sluggish speed of two to four miles per hour.

The art of low-water navigation consisted of devising ways of getting steamboats over river bars and into deeper water. Sandbars could be cut into and portions made wide enough for a boat to pass over, but this negated their damming effect, lowering the depth of surrounding pools and thus offsetting the benefits.

"Sparring off" was how to get a boat over a bar. The technique required an intricate knowledge of where to place the spars (great poles of straight-grained pine), in which direction to shove the bow, and whether to "walk her over" by setting the spars at a "fore and aft" angle (one at each side), and going straight ahead. By the 1850s, the procedure was accomplished by using tackle wound around the capstan and attached to the spars. A steam-driven hoisting-engine ("donkey") was turned on and the power used to maneuver the boat off the reef.

Falling Over the Ohio

The sudden appearance of rapidly moving water, or worse, a steep drop in elevation of the river presented unique and difficult challenges to navigation. These fell into three categories: ripples, rapids and falls.

Ripples or shoals: Water crossing large sand or gravel bars with a greater-than-ordinary slope. During low water, they assumed the character of rapids.

Rapids: The most formidable class of obstruction, being a series of more or less connected rock formations over which the river crossed at a great rate of speed. Rapids increased in number as a boat went upstream, and the severity of rapids determined the end point of steamboat navigation, beyond which a vessel could not pass. The four major rapids on western rivers were the Falls of Ohio, the Muscle Shoals on the Tennessee, and the Des Moines and Rock Island rapids on the Upper Mississippi.

Falls: The greatest obstacle on the Ohio River was the Falls of Ohio at Louisville, where the river fell twenty-two feet over a succession of rock ledges, covering a two-mile distance. The Louisville and Portland Canal opened in 1830 and provided relief for conveying steamboats over the falls, but within several years of its construction, technology had progressed so rapidly that the lock chambers were too small for the larger classes of vessels. Steamboat owners also considered the toll charges excessive, and often they were forced to hire local pilots to get their boats over the falls, incurring additional expenses.

Snag, You're It!

Permanent obstructions such as large rocks and boulders presented ongoing trials for steamboat navigation. These were charted and their appearance well known among river pilots. Other, less permanent obstacles fell into another category.

Hulks: Particularly in the early years of steamboating, many vessels were sunk in

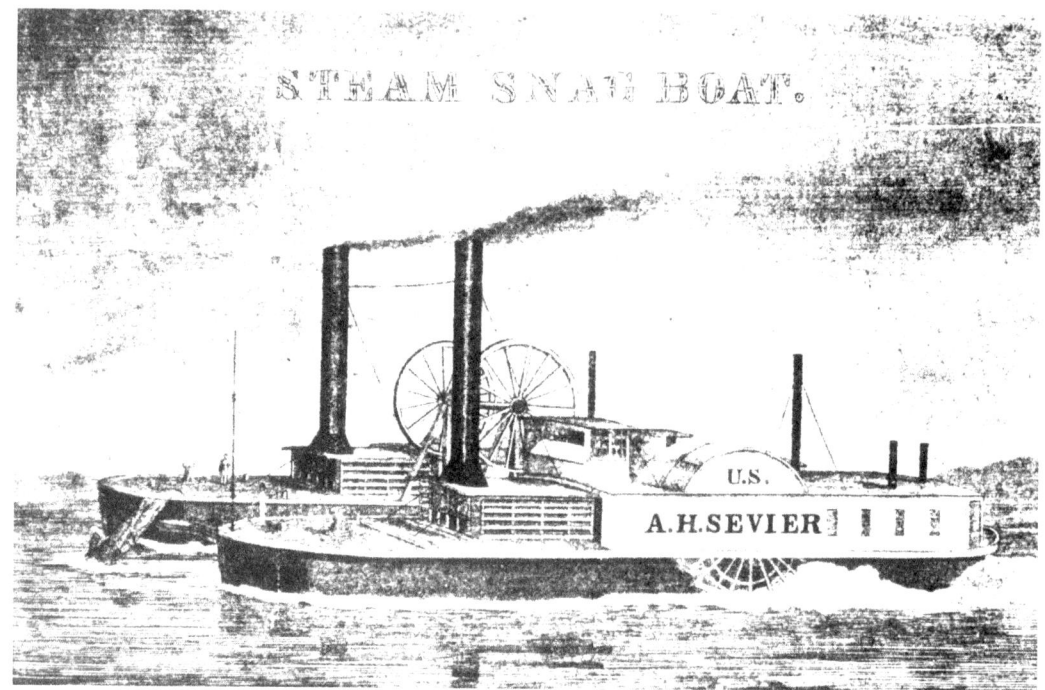

A steam snag boat, the U.S. *A.H. Sevier*. This is an early image demonstrating the unique construction of the vessel. A.H. Sevier was a delegate to Congress from Arkansas.

the rivers of the Lower Mississippi. Salvage operations were seldom done on these wrecks, and if they were, most involved only the removal of freight or the engines. The wooden carcass was abandoned. Time and current rendered most of the structure under water, where the solid frame became a serious danger to passing boats. A stretch of the Mississippi between St. Louis and Cairo became known in local parlance as "The Graveyard," due to the many steamboat hulks buried there. By 1875, there was an average of one sunken vessel per mile along that route.

Driftwood: Most likely encountered during periods of high water, these tree branches were found in the deepest areas of the river where the current was fastest. Their appearance was impossible to anticipate and harder to avoid. Over the course of a journey, a steamboat might strike any number of large limbs, causing extensive damage to the paddle-wheels. Driftwood was a prime factor in the short life of western steamboat hulls. The Sandusky *Clarion* of April 16, 1823, gives an account of a steamboat sunk by driftwood:

The steamboat Alexandria, in ascending the Mississippi on the night of the 1st of February, struck upon some drift wood, which caused her to leak so badly as to render it necessary to run her on a sand bar. She gained it in a sinking condition, and in five minutes later must have gone down without any hope of escape for those on board. An unfortunate circumstance occurred before the vessel was completely abandoned: One of the passengers, who remained behind to attend to some property he had on board, remarked to the mate that the vessel ought to be secured by cable, to which the mate replied that it was none of his (the passenger's) business — Some irritating words on both sides followed, till at last the passenger drew a sword from his cane and stabbed the mate in many places — it is not stated whether mortal or not.

Snags: Trees grew in abundance along the banks of western rivers. During flood periods, when water levels rose over their root systems, the soil washed away, destroying their foundations. This occurred with great frequency, causing hundreds of large, fully grown trees to fall into the river. After becoming waterlogged, the heavy trunks, stripped of smaller limbs and leaves, sank to the river bottom, where the heavy, gravel-encrusted root system became lodged. The lighter upper portions, buoyed on the rising water, emerged in the shape of a lance. If struck, the solid wood could easily pierce the hull and do great damage, including sinking the vessel. A snag, lying with the point downstream, was commonly referred to as a "sawyer," from the slow vibration, or sawing motion given it by the current. An article from the Madison (Wisconsin) *Express* of June 9, 1841, vividly describes such a situation:

Remarkable Incident

The steamboat Messenger arrived at our landing yesterday, from St. Louis. From Captain Beard, and from some of the passengers, we have learned the particulars of a most singular incident, uniting somewhat of the ludicrous with the serious.

On her passage up she was snagged near Shawneetown, Illinois, at 3 o'clock, last Saturday morning. The snag went through her starboard guards; forward of the wheel houses and near the main entrance to the cabin, up through her state rooms, three of which were demolished. It struck under the feet of two of the occupants, threw them out of their berths up against the door, which being locked inside, they could not move, after being thus abruptly aroused from their slumbers. The pantaloons of one, which had been thrown across his feet when he undressed, caught on the top of the snag; and when the boat stopped, were found dangling ten feet above the hurricane deck. At the time of the accident, a deck passenger fell, or jumped overboard in the fright, and being a good swimmer made for shore. Here in the darkness of the night, he found the banks too steep and slippery to land, and was obliged to swim one hundred yards or more down the stream, where he effected a landing; and in a few hours was taken on board the Wm. French. He overtook the Messenger at Louisville, his fellow passengers having given him up as drowned at the time of the accident.

The snag was a long black walnut log upwards of a foot in diameter. No damage having been done to the boat's machinery, she was under way again in a few minutes.

Snags were most often found in the numerous bends in the river where the current was rapid and erosion ate away the riverbanks. The Missouri, Lower Mississippi, Arkansas and Red rivers were unusually susceptible to the development of snags. This type of obstruction accounted for the vast majority of steamboat losses before 1826, and river obstructions in general caused three-fifths of all damage until 1849.

"White snags" were nearly invisible at night until a boat was directly upon them, and the same held true for "white logs." "Black logs," on the other hand, were more readily identifiable and easier to avoid.

Uncle Sam's Tooth-Pullers

Prior to 1829, there was nothing a steamboat pilot could do but keep a sharp lookout for snags and report new appearances to his fellow navigators. Snags remained an impediment until natural decay finally wore away the wood. In 1821, Henry Shreve (the famous riverman who ran his steamboat in the Battle of New Orleans and helped break the western Fulton-Livingston monopoly), began plans for a "snag-boat," a type of raft that he patented in 1838. His efforts culminated in 1829 with a finished prototype.[8]

Shreve called his vessel the *Heliopolis*. It was steam-driven and functioned like a battering ram. Striking a snag with the snag beam, the force tore the roots from their sand mooring, enabling the log to be hoisted from the water. Snags as heavy as 75 tons were thus removed from the paths of navigation by being dumped in a deep pool, or brought toward shore and deposited. Those snags too heavy or cumbersome for transport were chopped up by the crew and thus disposed of along the bank, out of harm's way.

Justifiably famous, the Red River rafts that Shreve developed were affectionately called "Uncle Sam's tooth-pullers." These snag boats were the only significant river improvement created until the post-bellum years. The *Adams Sentinel*, reprinted on June 20, 1836, an article from the *Western Telegraph*) that describes the system:

> *Uncle Sam's Tooth Puller*—This is the principal name which the boatmen have given to the public boat employed by Mr. Shreve in clearing the Ohio and Mississippi rivers of snags, sawyers and planters. The boat is of great strength; formed of two keels, between which the dental operation of snag pulling is performed. It is moved by an engine; the snags that are too firmly embedded, not to admit of extraction, are operated on with saws under water—This boat has been of immense service upon the river, removing the obstructions, dangers and risks of navigation; and it is not unaptly called, "Uncle Sam's Tooth Puller."

The Red River Raft

Of all western rivers, the Red was the most obstructed with debris. So great were the obstructions that they developed the name "raft," implying that the massive accumulations floated upon the river in a manner similar to a vessel of the same name. The Delaware *Patriot and American,* November 4, 1828, ran a fascinating excerpt from the Little Rock *Gazette,* written to Dr. Joseph Paxton of Hampstead County, Arkansas Territory, by the Hon. A. Sevier, delegate to Congress from Arkansas. In the text, the author describes the development of the infamous "rafts" which so plagued navigation on that inland waterway.

> At the time of the junction of the two rivers [the Red and the Mississippi, some three hundred years ago], which was probably when they were in flood, an eddy was made extending up the Red River, forty or fifty miles, as is now the case every spring, and the enormous quantities of drift wood which had been carried off by the current, were deposited at the head of still water, and the Red River, subsiding first or both together, no current was produced by the fall, and the drift timber was lodged in the bed of the river. By the arrival of another fresh the wood had become thoroughly soaked so that it would not float, and at the same time had gathered leaves and sediment so as to form a complete dam, impelling the water backwards still further, and arresting in its course the floating rubbish of another year.
>
> Such is the enormous quantity of brush, trunks of trees &c., that the deposits may be supposed to have gained at least one mile per annum, which it does at the present time. In its progress up the river, it frequently throws back the water upon the land for many miles, and makes a lake of what was before a prairie. The forests, too, which in some places line its banks, are often killed by the overflow of water, and after standing for a few years with their roots submerged, the trees become rotten and fall: the first fresh that

comes, they are propelled forward to the raft and thus aid in the destruction of other forests, which before were beyond the reach of the water.

Only thirty five years ago, the Bordeau Lake was a beautiful prairie, and probably several other large lakes in that vicinity were formed in the same manner. The raft has now reached a distance of three hundred miles from the place where it is supposed to have commenced. It is gratifying, however, to know that, as it is forming above, it is decaying below. While the memory of persons living in Nachitoches, the lower end of the raft was below that place, now it is sixty or seventy miles above. Almost every spring a portion of the lower end is carried away by the flood, all the timber except some of the more durable kinds have decayed and left the mass without any thing to hold it together. From these loose materials a new raft is frequently formed below in some narrow part or bend of the channel.—"This accounts," says Dr. Paxton, "for the rafts that choke up the old river below Nachitoches, for those that remain in the Rigolet de Bondieu, and also from those in the Achalalays. These immense floating rafts are frequently seen passing the settlements below, apparently in nearly the same situation as when they started. No one here is at a loss correctly to account for them, as it is well known no drift wood ever passes through the great raft."

The great raft is now eighty miles long. It will not be supposed that the river finds its way through or over this enormous mass of lumber. On the contrary, the river, when the water has risen to a level with the banks, finds a new channel at the next bend above, because the bends are nearly the margin of the valley which is generally lower than the bank of the river.—"As soon as an opening is made, the current stopped below runs thither with rapidity, accompanied with the driftwood which, while the opening is enlarging above, begins soon to lodge against and fill up the lower side, and even to extend itself diagonally across the river to the opposite bank. The new raft thus formed continues to accumulate, and becomes more impervious, until it again causes another opening above, and again in the same way, another raft so commenced." In this way the raft is constantly shooting off branches diagonally on both sides as it accumulates, and the boats that pass down the river find their way as they can through the side openings that occur, and through the lakes that are formed by the raft changing the outlet of tributary streams and occasionally empty into the Red River.

It is no wonder Henry Shreve's "tooth pullers" were sent to the Red River. That he had near-immediate success in breaking up the notorious rafts is revealed by the Huron *Reflector*, March 28, 1831, only two years after the snag boat was patented.

> *Remarkable change in the channel of the Mississippi.* — By the steamer Brandywine, arrived yesterday, we have the confirmation of a long expected cut off in this river.
>
> At the well known curve opposite the mouth of the Red river, the current burst across the neck of land by the aid of Capt. Shreve, and has made a channel four hundred yards wide and about four fathoms deep already. Flat boats and rafts have descended through it with tremendous velocity, and the Brandywine would have done the same but for business on the other route. The distance between this and Natchez is thus shortened about 30 miles by one small cut off; and should the same thing occur at the Turnica bend, (which is extremely probable) the distance will be lessened 100 miles! — *New Orleans Bee*.

An article from the *Adams Sentinel*, March 13, 1832, describes the raft problem on the Mississippi:

> *Immense Rafts on the Mississippi.* — One of the most interesting features of this river is the enormous rafts of drift timber it floats towards the sea, occasionally depositing them for a time, together with vast beds of mud and gravel, in some of its deserted channels. One of these rafts is described by Darby, in 1816, as *ten miles* in length, about two hundred and twenty yards wide, and eight feet deep. It is continually increasing by the addi-

Snagboat "AID", a few miles North of Shreveport, cleaning up remnants of the great Raft – 1873

This is an 1873 photograph of a typical snag boat. This is the next generation of snag boat developed by Henry Shreve that was used to such good effect on the Red River.

tion of fresh driftwood, and rises and falls with the water on which it floats — evidently waiting only an extraordinary flood to bear it off into the Gulf of Mexico, where far greater deposits of the same kind are in progress at the extremity of the delta.

Opposite the opening of the Mississippi large rafts of drift timber are met with, matted into a network, many yards in thickness, and stretching over *hundreds of square leagues!* They afterwards become covered with a fine mud, on which other layers of trees are deposited the year ensuing, until numerous alternations of earthly and vegetable matter are accumulated. The geologist will recognize in this relation of Darby the type of the formation of the ancient lignites and coal-fields.

Continued proof of the snag boat's success poured in. The Milwaukee *Sentinel,* February 13, 1838, reported:

Red River Raft.—A letter in the New Orleans Bulletin states that the snag boat passed Shreveport about the 29th ultimo, for the purpose of removing the remainder of the raft which will probably be effected in a few weeks. This will open an uninterrupted navigation above Shreveport on the Red River, of 15000 miles—on the Sulphur Fork, 200—on Little River 200—on the Blue River, 159—on the False Washita, 300-making 2300 miles of navigable waters above this place, through, probably, the finest country on the globe.

4. The Meschasipi — Or Is That the "Mississipi"?

Snagboat "AID", a few miles North of Shreveport, cleaning up remnants of the Great Raft – 1873

Snag boat at work, 1873. The debris in the view is typical of what a snag boat was required to break up in order for steamboats to have an unobstructed passageway. Snags were a primary source of danger along the western rivers, causing considerable damage and loss of life, particularly in the early years.

The *Republican Compiler,* May 1, 1838, followed up with:

RED RIVER RAFT. The New Orleans Courier mentions the receipt of a letter from Captain Shreve, dated 29th March, in which he states that the Raft was then cleared away, and the navigation easy and uninterrupted. He ascended the stream through the whole extent of the raft, a distance of 52 miles, in 9 hours. There is sufficient depth of water for any steamer that can navigate Red River. Ten feet are found in the shallowest places. *Baltimore American.*

Shreve performed a staggering improvement to western navigation by opening the Red River, but not all was completely well with the work of his snag boats, proving that no success ever satisfies everybody. The Bangor (Maine) article of October 25, 1838, notes: "The low state of the water has greatly increased the dangers of navigation on all the lower Mississippi when the stumps or snags cut off by the snag boats, and which generally are not perceivable on the surface, are calculated to do serious injury."

Henry Shreve was a monumental participant in the development of steamboats and consequently, river improvements in western waters. Appointed by President John Quincy Adams to the post of superintendent of western river improvements, he held the position for fourteen years. His greatest achievement during this period was the removal of driftwood that blocked 160 miles of the Red River in Louisiana. In 1835, he helped establish a port in that state, and four years later the town adopted the name Shreveport.[9]

His death in 1851 occasioned a fitting obituary, reprinted in the *Adams Sentinel and General Advertiser,* March 24, 1851:

> The St. Louis papers announce the death of Captain Henry M. Shreve, who, during the administrations of Adams, Jackson and Van Buren, fitted the important post of U.S. Superintendent of Western river improvements, and by the steam snag-boat, of which he was the inventor, contributed largely to the safety of Western commerce. To him, says the Republican, belongs the honor of demonstrating the practicality of navigating the Mississippi river with steamboats. He commanded the first steamer that ever ascended the river; and made several and valuable improvements both of the steam engine and of the hull and cabins of Western steamboats. Whilst the British were threatening New Orleans in 1814–15, he was employed by Gen. Jackson in several hazardous enterprises, and during the battle of the 8th of January, served one of the field pieces which destroyed the advancing column led by Gen. Kean.

Life Spans

Because of the hazardous conditions of western waters, the viable life span of a steamboat was greatly reduced, especially in comparison to other vessels.

Life Spans

Whaling ship	40 years
Sailing ship	20 years
Mississippi steamboat	4–5 years
Missouri steamboat	3 years

With attrition so high, there was usually no point salvaging any wooden parts from western steamboats. Even the boilers and engines suffered from wear and tear, although that seldom prevented them from being saved whenever possible and the engines, especially, placed in new boats.

Whether a newcomer to the western waters considered steamboats sights of inspiration or harbingers of death, Americans and foreigners by the thousands traveled aboard these vessels. Some found new lands to farm; others took jobs in the river towns. Many perished from accident, whether by unseen obstructions or boiler explosions. Disease swept up the rivers; economic conditions made millionaires overnight and bankrupted the rich and poor alike with equal disdain. Yet nothing stemmed the westward tide, and the trust the settlers placed in pilots to navigate their steamboats around obstructions, over rapids and across channels spoke of the faith that characterized the vast, unprecedented western migration.

5

"St. Looy" and the Upper Mississippi

St. Louis, established by Pierre Laclede Liquest in 1764 as a base for French fur trading operations, sits on the Mississippi near the confluence of the Missouri River. On April 9, 1804, Spain transferred the Upper Louisiana Territory to France, and a day later, France turned the land over to the United States. By 1808, this American city had a population of nearly 1,000, most of whom were involved in the lucrative fur trade. Fur trading proved a significant incentive for the development of Upper Mississippi River navigation because early settlers and trappers required the protection of the government from hostile Indian tribes — and vice versa.

In order to develop a more complete understanding of the Mississippi River and its tributaries, which were vital to the fur trade, Zebulon Pike set out from St. Louis in 1805, attempting to discover the headwaters of the great river. He reached as far as Leech Lake, in northern Minnesota, on January 31, 1806, before turning back. He explored farther than any other adventurer but found, as others had before him, that river passage upstream on keelboats (and later steamboats) was extremely difficult due to stiff currents and dangerous, if not deadly obstructions.

Not until 1820 was another attempt made. Lewis Cass, soldier and politician (he was appointed governor of the Michigan Territory in 1813 by President James Madison) and a group of thirty-eight traveled upriver with the same goal as Zebulon Pike. More significant than exploration for the sake of knowledge, their task was to define the then unidentified border between the United States and British North America.[1] Cass and his party, which included geologist Henry Schoolcraft, entered what became known as Cass Lake, incorrectly identifying that area as the source of the river.

It took another decade and a more pressing reason before Henry Schoolcraft, now an Indian agent, was sent by the United States government "to proceed into the Chippewa country, to endeavor to put an end to the hostilities between the Chippewa and Sioux."[2] Low water prevented Schoolcraft from getting beyond the St. Croix River.

In the spring of 1832, the government again ordered Schoolcraft to "proceed to the country upon the heads of the Mississippi, and visit as many of the Indians in that, and the intermediate region, as circumstances will permit." His party entered the Mississippi from Lake Superior via the St. Louis and Sandy Lake River. At Cass Lake, a Chippewa guide directed a select party up the east fork of the Mississippi, where they finally arrived at Itasca Lake — the source of the Mississippi."[3]

It well behooved the United States government to make treaties with the Native Americans, for the opportunities that white immigrants saw in the territory drew them forth in unprecedented numbers. John Jacob Astor was one who took advantage of the new land. In 1811, he sent his representative, Wilson Price Hunt, from St. Louis to establish Astoria, while the South West Company had already established fur trading posts on the Upper Mississippi. After 1822, this company, together with a rival organization, dominated the commerce of nearly the entire Indian region.

The United States' relationship with the Native Americans was complex and fraught with difficulties, if not outright deception. Numerous treaties were made for the rights to the land, many under protest. As payment, the government promised regular shipments of foodstuffs, merchandise and other necessities. In order to provide these goods, a series of congressional legislations were passed, called the Indian Intercourse Acts. The Factor System, already established by the Spanish and French, was officially sanctioned, the original intent being to protect Native Americans from exploitation.

The term "factory" comes from a Latin word meaning "trading post." The head of such a post was called a "factor." Factories sometimes served as forts, defending Indians and colonists from attack. They were also settlements that included a blacksmith to forge and repair farming utensils and a milling operation to make flour from locally grown wheat.[4]

Factories were usually constructed along waterways, and were served by keelboat. It was backbreaking work hauling tons of supplies to out-of-the-way factories, and the arrival of the *Zebulon M. Pike,* the first steamboat to reach St. Louis, on August 2, 1817, was hailed with great enthusiasm. While steamboating had already become a proven method of transportation on select portions of the Lower Mississippi, their late appearance as far north as St. Louis was occasioned by a number of important factors.

The "Great American Invention"

Prior to the breaking of the Fulton-Livingston monopoly, running an independent steamboat ran afoul of the law. Such boats were also extremely expensive. Although boatyards for the further development and construction of steam-powered vessels were built along the Ohio, fewer than ten boats had actually been put into service by 1816. Those vessels were made on a more or less experimental basis and suffered a staggering rate of attrition from mishaps with snags and boiler explosions. Further complicating matters, any boat constructed at Pittsburg or Brownsville had to steam down the Ohio River and pass the Falls of Ohio before reaching the Mississippi and turning upstream. Steamboat construction and usage was also hampered by the War of 1812, and it was not until the cease of hostilities that capital was freed for investment.

The Zebulon M. Pike

Built: 1815 in Henderson, Kentucky
Tonnage: 31.76
Hull: Wood

Engine: Low pressure
Construction: A little scow, built on the model of a barge. Next to the *Comet*, she was the smallest steamboat documented in the Mississippi Valley
Design: The cabin was on the lower deck; the paddle-wheels had no housing

Since the low-pressure engine (originally introduced by Fulton for use on the placid eastern waters) had difficulty working against the strong current of the Mississippi, the crew of the *Zebulon M. Pike* were compelled to resort to poling, the technique developed by keelboaters. It took the *Pike* six weeks to go from Louisville to St. Louis.

In October 1817, a second steamboat, the *Constitution*, reached St. Louis, and thereafter, the trend continued. By 1819, the city had become a major port of call. Ideally situated in a position to serve river traffic from the Upper and Lower Mississippi, the Missouri and Illinois rivers, St. Louis soon established itself as a hub of western water commerce.

In 1819, The Missouri *Gazette* proudly declared that the steamboat, a "great American invention," would open up a "new arena" for state merchants.

> A steamboat ... has started from St. Louis for Franklin, two hundred miles up the Missouri, and two others are now here destined for the Yellowstone. The time is fast approaching when a journey to the Pacific will become as familiar, and indeed more so, than it was fifteen or twenty years ago to Kentucky or Ohio.

Word spread quickly. The *Adams Centineal* reported on July 21, 1819:

YELLOW-STONE EXPEDITION

> The St. Louis Gazette, of the 26th of May [1819] states that the steamboat Johnson passed that place on the 19th ult. with troops, &c. for the Yellow-Stone.

The first steamboat destined to reach Yellowstone on the Missouri River was called the *Western Engineer,* and she had been built by the United States government.

The Government Gets Its Feet Wet

No discussion of steamboating on the western waters is complete without detailing the involvement of the federal government. Almost as soon as the *Clermont* proved her worth on the waters of the Hudson, Robert Fulton pitched the value of steam-powered vessels as military weapons. Having considerable interest in the subject, as demonstrated by his previous work on submarines in France and England, Fulton submitted plans for a steam warship carrying 30 long guns (32-pounders) to the Coast and Harbor Defense Committee in 1814. He succeeded in obtaining a contract and, in October of the same year, launched the *Demologos*.

The Demologos (*also known as* Fulton the First)

Built: 1814 in New York at the shipyard of Adam and Noah Browne
Length: 167 feet
Boilers: Copper
Depth of beam: 56 feet
Depth of hold: 12 feet
Height of gun deck: 8 feet

Thickness of sides: 5 feet
Draws: 9 feet, 2 inches fully loaded
Power: 120 horses
Estimate of cost (engine and hull): $150,000
Estimate of total cost: $235,000–240,000

On September 11, the *Demologos* made a trial run, loaded with 26 guns and armament, achieving the speed of 5.5 miles per hour.

Word traveled fast. On August 29, 1815, the London *Times* reported the following:

> An American gentleman who is lately arrived from New York, states that there is just completed in that harbour, a steam frigate, the length of which is 100 yards, and breadth 200 feet; her sides, which are alternately composed of oak, plank and cork wood, are 13 feet thick. She carries 44 guns, four of which are of very large bore, the others 42 pounders; in case of being boarded, she is enabled by machinery to discharge 100 gallons of boiling water on her enemies per minute, and at the same time 300 cutlasses branch over her gunwales, and an equal number of pikes dart out from her sides.

The *Demologos* was the first war steamer. The boat was intended to serve in harbor defense at New York Bay, but the War of 1812 ended with the Treaty of Ghent, December 24, 1814, before she was put to the test. The vessel was subsequently laid up in the Navy Yard in Brooklyn, where she was used as a receiving boat. On June 4, 1829, an accidental explosion completely destroyed her. Twenty-five people were killed and nineteen wounded in the disaster.[5]

Original government interest in western steamboating had an entirely different character. The necessity of maintaining a military presence to deal with Indian matters and protect early settlers prompted Secretary of War John C. Calhoun to order the establishment of Fort St. Anthony (renamed Fort Snelling in 1824), above Prairie du Chien on the Upper Mississippi, and Fort Atkinson on the Missouri. To supply Fort Atkinson, Calhoun contracted with Colonel James Johnson of Kentucky to charter a steamboat for that purpose.

Calhoun was roundly criticized for risking the success of the expedition by the use of an untried mode of transportation, and his critics were ultimately proven correct. Johnson was unable to get the supplies up the Missouri, but that did not prevent him from billing the government for the staggering sum of $258,818.15. The fee was ultimately paid by Quartermaster General Thomas S. Jessup, but prompted a congressional committee to investigate the scandal.[6]

The *Adams Centinal*, December 13, 1820, provided readers with a summary:

> HOUSE OF REPRESENTATIVES Mr. Cocke, of Tennessee, offered the following resolutions: Resolved, That the Secretary of War be directed to communicate to this House what sums of money have been actually paid to Colonel James Johnson, on account of transportation furnished the expedition ordered up the Missouri river; and also what sums have been paid him for detention of Steam boats or other incidental charges; whether any difference of opinion existed between the Department of War and said Colonel J. Johnston* relative to the value of transportation or other charges exhibited by him against the United States; if any difference existed, how were they adjusted; if by reference, who were the referrers, what was their award, and what evidence was submitted to them, on which they formed their award.

---

*Both the "Johnson" and "Johnston" were used in the article.

On a motion of Mr. Rich, with the consent of Mr. Cocke, the resolution was amended so as to require an account also of the causes of the detention of the Steam boats.

Interestingly, an addendum to the newspaper article on government resolutions noted that:

> Mr. Lowndes gave notice, that he should, on Wednesday next, move for the consideration of the resolution declaring the admission of the State of Missouri into the Union.

While this was transpiring, it was decided that the United States government would go into the steamboat construction business. In 1818, work on a boat was begun at the arsenal on the Allegheny River near Pittsburg. Christened the *Western Engineer* (in honor of the engineering corps which designed the boat and for her destination), the steamer was launched on March 28, 1819.

The Western Engineer

Built: 1818 at the United States Arsenal near Pittsburg
Tonnage: 30
Draw: 19 inches of water when unloaded
Description: A dingy-looking craft carrying artillery, the flag of the United States, portraits of a white man and an Indian shaking hands, the calumet of peace and the depiction of a sword.

The *Western Engineer* was further described as a "black, scaly serpent, rising out of the water, with waste steam escaping from her sculptured figurehead," and as an "apparent monster with a painted vessel on his back, the sides gaping with port-holes, and bristling with guns."[7] The boat left Pittsburg on May 5, 1819, and reached St. Louis on June 9. She continued her trip and reached as far as Fort Lisa, a trading post of the Missouri Fur Company near present-day Omaha, on September 17, 1819. With freezing weather on the horizon, the boat was put into winter quarters, having ascended the Missouri River farther than any other steamboat.

Subsequent voyages took the *Western Engineer* up the Mississippi as far as Keokuk, in August 1820. This passage proved a steamboat could travel farther up the river than had previously been thought possible and opened the door for further experiments.

The *Republican Compiler,* August 23, 1820, provides details:

> Franklin, Missouri, June 24.
> The steam boat Western Engineer arrived at this place on Saturday the 17th inst. in 4 running days from the Council Bluffs (500 miles). Captain Perkins, Lieut. Graham, and Mr. Ranney, came passengers. The Engineer has since progressed to St. Louis, where she is to await the arrival of Major Long and party, from their exploring expedition.
> Major Long, accompanied by Capt. Bell, of the corps of artillery, Lieut. Swift, Doctors Say and James, and Mr. Peale, with others, the whole party consisting of 21, left the Council Bluffs on the 6th inst. They will explore the La Platte by land, then cross over to the head waters of the Arkansas; part of the company to descend that river, the other part continue to the head waters of Red River, and descend that river. The parties are expected to complete their tours, and arrive at the Mississippi in September.
> The sickness at the Bluffs has entirely subsided. The Expedition, steam boat, with provisions for the Council Bluffs, arrived at Fort Oswego on the 10th inst. and has progressed well.

That the steamboat had proven an acceptable means of transportation was expressed in the *Adams Centineal*, March 24, 1819:

Astonishing Facts

In the year 1811, a steam boat, to navigate the western waters, was launched at Pittsburg, Pa. There are now, in full tide of success, on the Mississippi and its tributary streams, *thirty-one* steam boats, and thirty more are building, and nearly completed, for the same navigation. Allowing each boat to make three voyages a year to New Orleans, at the present rate of freight and passage, the income of sixty one boats is estimated at the enormous sum of $2,256.90 per annum! What a world of industry, enterprise activity and productiveness!

Had steam navigation reached a point of equal success across the Atlantic or the opposite? The answer may be surmised from a number of adverts running in the London *Times* of March 27, 1816, when the steamboats *Princess Charlotte, Prince of Orange* and the *Duke of Wellington* were put up for public auction. The *Duke* boasted a steam engine of six horsepower and a boiler of ten; the whole of the machinery was entirely new and of "very superior construction." The *Princess* and *Prince* (running the Clyde between Glasgow and Greeneck) were described as being light and airy, with neatly fitted-up cabins for ladies and children, a steward's room and well furnished cabins, where "no expense has been spared" on elegant and comfortable accommodations. The boats drew only 3 to 4 feet of water and measured about 33 tons.

The Indian Question

White Americans' relationship with Native Americans was fraught with problems from the beginning. With white settlers' acquisition of tribal land as the ultimate goal, it is no wonder hostilities arose between the two factions. While a discussion of the evolving tensions and outright warfare exceeds the scope of this book, two newspaper articles from the time sum up some of the issues. The first is from the *Republican Compiler* (Gettysburg), August 1, 1827:

Wheeling, (Va) July 12, 1827
Indian Hostilities — About the first of this month, as the keel boat O.H. Perry, owned by Mr. Robert P. Clarke of this place, was returning from Fort Snelling, whither she had been conveying military stores, the crew were twice attacked by a party of Winnebago Indians; at the second attack the Indians got possession of the boat; but the crew afterwards recaptured her. In these several engagements a number were killed on both sides. The clerk of the steamboat Mexico, Benjamin Thaw, formerly of Pittsburg, after killing three Indians, was slightly wounded; he is now at Fevre River under the care of physicians. The men working the lead mines in the vicinity of Fevre River have collected at Galma and are erecting fortifications.
"The foregoing particulars were communicated by Gov. Cass at St. Louis, to a gentleman who passed through this place on Wednesday last."

The second report is from the Huron (Ohio) *Reflector,* August 15, 1837. While a bit self-serving, it adds a new light to Indian and settler (and government) relations.

Uncle Sam Turned Pedlar*
Start not, gentle reader, at this not very euphonical caption — there is more truth than

---
*Spelling copied from the original.

poetry in it. Yes, be it known, that the United States Government, not being able to procure a few thousand dollars of specie or from some other cause, has determined to pay its honest debts to the Menomonee Indians in this vicinity — in what? in trinkets and calicoes, and this, too, in opposition to express treaty stipulations. Alas for our long-cherished visions of gold eagles and silver dollars — alas! for the credit and honor of the government — and alas! for the *better currency* which we have all along been promising ourselves! Several bales of goods, marked "U.S." from Suydam, Jackson & Co. New York, have already arrived, and we understand that a steamboat has been chartered expressly for the purpose of bringing out to the north-western Indians their annuities, in this novel and new kind of currency.

But to be *very* serious. We cannot view this step on the part of the government in any other light than as an unjustifiable and flagrant infraction of its treaty, a heartless disregard of the interests of the aborigines, and of our citizens. Our merchants, anticipating the demand that would naturally arise upon the payment of the Indians under the treaty stipulations, provided themselves with large quantities of goods, suitable to the market. This unnecessary interference, therefore, of the government (from which protection rather than opposition should be expected) with their trade, cannot but be attended with disastrous consequences to those, who, relying upon the good faith of the government, have invested largely, some of them their all, for the supply of the Indian trade.

But will the Indians receive these goods in payment of their dues? We confidently hope and believe not. And after Uncle Sam finds the occupation of *pedlar* at the west rather unprofitable, we presume he will take his trash back to the east, and turn *auctioneer. Sic transit Gloria mundi.* (Wisconsin Dem.)

In 1832, the "Indian question" caught national attention when the Sauk and Fox tribes murdered twenty-eight Menominee Indians at Prairie du Chien. Not only were Indian wars of concern to the United States government, the country owed the Menominee a debt of gratitude. Other tribes had joined the British during the War of 1812, but the Menominee had steadfastly refused. They were considered a hardworking and friendly people, despite the fact they were reported to have peculiar culinary habits, such as serving human flesh to honored guests.[8]

The Sauk and Fox tribes were allies, resembling each other in features, dress, customs, language and weapons. The largest Sauk village in the Mississippi Valley was called Saukenuk, situated three miles to the southeast of the Rock River. Both tribes were highly skilled with the bow and arrow, and archers could easily strike coins affixed to trees at twenty-five paces. Their leader at the time of the Menominee massacre was Black Hawk.

In April 1832, the Sixth Regiment of the United States Infantry left Jefferson Barracks (outside St. Louis) aboard the steamboats *Enterprise* (the same boat which saw service in New Orleans during the War of 1812) and the *Chieftain.* Their object was to make the tribes give up the warriors who had murdered the Menominee. Ignoring the military threat, Black Hawk defeated a group of militia at Stillman's Run. This prompted Governor John Reynolds to issue a proclamation to the citizens of Illinois. Word of his call was quickly disseminated throughout the states, and *The Mail,* receiving the news from the steamboat *Herald* out of St. Louis published it in their June 1, 1832, edition.

To the Militia of the State of Illinois: It becomes my duty again to call on you for your service in defense of your country. The State is not only invaded by the hostile Indians, but many of your citizens have been slain in battle. A detachment of the mounted volunteers commanded by Major Stillman of about 272 in number, were overpowered by the hostile Indians on Sycamore creek, distance from this place about thirty miles, and a considerable number of them killed. This is an act of hostility which cannot be misconstrued.

I am of the opinion that the Pottawatamies and Winnebagoes have joined the Sacks and Foxes, and all may be considered as warring against the United States.

To subdue these Indians and drive them out of the State, it will require a force of at least two thousand mounted volunteers more, in addition to the troops already in the field. I have made the necessary requisitions on the proper officers for the above number of mounted men, and have no doubt the citizen soldiers of the State will obey the call of their country. They will meet me at Hinepin, on the Illinois river, in companies of 50 men each on the 10th of June next, to be organized into a Brigade. JOHN REYNOLDS, May 15, Commander in Chief.

Before 1848, when telegraph lines had been run as far west as St. Louis, steamboats were the only reliable source of news, and captains arriving from the Upper Mississippi were eagerly greeted for information. Steamboating was particularly dangerous during hostilities. Fewer boats attempted voyages up the river in 1832 than in previous years, so the crews of those that did were hailed as heroes.

Among the most notable steamboat captains of the era was Joseph Throckmorton. With four years of experience to his credit, he brought his boat, the *Warrior*, to St. Louis in midsummer of 1832 and determined to go upsteam.

The Warrior

Built: 1831–32 in Pittsburg
Owned by: Joseph Throckmorton and William Hempstead of Galena
Style: Side-wheeler
Tonnage: 100 tons
Engine: High pressure
Boilers: Three
Measured: 111 feet, 5 inches
Beam: 19 feet
Hold: 5 feet
Description: One deck, a transom stern, a cabin above deck for officers
 and crew, and a figurehead.

Throckmorton arrived at Prairie du Chien and received orders from a United States major to patrol the river above Fort Crawford to prevent Black Hawk from crossing. With the aid of the friendly Winnebago Indians, the boat succeeded in accosting Black Hawk's group just as they were attempting to cross. After failed peace talks, the crew of the *Warrior* and the Indians exchanged hundreds of shots. One white man was wounded in the engagement, and the battle ended after four blasts of grapeshot were discharged from the vessel.

This contest delayed Black Hawk long enough for soldiers to arrive, whereupon they routed the Indians at Bad Axe. Soon after, the Winnebago captured Black Hawk and brought him to Prairie du Chien. He and eleven of his party were taken to Jefferson Barracks, appropriately enough, aboard the steamer *Winnebago*. While the incident was hardly a "war" in the true sense of the word, a number of men who would later participate in greater challenges were involved: Colonel Zachary Taylor, Lieutenant Jefferson Davis and General Winfield Scott.

The *Warrior* was one of few Upper Mississippi steamboats to tow a safety barge when working the freight and passenger trade. The barge served as a protection for passengers in case of explosion, and as a means of decreasing the draft to ease passing over rapids. As the *Warrior* had no staterooms, the barge in this case was used for passenger accommodations.

Safety Barge

Tonnage: 85 tons
Length: 111 feet, 8 inches
Beam: 16 feet
Hold: 4 feet, 8 inches
Draw: 20 inches of water
Description: 52 berths, three cabins

Captain Shallcross of the steamboat *St. Louis and Galena Packet* was the first to use a safety barge in 1827.[9]

A significant source of revenue for Upper Mississippi steamboat captains was the task of removing Native Americans from their lands. This black mark on American history was preordained by greed, a swelling population and misguided policy dictated by the federal government, which, in some instances, can hardly be considered anything less than genocide.

Not all Indians were removed by steamboat; many were forced to march overland to their new "homes." The list of those removed is depressing, seems almost never-ending, and included nearly every tribe. In 1832, over two thousand Choctaw were transported to Rock Roe on the White River. The same year, Seminoles reached Memphis on the *Little Rock,* and from there were marched to Fort Gibson. In 1846, after the Winnebago gave up their land, it was determined to move two thousand of them to the Crow Wing River aboard the *Dr. Franklin*. After much discussion and a consultation with neighboring Sioux, whom the Winnebago feared would object to their intrusion into Sioux land, some, but not all the Indians were moved. Not until the following year did the one hundred left behind agree to leave.

When tribes did agree to be transported from their homelands, many resisted travel by steamboat, being fearful that their children might be drowned and that the young and old would be in danger of being scalded to death, "like the white man cleans his hog."[10] Their suspicions were well-founded, for explosions were common and always deadly. (See Chapter 10 for a disaster with Indian casualties.)

Transporting soldiers (by military order, troops were moved from fort to fort on an annual basis), supplies, armament, equipment and foodstuffs for the troops provided an irregular but highly lucrative trade for steamboat captains. According to William Watkins, the secretary of war, the public property carried yearly on the Upper Mississippi was worth $272,213.90. In 1853, for example, the cost of government supplies at St. Paul surpassed that of civilian goods. It is estimated that from this transport, steamboat captains earned the immense sum of $1,500,000. By the dawn of Civil War, Upper Mississippi captains had garnered almost $3,000,000. Although not as profitable as carrying Indian subsidies and furs, the importance of this source of revenue cannot be overestimated.[11]

Get the Lead Out

Perhaps the single greatest boon to Upper Mississippi steamboating was the lead mines. The presence of lead was found in the territories of what would become northwestern Illinois, southwestern Wisconsin and eastern Iowa. The Fox Indians, who controlled the land around the Fever River, were well aware of the precious metal, and mined it at Dubuque for their own use. Carrying the newly mined ore to a hole dug in the rock, they melted it and reduced it to manageable size.

Nicholas Perrot, a French soldier, was the first white man known to mine lead in the Upper Mississippi Valley as early as 1685. Over a century later, in 1788, Julien Dubuque received permission from the Fox to mine lead west of the Mississippi. By 1810, the Fox (who owned the land), were melting over 400,000 pounds annually. That year, Henry Shreve became the first American to take a barge of lead out of the Fever River. (With the net profit of $11,000 from this venture he opened a barge business in Pittsburg, and later achieved fame with his challenge to the Fulton-Livingston line.)

Shreve's success officially opened the lead market, and miners and investors invaded the lead-mining area around Galena. By 1828, the government paid the Native Americans $20,000 for the lead district and began leasing 160-acre plots, on the stipulation it receive one-tenth of all the mineral mined. (In 1830, the fee was reduced to 5 percent and in 1846, the government simply offered the lots on a straight sale basis.)

Galena Gets Mad at St. Louis

As steamboats became more and more efficient, these vessels replaced keelboats in the removal of lead. Lead became the single most lucrative freight a steamboat could carry, and many a captain covered the cost of his boat in one successful season. But all was not perfect with this highly profitable trade. By March 1840, an editorialist in the Galena *Democrat* bitterly complained about "the exorbitant and unjustifiable charges adopted by the monopolists" (the owners of St. Louis boats trading at Galena), that have "a tendency to retard, rather than increase the growth and prosperity of Galena, and the mining region generally." For example, he gives the following statistics:

The steamboat *Demone,* worth about $6,000, cleared upwards of $12,000 during 1839

The *Ione,* in two trips brought up from the Rapids six or eight hundred tons of freight, charging $1 per 100 lbs; returning with three keelboats in tow, she took 20,000 pigs of lead, weighing 1,400,000 pounds, for which she was paid 75 cents per 100 lbs, amounting to $11,500. Supposing her trips up, to be worth two-thirds that amount, she must have been paid over $24,000. Deduct for expenses one third of the whole amount and the boat cleared in two trips $10,000

The *Illinois* charged $1.50 per 100 lbs, in a single trip received $9,000. Allowing her expenses to be $3,000, she cleared $6,000

The steamboat *Rapids* (running from August to November), cleared upwards of $10,000

The writer estimated that the amount paid by Galena and other merchants in 1839 was over $50,000. Interestingly, he goes on to compare prices in St. Louis and Galena:

	St. Louis	Galena
Flour	$5–6 barrel	$10
Sugar	5½–6½ cents/lb	12½ cents/lb
Molasses	33 cents	$1.00

5. "St. Looy" and the Upper Mississippi

The *Amaranth* was a steamer working out of Galena, Illinois, in the lead trade. She is mentioned as working as early as 1843. This image was taken outside St. Louis in 1848.

To remedy the situation (borne by merchants and consumers alike), the suggestion was made to Galena importers that they should "set their faces against such extravagant charges and demand a return to the prices of 1837 and 1838. There are many boats on the Ohio and Lower Mississippi that would gladly enter into competition with the "St. Louis monopolists" if they received encouragement. These boat captains would "engage for the convenience of freights at *half the price charged last season,* and clear as much, if not more, than by running in any other trade."

The conclusion may be construed as a warning to greedy St. Louis boat owners:

> We trust our suggestions may have some weight with those most interested; and for the present we shall leave the subject, with the remark that, in the coming season we shall take such measures as will make known to the community such boats and owners of boats as are friendly to the interests of the mining country.

Whether or not the newspaper succeeded in its demands, business continued at a brisk pace, and profits soared, particularly in cargoes taken from Galena and the upper rivers.

In the period 1841–43, the fees garnered by twenty-two regular boats and eleven transients amounted to $620,000.

Breakdown of Revenue

$236,000	from lead
$156,000	from freight
$228,000	from passengers[12]

During the first twenty-five years of lead mining, approximately 472,000,000 pounds (6,728,000 pigs) of lead was mined and shipped down the Mississippi.

Values of Cargo

Lead	
1841–48	$8,676,647.39
1847	$1,654,077.60
Quarter century	$14,178,000
Furs	
1848	$300,000
Produce of the Santa Fe Trail	
1848	$500,000[13]

The remuneration for the steamboat captain depended on a number of factors, including the season, the state of the economy and the competition of rival steamboat owners. In 1843, the *Amaranth* left Galena with 13,000 pigs of lead (455 tons). The captain earned $1,265 upon delivery. Two months later, during a period of aggressive competition, the same trip would have earned him a scant $500, and during a period of low water, he could have earned as much as $4,550.[14]

The Development of Steamboats on the Upper Western Waters

Steamboating on the Upper Mississippi past St. Louis and up the Missouri presented unique difficulties: the preponderance of rock bars, narrow streams, rapids and, most particularly, weather. During the years 1818 and 1819, over sixty steamboats were constructed for western commerce, but of the total freight shipped, only one-third of that actually went by steamer. The remainder was still carried on barges and keelboats.

Advancements, however, were steady, and with the development of the high-powered engine (discussed in chapter 6), boats had the power needed to ascend the rapids. The principal development of trade concentrated on the St. Louis-to-Galena route. Supplies for the settlers were the primary cargo upriver, and lead was taken on the return trip. A typical greengrocer's order might include

> Peaches, dry apples, line, barrels of whiskey, crackers, smoking and chewing tobacco, bags of oats, corn and wheat, flour, coffee, lard, molasses, salt, soap, butter, barrels of cider, dried fruit, onions, beans, bacon, pork, and the occasional "ham on the hoof."

Items of a more substantial nature (and weight) included

> Axes, grindstones, picks, plows, reapers, saws, scythes, shovels, andirons, candles, guns, ovens, pots and pans, powder, shovels, and rolls of line.

The apple trade by itself was an enormously profitable seasonal cargo, and occasionally, its delivery was mentioned of in the newspaper, albeit with a slight prejudice. The Milwaukie *Commercial Herald,* November 2, 1843, gave a nod of thanks:

> The Propeller Samson, arrived at Milwaukie with a fine lot of freight, a part of which (a barrel of Apples) was consigned to (us) by the Captain. We shall remember the Captain for the consignment, especially when partaking of the treat.

The transport of glassware, hardware and coal from Ohio River cities comprised other valuable cargoes. Even ice was a desirable commodity, and boats specially constructed to carry it from Minnesota were brought downriver. Clearly, this trade was the inspiration for Ambrose Bierce's brilliantly witty short story, "The Failure of Hope and Wandell."

Lumber also made up a significant amount of revenue for steamboat captains. The Milwaukee *Daily Sentinel* (June 12, 1854) reported:

> The huge pile of lumber and shingles that have been heaped upon our docks on both sides of the river are now rapidly disappearing. Every steamboat and vessel that arrives in our port takes away a heavy load.... All the mills in this vicinity are worked to their utmost capacity, and yet our dealers find it impossible to fill the orders they are constantly receiving from abroad.

This type of lucrative commerce invariably led to the transfer and storage business, which also proved highly profitable for riverside merchants. Cargo waiting to be shipped out was stored in warehouses or on the levee, covered with tarpaulins. Regardless of the method of storage, a fee was charged for the service, and a guard was always kept posted to prevent keelboatmen from pilfering the goods onto their rafts. Transfer agents paid the freight on goods from the lower river, and a commission of 5 percent provided a substantial remuneration for the business owner.[15]

Freight fares typically ranged from 25 cents to 75 cents per hundred pounds. The longer the trip, the less expensive the fare. For example, if the cargo were taken from St. Louis as far as the boat went, the shipper paid the lesser price; if the cargo were to be unloaded along the way, the shipper paid a higher fare.

Standard operating practice dictated that a boat left port when the captain had enough freight on board to make a profitable trip. That meant passengers were often stranded in town, despite promises of departure by a certain date. Throughout the history of steamboating, departure and arrival dates were highly speculative, and boats seldom, if ever, kept to schedule. A passenger buying "through passage" (all the way to the end of the line) often discovered his trip cut short when the captain cancelled the remaining voyage. In such cases, fares were often difficult, if not impossible, to recover, leaving the unlucky person to sue for his money.

Charges from Galena upsteam, where navigation was more treacherous, were usually 50 cents per pound, if not higher, and kegs of powder from that same port to Fort Snelling cost $1.00 per keg.

Fees were determined by four factors: competition, the stage of water, the available freight and the season of the year. When freight was plentiful, profits on an uneventful trip were high. In 1844, the steamer *Lynx* netted a scant $161.04 after paying for repairs. The following season, the boat earned $11,144.73, which was the average earning for an Upper Mississippi boat.[16]

Competition was always the order of the day and many captains were ruined by cutthroat underbidding. Too often, they were compelled to reduce rates in order to secure cargo, reaching a point where they sacrificed profitability. The shippers benefited from this generous circumstance, but operators who did not keep a sharp eye on their books, worked at below cost, hoping for a profitable return trip. If that did not happen, they occasionally lost their boat to creditors.

In April 1843, for example, the cost of cargo averaged 18 cents per one hundred pounds, but by June, with a surplus of available transport, the price fell to 6¼ cents. Owners who had money in the bank to tide them through hard times survived to try again the next year. Those less fortunate were compelled to leave the trade. Occasionally, feuds would break out between competing captains, and blood would be spilled.

Ice and Steamboating

Steamboating was entirely dependent upon the season, for the weather dictated the rise and fall of water levels. In the Upper Mississippi trade (beyond St. Louis), where the rivers froze for four months out of the year, that meant only eight months of navigation. And even that could not be counted on if the winters were long or the waters were late rising. (In and below St. Louis, frozen water was not nearly so much of a problem. Ice closed the river only twenty-nine days on average, and occasionally the river was navigable year round.)

Ice also posed a problem for cities in the northern parts of the country. The rise of water and ice breaking up in Pittsburg in February 1840 caused serious damage to boats in the coal trade. Fourteen boats and cargo were crushed and sunk. Two steamboats, the *Beaver* and *Ontario,* were also sunk. In all, the water for that early period rose seven feet.[17]

As soon as the ice broke, steamboats set out. Leaving too early could mean damage to the vessel by floating bergs, but captains often took that chance, for there were always ample amounts of lead that had been mined during the winter ready for pickup. The first boats earned the greatest reward because miners were eager to get the mineral to market.

Once the spring thaw had completely melted the ice, prices slowly dropped as more and more boats entered into competition. Melting ice also meant a rise in the rivers, which made navigation much easier. During the heat of the summer, water levels steadily dropped, so that by August and September, sandbars and snags presented a serious problem, while the rapids became harder to negotiate, sending rates steadily up. A second rise in water level occurred around the middle of October that temporarily lowered rates, but as the season progressed and navigation became more difficult, they rose to their highest levels. The unofficial end of the fall season came around the middle of November.

Ice was a significant factor toward the end of the season. The early and unanticipated appearance of ice could damage a boat, or worse, freezing water could close up around the hull and literally trap the vessel in a matter of hours. In that case, the boat could not be extracted and was thus compelled to stay in the middle of the river until spring thaw. If that happened, the cargo was either left aboard with a guard, or manually hauled inland and retrieved by horse and sled. In either case, profits were greatly diminished or lost altogether, especially if the ice shifted during breakup and crushed the hull.

The second problem a steamboat captain faced by attempting a late trip was the extravagant wages charged by the crew. Fully aware that if the boat were stuck in ice they would have to work their way to the nearest town by foot, an unpleasant and often dangerous proposition due to the threat of Indian attack or freezing to death, the demand was not unreasonable. Given those circumstances, most captains opted not to take chances. This reticence, however, often left miners and distant pioneers stranded in their little waysides, some even starving to death before spring thaw brought a boat laden with supplies.

To protect steamboats from being crushed by tons of floating ice, the vessels were laid up during the impassable months. It was too dangerous to dock them along the wharfs and piers of river towns because there was no predicting how fast or how high the spring floods might be. A boat caught in the swelling tides could be washed inland or smashed between bergs and rendered into tinder. For example, in the winter of 1832, 34 inches of ice on the river at Keokuk was shattered by the force of floodwaters rising fourteen feet in a single hour. Five thousand pigs of lead were buried in the mud and not recovered until June.

Lagoons (sloughs) or small tributaries were chosen for winter quarters, but these were often difficult to reach and protected havens were hard to find. During the heyday of Upper Mississippi steamboating, St. Louis represented the southernmost point boats reached. The west bank of the Mississippi near Arsenal Island was a favorite spot for laying up. Typically, a guard was left to protect the owners' property. Before the thaw, carpenters were sent to ready the boat for spring work.

Fare and Not So Fair

Before George Catlin made the "Fashionable Tour" popular in the late 1830s (see chapter 16), little attention was paid to passenger accommodations on Upper Mississippi steamboats. Cabins were plain but well furnished; there were no staterooms, only tiers of bunks running along the sides of the boat, with curtains separating sleeping chambers from the saloon.

In the 1820s, the frontier had reached as far as St. Louis. By 1830, it had been pushed up to Keokuk. A decade later, miners populated Galena and Dubuque and by the 1850s, civilization reached the northern boundary of Iowa. Into that decade and beyond, pioneers settled as far as St. Paul. As the United States expanded and opportunities arose for inexpensive farmland and new, well-paying jobs, most families traveled west by wagon through the Cumberland Road. This famous thoroughfare reached as far as Columbus, Ohio, by 1833 and stretched to Illinois by the mid-1840s.

A more roundabout route that provided the security of one major river passage took travelers from New Orleans as far north as the Falls of St. Anthony. Steamboat passengers could consign their household goods to the rivermen and concentrate on enjoying

A particularly sharp image of the *Tom Greene*, left, at St. Paul beside an unnamed boat. The boat flies both the flag of the United States and a private, checkered flag, likely denoting the line to which it belonged.

what they could of the journey. Prior to 1845, a trip by water from New Orleans to St. Louis required about fourteen days. From there, most continued northward as far as their finances and perseverance allowed.

Cabin fare from New Orleans to St. Louis was $25.00 and freight costs averaged 62½ cents per one hundred pounds. Deck passengers were charged $3–4.00 and children went for half fare, with each person being allowed one hundred pounds of baggage. Deckers, as they were commonly known, lived in the open without appreciable shelter and had to provide their own food. When compelled to go ashore to replenish their stores, they might find deer meat, Indian corn cakes, buckwheat cakes, pancakes, wheaten bread and skimmed milk or buttermilk for sale.

Fares in 1833*

Cabin fare from Pittsburg to St. Louis	$24.00
Deck fare from Pittsburg to St. Louis	$8.00
Cabin fare from St. Louis to Galena	$15.00
Deck fare from St. Louis to Galena	$5.00

*Passengers could buy meals for 25 cents.[18]

To reach St. Louis from Philadelphia, a traveler went by stage and steamboat, with a one-way fare costing $55.00.

Fares in 1834*

Cabin fare from New Orleans to St. Louis	$25.00
Fare from St. Louis to Quincy	$6.00
Cabin fare from St. Louis to Galena	$12.00

Fares in the 1840s*

Cabin fare from St. Louis to Galena	$12.00
Deck fare from St. Louis to Galena	$6.00
Cabin fare from Galena to St. Paul	$8.00
Deck fare from Galena to St. Paul	$4.00*

In a very short span of time, steamboats came to dominate the freight and passenger trade. From 1817, when the first boat arrived in St. Louis, until the outbreak of Civil War, thousands of people and hundreds of tons of cargo passed over the western waterways. In the beginning, travel centered on the fur trade and then steadily developed into the transportation of lead. Soon, the delivery of Indian annuities, the transport of soldiers, and the heavy manufacturing equipment of industrial cities such as Pittsburg, helped rivermen make hefty profits.

Largely ignoring the loss of life from accident, explosion and disease, for the possibility of death was an accepted risk of frontier existence, the opening of the country was made possible by the once terrifying mode of steamboat travel. Competition, harsh winters, devastating floods, rapids, low water, rapidly changing fees and the threat of attack hindered but rarely stopped the steamboats. The ultimate threat to success would not come from natural disaster or interior bickering, but from an entirely different quarter: the railroads.

Passengers could buy meals for 25 cents.

6

A World Unto Their Own: The Mississippi and Ohio Rivers

The beginning of the steamboat era on the Mississippi and Ohio waterways might fairly be said to have started after the breaking of the Fulton-Livingston monopoly in 1818. Thereafter, there was no stopping the development, construction, ownership and travel of this new mode of transportation.

The High and the Low Roads

Ironically, the two developments which would have the most impact on steamboat advancement on the western waters actually began prior to 1818 by Henry Shreve and his Brownsville group. Convinced the low-pressure Boulton and Watt engine used by Robert Fulton was incapable of providing enough power to handle stiff currents or offer the maneuverability to navigate around or over the numerous river obstructions, Shreve developed the French oscillating (or high-powered) engine. These were placed in the first three of Shreve's boats: the *Comet* (1813), the *Despatch* (1814) and the *Enterprise* (1814), the last-named of which played a significant role in the defense of New Orleans.

The fourth of Shreve's boats, the *Washington*, had an almost horizontal cylinder that was directly connected to the paddle-wheel through a pitman, using neither beam nor flywheel. Her success proved in practice what Shreve held in theory, and set the standard for all later boats.

The Washington

Built: 1816 in Brownsville
Tonnage: 403 tons
Type: Stern-wheel
Engine: Double, high-pressure
Length: 139.9 feet
Breadth: 24.7 feet
Depth: 12.3 feet
Design: The first "two-decker." The cabin was placed between decks, the main

cabin being 60 feet long. She had three "handsome private rooms" and a commodious barroom. The boiler was placed on the main deck instead of in the hold.

At the time of her construction, the *Washington* was the largest steamboat ever built on the western waterways.

By 1825, the vast majority of boats utilized the high-pressure, noncondensing engine. Its advantages were noteworthy: the high-pressure engine took up less space, weighed less and generated far more horsepower. The high-pressure engine (although having a reputation as being more susceptible to explosion) was a less complex machine, making it cheaper and easier to repair, "having less than half as many moving parts" as a low-pressure engine.[1]

By contrast, low-pressure engines lacked flexibility. Equally significant, they lacked the reserve power required for western navigation. In 1834, commenting on the advantages of high-pressure engines, one writer explained of their reserve power, "It might almost be said that the only limit upon its power was that imposed by the limits of strength of boilers, steam pipes, and cylinder."[2] Customary running pressures might be exceeded when occasion or necessity required by dint of hard firing and the use of highly combustible materials such as resin, oil or pine knots. This "wad of steam" was highly appreciated by captains, engineers, and passengers alike, most notably during steamboat races.

Additionally, low-pressure engines used river water for generating steam and condensing it. Because the Mississippi and particularly the Missouri rivers were laden with silt, the grit wore out valve leathers and lowered the efficiency of the condenser. Because of the large amounts of water consumed, it was not possible to carry clean water for use in the boilers, and the development of a filtration system never reached the point of high necessity. The heavy accumulation of mud, therefore, necessitated the unpleasant and occasionally dangerous task of frequent cleaning.

To clean a boiler, the boat was made fast, the mudvalves were opened, the fires drawn down and the water let out. The entire head then had to be removed in order to allow a crewman inside. The task was usually performed by the cub engineer, who dropped through a narrow manhole into the body of the boiler. He "scaled" the walls with a hammer and sharp-linked chain, pounding on the two large flues and the sides with the hammer, and then sawing the chain until the mud and sediment was dislodged. The task was physically demanding and the work carried out in extreme heat.[3]

After the scaling was finished, the residue was washed out with water or carried out in buckets. Carelessness in leaving behind a cleaning rag was often a cause for explosion. By the 1850s, the blowout valve and the introduction of mud drums to act as settling basins underneath the boilers greatly simplified the problem of boiler maintenance.

The low-pressure engine did, however, have two substantial advantages: safety and efficiency. Safety was much studied and written about. Both low- and high-pressure systems had advocates. But it was well documented that when a high-pressure engine blew up, there was far greater damage to the boat and a greater loss of life. Additionally, low-pressure engines were said to use 30–40 percent less fuel than high-pressure engines. Steamboats seldom carried more wood than was necessary and stopovers for fuel were frequent. However, as wood was plentiful and cheap in the early years of steamboating, this factor did not tip the scales in favor of the low-pressure engine.

An article from the Madison *Express,* September 19, 1840, provides some pertinent statistics:

> The Buffalo Journal states there are about fifty-three steamboats now navigating Lake Erie. Their cost is between $24,000 and $120,000 each. Between buffalo and Detroit, the fuel consumed is about 100 cords, at $1.75 per cord. The cost of outfit between the high and low pressure boats is significantly great. The Missouri, a high pressure boat of 600 tons, when ready for service cost $80,000. Her engine is horizontal, and $28,000 was paid for it. The Cleveland, a low pressure boat of 570 tons, cost over $70,000, her engines $45,000. She consumes 600 cords of wood between Chicago and back, while a high pressure boat of the same dimensions, will burn about 364 cords in running the same distance.

The Stern- versus the Side-Wheeler

It has been speculated that Henry Shreve developed the stern-wheel paddleboat (one wheel at the rear of the boat, as opposed to two opposing paddlewheels on the sides) as an early attempt to thwart the Fulton-Livingston patent. All his early boats were created following this pattern, with the *Washington* being the most famous. In 1817, when this boat steamed from New Orleans to Louisville in a record 21 days, it proved the viability of upstream navigation and brought great credit to its captain.

The success of the *Washington* did not establish the superiority of the stern-wheeled boat, however, and that style soon fell into disfavor. During the period before the Civil War, side-wheelers were by far the most dominant design. There were a number of reasons for this, the most apparent being the breaking of the patent, which permitted boat builders to copy Fulton's style. Second was the issue of maneuverability. When pulling into the wharf, steamboats went in headfirst. That required the boat to depart by backing out. Having the ability to work the side-wheels in opposite directions greatly facilitated this often tight and delicate procedure.

When going around a sharp bend in the river, the side-wheeler used one wheel working forward and the other operating in reverse. A stern-wheeled boat, on the other hand, was required to "flank" around the elbow, or tight angle, by backing against the point and allowing the current to swing the bow around the curve.

Paddle-wheels placed on the sides also had the added advantage of better weight distribution, a crucial factor in preserving an even keel. This was particularly important in maintaining even water levels in the boiler. If the boat tipped, even for a brief period, unequal water distribution exposed portions of the boiler, and was a prime cause of explosions. On the river, dual paddle-wheels facilitated the steering process, enabling pilots to avoid obstacles by shutting down one side while working the other.

Stern-wheelers carried most of their weight in the back and thus were susceptible to heavy winds. Going downstream while the wind blew upstream created a hazardous condition that rendered the boat nearly helpless. The current acted on the stern and the partially submerged paddle-wheel, pulling the boat downstream, while the wind blowing against the tall chimneys and pilothouse pushed the bow upstream. It often required hours to reverse the process and straighten the boat. The technique required the pilot to turn the wheel hard over, throwing the rudders as far to one side as possible and then back strongly against them. If the boat could not be backed up fast enough, the engines had to be stopped and the vessel swung to a right angle with the river, permitting the current to straighten her. This type of maneuvering might have to be repeated twenty times before the boat would get underway in the proper direction.

If all efforts were unsuccessful, the boat had to be run ashore, a line attached to a sturdy tree and the boat backed out so that her bow pointed downstream. A side-wheeler under similar conditions would have come ahead on one wheel while backing on the other and in a manner of minutes been straightened around.[4]

Because of these difficulties, insurance agents charged higher premiums on stern-wheeled boats. It was only in the later years, as technology advanced and main decks became wider, that the placement of freight toward the front of the boat helped correct the unequal weight distribution and ease navigational problems.

In favor of stern-wheelers, the position of the wheel at the back gave it significant protection from floating debris. Half-submerged logs or chunks of ice that frequently struck the side-positioned wheels, breaking the paddles or becoming lodged in the mechanism, were either avoided entirely by stern-wheelers, or the driftwood hit the more protected hull. There were far fewer accidents of this type on stern-wheeled boats.

When navigating through narrow tributaries, the stern-wheeler's trimmer configuration helped it get by, while the significantly wider side-wheelers had trouble. The position of the wheel in the back also facilitated getting the boat off a sandbar, as power generated from behind more easily propelled it forward.

On a more emotional level, the side-wheeler presented a picturesque beauty of balance and speed, appealing to the eye of both rivermen and the public. And what the paying customers liked, the paying customers generally got.

By 1830, stern-wheelers had nearly vanished, relegated primarily to boats of small tonnage working the shallow tributaries and performing harbor duty. Smaller stern-wheelers, known collectively as the "mosquito fleet," were capable of working through all four seasons, while the larger side-wheelers were laid up during the winter and at periods of low water. During these bad river conditions, however, freight was too heavy to carry and only passengers were transported. It was not until the post-bellum years that stern-wheelers came into their own, but these vessels were primarily used for the transport of cargo, inasmuch as the passenger trade had all but been consumed by the railroads.

Technology Races Ahead

Power on a steamboat came from the paddle-wheels. The ability to increase power, without altering the depth of immersion or revolutions per minute was obtained by constructing longer and wider buckets (the paddle-wheel slats) and by raising the level of the shaft as the diameter of the paddle-wheel increased. It did not take long before travel times radically improved, and by 1825, many steamboats made over 100 miles a day.

Travel Tables

New Orleans to Louisville (1,350 miles)

1821	17 days, 17 hours
1827	8 days
1833	7 days, 6 hours

By the mid–1850s, a steamboat made about 10 miles per hour upstream, and 14–16 miles an hour downstream.

Cincinnati to Louisville (135 miles)

1819–20	upsteam 45–50 hours
	downstream 15–16 hours
1850s	upsteam 11–12 hours
	downstream 8½–9 hours

Pittsburg to Cincinnati

1830s	3½ days
1850	48 hours

St. Louis to Galena

1836	3 days, 6 hours (5 mph)
1845	43 hours, 45 minutes (9 mph)[5]

On May 11, 1830, The *Adams Sentinel* reported on the staggering achievements achieved by steamboats:

RAPIDITY OF INTERCOURSE

The facilities offered of our commercial transactions and intercourse, by the improvements of our steamers and the navigation of the Ohio, are truly astonishing. Passengers arrived in this city [Pittsburg] on Friday last, in the short space of *fifteen days,* by steamboats from New Orleans.—The distance is 2400 miles!

A gentleman who departed in the steamer Trenton, writes his friend in this city as follows: "I left Pittsburg in the steamboat Trenton, (Capt. Jesse Hart, master,) on the 28th ult. and arrived at St. Louis on the 5th inst. having completed the trip from Pittsburg to St. Louis in *eight* days. The Trenton ran from Louisville to St. Louis in three days and nine hours. This was considered to be unusually quick, as she was so heavily loaded that her guards were literally under water." *Gaz.*

Eight years later, the *Adams Sentinel* of April 2, 1838, records another letter originally published in the New York *Star*:

We received here yesterday express slips from New Orleans of March 17th, distance 1500 miles, which contain dates from St. Louis, 1000 miles distant, of March 7th. by steamboat down the Mississippi. This is a space of 2500 miles, nearly two-thirds of the circuit of the U. States, has been run over in eighteen days.

Wooding Up

The typical steamboat took on wood ("wooded up") approximately every thirty miles. It was not economically feasible to carry more wood than necessary, for it took up an inordinate amount of space. Of greater concern, however, was weight: carrying more than one day's supply kept the boat low in the water and increased the amount of energy required to maintain good time. That meant frequent stops, and supplying steamboats with wood quickly developed as a primary means of seasonal employment for hundreds of men living along the shorelines. When a young Thomas J. Jackson (better known as "Stonewall"), and his older brother ran away from their home in western Virginia, the pair worked down the Allegheny River and became wooders for a time.

Western steamboats were notorious for their inefficiency, and burned prodigious

An 1870 image of the steamboat *Arkansas*. Docked beside it is a wood flat, waiting to be loaded. Note the fancy embellishment of the name on the pilot house.

amounts of wood. As it was inexpensive and plentiful, however, this was never considered a problem. Nor did the frequent wooding stops unnecessarily delay a trip, for crews became adept at carrying wood aboard in a very short time. Frequent stops also permitted captains to pick up passengers or whatever cargo happened to be on the wharf. Officers often delivered mail from point to point (despite the federal government's prohibition of such practice), and the stops also gave cabin passengers a chance to stretch their legs. Deck passengers, too, took advantage of these brief stops to purchase, at exaggerated prices, food and perishables to sustain them for whatever remained of the voyage.

Wood supplies varied depending on which part of the river the boat worked. Resinous pine was considered the best fuel, followed by oak, beech, ash and chestnut. Cottonwood was readily available, and it burned well but too quickly. Always, seasoned wood was preferred over recently chopped trees. When green timber or rotten driftwood was substituted, efficiency suffered. From Cairo to New Orleans, wood was less plentiful and sold for one to three-quarters more than on the Upper Mississippi. On the Upper Mississippi, cottonwood was the chief fuel, but the farther up the river a boat steamed, the more expensive the price. Toward the headwaters, several times as much was paid.

At woodyards, the method of stacking was as different as the types of wood offered. Typically, ranks contained twenty cords of wood, measuring eight feet high and eighty feet across. Disreputable dealers stacked rotten or green pieces in the middle, much the same as chandlers sold kegs of pork to sailing ships with bone and gristle buried at the bottom. By the time the worthless offal was reached, the vessel was far out to sea, too late to claim redress. For the woodman, however, the comeuppance came as soon as the return trip. A clerk cheated once took extra precautions when buying from the same seller, haggling down the price in order to redeem himself in the captain's good graces. If he did not obtain suitable discount, he refused to buy, and had the mate, with his colorful vocabulary and bulging muscles, stand ready to protect him from the less-than-righteous outrage of the seller.[6]

Woodyards were either temporary affairs, where squatters settled on federal lands and chopped trees until ushered off by agents, or those of a more permanent nature. Farmers who owned land reaching to the river often supplemented their income by felling trees and shaping the logs to the proper length for sale to passing boats. Those who provided the best service and offered fair prices were the most frequented. Often, captains would contract for all the wood of a certain yard and stop there throughout the season.

In later years, flatboats filled with cordwood were taken in tow and captains paid a higher price for this wood, it being more convenient to unload from a raft than send deck hands ashore to do the heavy labor. It was a matter of no consequence to bring the flatboat alongside and rapidly take aboard the needed fuel. When empty, the flatboats were turned loose to float back to the boatyard for restocking and use by the next steamboat. This arrangement was not used when a steamer was going downstream as there was no way to return the woodboat.

The danger in this procedure came from an excessive use of speed by the steamboat, which continued upriver during the unloading process. If the pilot were not careful, the raft would be swept under the boat. To avoid this, wood was always taken first from the bow, and pilots were required to modulate forward progression until the flatboat was set free.

Deck passengers lacking money, particularly emigrants, often traded their services as wooders in exchange for reduced fares. It may well be said they earned their way, for the task was onerous and sometimes dangerous. The wooder was required to carry six or seven logs, each 4–5 feet in length, over his shoulders. In bad weather when the boarding ramps were slippery, the chances of falling into the churning river were high and dangerous. If a passenger-wooder were injured, responsibility for care fell to the individual, for he was not considered an official part of the crew and therefore was not covered by maritime law.

When boatyards were scarce or there was no wood waiting, crews were dispatched to chop their own. Steamboats of the smaller classes burned 12–24 cords of wood every twenty-four hours, and the larger boats of the 1850s burned 50–70 cords per day. When wooding up at night, large iron baskets were filled with fatwood and suspended over the water. The light illuminated the shore for several hundred yards, enabling the laborers to see while they worked. Adding pulverized rosin to the "jack" lamp caused the fire to flare and burn fiercely for a time.[7]

To increase speed or to augment the properties of green wood, auxiliary fuels such as resin, turpentine, oil and lard were used. These also substituted in case of emergency, and were liberally utilized during races.

Interestingly, on her groundbreaking voyage in 1811–12, the *New Orleans* burned both coal[8] and wood.[9] Although readily available at the upper Ohio River towns such as Pittsburg, coal was far scarcer lower on the Mississippi. Since a boat typically carried 150 bushels (only enough for one day), supply became a problem, and it was not until the 1840s that the availability of coal made it viable as fuel.

In the early years, engineers felt wood was easier on the boilers, being less liable to injure them by excessive heat. Wood burned until almost entirely consumed and the flues remained cleaner longer, consequently requiring less maintenance than if coal were used.

By the late 1840s, most eastern steamboats on the Hudson River and Long Island Sound had abandoned the use of wood and adapted their machinery for the use of anthracite coal. Coal in the Atlantic states was less expensive and, by this period, more readily available than wood. Coal also took up less space, always a major consideration. Toward the end of the decade, the use of coal on western boats had become more common, and mine owners aggressively marketed it to steamboat captains. Coaling stations were quickly established, and by 1848, many were in business along the Cincinnati River.

The transition from wood to coal had its cost. Fireboxes and grates were made for wood and were not easily adapted. Coal required a more shallow and smaller firebox and a large grate opening. In the early 1850s, engineers working the trades along the Ohio and Mississippi rivers above Cairo developed the practice of burning both wood and coal in equal amounts. This created a hotter fire, and experimentation soon proved that when wood logs were placed between the coals, the result provided for better airflow than if coal alone were used. Steamboats working the Lower Mississippi and Louisiana trades used only wood, as coal costs there were prohibitive.

The real coal consumers in the 19th century were not steamboats at all but railroads. Early trains used wood for fuel, but with the rapidly expanding technology, engines were soon adapted for coal. This insatiable need was a boon for the coal mine operators of the midwest. Those who had once looked to river transportation as the means of their profitability soon had bigger clients, and the coal industry grew in proportion.

What the coal-eating Iron Horse did for steamboat transportation is another story, entirely — one with a far less happy ending.

Stacking the Decks

Robert Fulton brought the steamboat west. What he did not fully comprehend was that the steamboat was a work in progress, barely set out upon a journey that would take the original "engine on a keelboat" design well into the realm of "Steamboat Gothic."

Along the interior eastern waterways, freight was transported by sailing ships. Barges were used for heavier materials, while the rapidly developing steam-towing system took portions from each. Unable to compete in that quarter, Hudson steamboats were conceived and designed primarily as passenger transport. As the decade of the *Clermont* advanced, Fulton and his (illegal) competitors sought to develop the New York City-to-Albany trade and the basic ferry routes between the northeastern states. For this purpose, a low-pressure engine was adept and more than adequate. Although improvements were made to the internal workings, an equal or greater concentration centered around structural integrity and creature comforts.

On the opposite end of the spectrum, there were not enough people on the frontier to make passenger-only trade practical. Only the Cincinnati-Louisville passenger trade came close to approximating that on the Hudson. Immigrants traveling as deck passengers provided some money, but it was readily apparent that for early western steamboat ventures to succeed, their concentration had to be on the delivery of cargo: either produce and manufacturing from the Ohio Valley, or cotton, sugar and foreign goods from New Orleans. This set the pattern for all future development, and few if any western boats ever provided passenger-only service. With competition stiff and ever growing, the idea of making a profit by any and all means assumed precedence.

Mississippi pilots required a vessel capable of withstanding the rigors of mud and debris-choked rivers, variable currents, and a superstructure able to withstand tremendous weight. In order to accommodate both well-paying travelers and many tons of cargo, the cabin quarters were shifted to the upper deck, freeing the lower, more accessible main deck for cargo.

The prospect for making money was already established by 1817, when an operator could hope to net 40 percent of the cost of the steamboat in a single season. Working just the New Orleans to Natchez route (270 miles), hauling both cargo and passengers on a single trip, a captain might net $4,000, a staggering return on his initial investment. The following year (after the Fulton-Livingston monopoly was broken), owners reported making a capital gain of $40,000 for the season, after deducting expenses, repairs and interest on loans. This was superior to any other type of monetary investment in the United States.[10]

Quick-witted westerners rapidly determined what needed to be done to make the prototype more profitable still. Although early boats of the era still appeared as an odd combination of rafts and sailing ships (as late as 1828, more than one-third of all steamers still had figureheads on their prows), the innovators eagerly set themselves to the tasks at hand.

In the 1830s, the typical design still presented the top-heavy, boxlike, and rectangular paddle-wheel housing (the wooden structure built around the buckets for protection), and the forward, semi-detached position of the chimneys. Using trial-and-error as a guide (and seldom, if ever, bothering to patent their improvements, leaving that which worked available for anyone who chose to copy them), individual engineers and designers tackled sundry problems.

In twenty short years, the paddle-wheels became enclosed in circular housing and were increased in size, reaching to and rising above the line of the hurricane deck. Chimneys passed up through the deck, instead of being placed directly forward of the main structure. The pilothouse also evolved, being repositioned somewhat aft in the boat, streamlining the silhouette. Hulls were made lighter (with an eye toward the moment, rather than longevity), and high-pressure engines were placed in newly constructed vessels. Again with an eye toward weight, the superstructure (that above the deck) was constructed of flimsy material, also reducing draft and the problem of top-heaviness.

These adaptations succeeded in lessening the overall weight, but tended to augment the tremendous vibrations of the machinery that shook the boat from stern to bow, a fact well documented in contemporary accounts. These solutions also occasioned considerable noise, heat, and danger from fire, but engineers, if not passengers, accepted the benefits over the drawbacks, and little was done to address these concerns.

A loaded steamboat sat nearly four-fifths out of water, and when empty, less than four feet was actually submerged. Given the above changes and the fact most of the heavy equipment aboard, including the machinery, wood and freight, was on the main deck, the appearance of being poorly proportioned was deceiving. By 1847, only 24 inches of water was required for a steamboat to carry 50–100 tons of cargo. This fact alone extended the ability of the boats to operate during periods of very low water and extended the amount of days per year they worked.

Most of the reconfiguration was done to provide more room for freight, the primary and most profitable concern of owners. As the hull decreased in size, new decks were added. The upper between-decks was a characteristic of western steamboats. It was used solely for cabin passengers, and the hurricane deck, which formed its roof, served as an open promenade.

The lower between-decks used for machinery, cargo and deck passengers. This deck was closer to the water level facilitating the transfer of heavy freight to and from the wharf. This reduced the capacity of the hold, eventually making it impractical in smaller-tonnage boats, where a lack of headroom prohibited easy transfer. The height was 11–12 feet, reaching 15–20 feet on the largest vessels. When the hull depth fell below six feet, most of the cargo was stored on the main decks. When more space was needed, designers increased the height of the boiler deck, so more crates, barrels and bales could be piled atop each other.

As the steamboat gradually developed her distinctive and readily identifiable shape, other alterations were made. The fully developed boat had three primary decks: the main (lower deck), the boiler deck (which came into use as a promenade deck) and the hurricane deck, from which the Texas deck rose. Atop sat the pilothouse.

An addition to the upper works was more narrow than the main cabin deck and shorter in length. It became known as the "Texas" (sometimes spelled with a lowercase "t"). Its earliest use was to give greater elevation to the pilothouse, as well as provide living quarters for the officers. The Texas dates from the mid–1840s, but the origin of the name is more obscure. An editor for the Cincinnati *Gazette* (October 11, 1846) speculated that it came from the fact the state of Texas had been annexed. Another explanation offered was that staterooms were named for states, and that when the new deck was added, only one state name — Texas — remained available.

Thereafter, "guards" were added as extensions of the main deck beyond the line of the hull. They were developed to protect the projecting paddle-wheels and to provide bracing and support to the outer ends of the paddle-wheel shafts. The guards also provided additional room for cargo and wood, and served as a passageway between various quarters of the boat. When not fully loaded, they also provided pleasant walkways for cabin passengers.

Mechanical Developments: A Work in Progress

In order to cope with seasonal water depths that fluctuated up to fifty feet, as well as varying currents, the high-pressure engine became the standard on western steamboats. In the early years, these engines typically created a pressure of 100 pounds; later, 120 pounds became the norm. Hard firing by use of resinous materials increased pressure to over 150 pounds. Steam was channeled directly to drive the pistons, then exhausted into the air through the chimneys. (Chimneys were typically thirty feet above the water, reaching 50 feet by the 1850s. Thereafter, larger classes of steamboat carried ones as high as 90 feet, which seemed to have been more for show than necessity.) Discharged steam was released at high pressures. Black clouds of smoke and cinders were indicative of great fuel waste.

Hard firing and supplemental fuel resulted in greater speed, but to increase power even further, engineers added a second engine to the side-wheeler. That permitted each paddle-wheel to move independently, often in opposite directions to increase ease of maneuverability and turning power.

Steamboat boilers were often blamed for explosions and for good reason. The iron plates used were half the thickness required by law in France, and were habitually badly joined. Cast-iron boiler heads were carelessly fastened and internal flues were without bracing. Complicating matters, safety devices were slow to develop and inspection of finished boilers at the factory were inadequate. Boilers "panting"—visibly expanding and contracting—were evidence of the opening and closing of engine valves.

Inefficient and badly proportioned grate and boiler surfaces led to the practice of driving the furnaces at a furious rate, continually adding wood until the firebox was literally choked with ash. Cylinders, steam lines and tops and sides of boilers were seldom insulated, and the practice of leaving furnace doors open to augment the draft coming in off the river, or the use of fan blowers, added to heat loss from condensation.

Importantly, water was supplied to the boilers only when the engine was running. Steam, however, continued to be made as long as the fires were allowed to burn. Unless care was taken to dampen the fires every time a boat stopped, water levels would fall to dangerous levels. This weakened the boilers and allowed steam pressure to build, and explained why most explosions happened just after a boat left the landing.

The remedy for steam buildup was to keep the paddle-wheels slowly rotating while at landings, or to actually keep the boat moving in a circle while passengers and freight were carried aboard or discharged via yawls or flatboats. Since this added to delays at the wharf and proved irksome, few captains resorted to these safety measures. (If the vessel happened to be single-engined, the paddle-wheel could be unshipped. This permitted the engine and thus the supply pump to keep running and thus avoided a buildup of steam.)

In the early 1840s, a small auxiliary engine was introduced. Called "the doctor" because it cured many ills, its principal use was to supply the water pump independently of the main engine. Most larger classes of boat utilized this improvement by the 1850s. It hardly proved a cure-all, but it minimized danger and lessened engine and boiler wear. Later uses for the "doctor" (used primarily on double-engine boats) included running the bilge pump and fire engine. It also shaved time at landings, saving the engineer from wetting down the fires and then having to work for half an hour to get them hot again.

In reference to fire engines, a note in the *Adams Sentinel,* April 13, 1840, reports:

A Fire Engine for Steamboats has been made in Boston by Mr. Creed. It can be worked by the steam engine at the rate of 400 gallons per minute, [compared to] 200 by sixteen men. It may be put into operation five minutes after the discovery of a fire. Should a boat strike a snag and leak, the engine will also pump out the water at the same rate.

As late as the 1850s, few western steamboats had steam gauges. Engineers depended upon the sound of the exhaust, the appearance of the engines and a visual and auditory appraisal of steam issuing through the cocks. That was soon to change.

Steamboat Acts: From Self-Regulation to Government Oversight

In 1830, the House of Representatives began debating a bill put forth by Mr. Wickliffe, from the select committee appointed "in pursuance of a resolution of the House upon the subject of preventing accidents in steamboats under certain penalties, for the strength and fitness of the machinery (and in particular of the boilers) attached to steam vessels, and respecting the skill and experience of the engineers entrusted with their management:

Resolved: That the Secretary of the Treasury be directed to collect and communicate to this House, at the next session of Congress, such information (and report his views on the same) as in his opinion may be useful and important to Congress in enacting regulations for the navigation of steamboats or steam-vessels, with a view to guard against the dangers arising from the bursting of boilers.[11]

Three days later, the *Adams Sentinel* elaborated on its previous story, so near to the hearts and purse strings of passengers and merchants.

Mr. W [Wickliffe] said, he offered this resolution because he believed the House possessed the power, and might, by a proper enactment, greatly diminish, if not prevent the reoccurrence of the distressing accidents which so often take place on board of steam boats — Whoever, he said, had read the account of the late dreadful calamity on board the Helen Macgregor,* must be satisfied that it was owing to the negligence of some officer on the boat.

Mr. Wickliffe went on to call for any interested persons who might assist in framing a bill on the subject. The newspaper editorialist then proposed that whenever a steamboat stopped for any purpose, the safety valve should be raised: "At present it was the practice, for the purpose of saving fuel, to refrain from letting off the steam, regardless of human life."

Amid continued public outcries over steamboat accidents, the United States Congress first enacted legislation in 1832. While intended to provide for "the better security of the lives of passengers," it had little effect in that area. Provisions called for licenses on all steamboats that would be granted only when a boat met certain criteria. Hulls, boilers and machinery had to undergo periodic inspection and be certified as sound.

---

*The Helen Macgregor *exploded near Memphis from a burst boiler in 1830, just as the boat was leaving the dock. Over forty persons died and many more were extremely scalded, all of whom were deck passengers. By 1830, it was becoming apparent that most of the steamboat accidents stemming from burst boilers happened when the boat was just leaving the wharf.*

Owners and masters were required to employ "a competent number of experienced and skilled engineers," and when the boat was not in motion, the master was required to open the safety valve to keep steam pressure low.

All steamboats were required to use iron rods or chains in place of tiller ropes and to carry navigation lights at night. Additionally, the act stated that all explosions were to be accepted as being the result of negligence until evidence to the contrary was presented.

Before passage, the bill was stripped of its workable specifications and watered down by vague language and impractical provisions. Most captains considered the act more a nuisance than an order and ignored it. Equally as bad, selection of inspectors was given to federal district judges who knew nothing of steamboats, making it difficult for them to select competent men. The latter problem basically resolved itself as the remuneration offered was inadequate to the work expected, and few applied.

The *Racine Advocate,* April 5, 1843, offered details of another attempt at steamboat regulation:

> SAFETY OF STEAMBOAT PASSENGERS—The Act passed at the recent session of Congress, provides that every vessel propelled in whole or in part by steam, shall be furnished with additional steering apparatus, to be located in such part of the vessel as the inspector may deem best to enable the officers and crew to steer and control it, in case the pilot or man at the wheel is driven from the same by fire; and it makes the filing of a certificate with the inspector, that such apparatus has been provided, a condition precedent to the registry of any vessel exclusively propelled by steam.
>
> Steam vessels so furnished are permitted to use wheel or tiller ropes, composed of hemp or other good and sufficient material, around the barrel or axel of the wheel, and to a distance not exceeding twenty two feet therefrom, and also in connecting the tiller or rudder yoke with iron rods or chains used for working the rudder. *Provided,* that no more rope for this purpose shall be used than is sufficient to extend from the connecting points of the tiller or rudder yolk placed in any working position beyond the nearest blocks or rollers. *And provided further,* That there shall be chains extending the whole distance of the ropes, so connected with the tiller or rudder yolk, and attached or fastened to the tiller or rudder yolk, and the iron chains or rods extending toward the wheel, in such a manner as will take immediate effect, and work the rudder in case the ropes are burnt or otherwise rendered useless.
>
> Courts before which any suit is pending for neglect to use iron rods or chains, instead of wheel and tiller ropes, are required to order the discontinuance of such suit, on such terms as to costs as to the Judge seems reasonable, if it shall appear that ropes were used from a well grounded apprehension that such rods or chains could not be employed for the purpose aforesaid with safety.
>
> The act further requires the Secretary of the Navy to appoint a board of examiners, consisting of three persons, of thorough knowledge as to the structure and use of the steam engine, whose duty it shall be to make experimental trials of inventions and plans designed to prevent the explosion of steam boilers and collapsing of flues as they may deem worthy of examination, and report the result of their experiments and efficacy of such inventions and plans, which report the Secretary shall cause to be laid before Congress at its next session. It shall also be the duty of said examiners to examine and report the relative strength of copper and iron boilers of equal thickness, and what amount of steam to the square inch, when sound, is capable of working with safety; and whether hydrostatic pressure, or what other plan, is best for testing the strength of boilers under the inspection laws; and what limitations as to the force of or pressure of steam to the square inch, in proportion to the ascertained capacity of a boiler to resist, it would be proper to establish by law for the more certain prevention of explosions.—*Jour. Com.*

Problems persisted and in 1858 a stronger steamboat act was passed, although not without opposition. Two sides quickly — and bitterly — developed. Senator Robert F. Stockton of New Jersey defended the laissez-faire system with an extraordinary statement, asserting that a higher principle than "liberty and equal rights" existed: one pertaining to the protection of steamboat owners who had the right to do with their property as they pleased. This "right," Stockton continued, surpassed human life, which was merely "transient and evanescent."[12]

His comments, although reflecting the temper of the times, seem particularly ill-suited for a career seaman who participated in the War of 1812 and later, while stationed off the coast of West Africa, helped negotiate a treaty that led to the founding of Liberia. In 1841, Stockton was offered the post of secretary of the navy by President John Tyler, but he refused. Stockton successfully campaigned for the development of the Navy's first screw-propelled steamer and eventually assumed command of this boat, named the USS *Princeton*, which operated between 1843–1849.[13]

In an 1840 workup to what became the Steamboat Act of 1852, a senate committee noted, "We have had too many proofs of the futility of relying alone on the self interest of proprietors, or the sense of personal hazard of engineers, for security against steamboat disasters. Legislative regulations and penalties must, therefore, interpose their protection."[14]

One of the first specifications in the act of 1852 was the requirement that all steamboats be equipped with "suitable steam gauges." This brought into general use the Bourdon bent tube gauge, invented in 1845. The Steamboat Act additionally called for the annual inspection of steamboat boilers, established standards for the dimensions and methods of construction of parts, and called for the inspection and stamping of the boiler plate during manufacture.

A licensing system was also established for pilots, with licenses typically being granted only for those parts of the rivers where the pilot was capable of demonstrating his competence. Supervising inspectors were authorized to create regulations for the passing of boats on the river, which common practice had failed to standardize. One rule in particular proved effective: the substitution of the steamboat whistle for the bell when communicating intentions to another vessel. This rule was credited with a marked reduction of collisions from contrary or poorly heard signals.[15]

This legislation also provided for adequate means of escape for deck passengers in case of fire or explosion. Numbers were determined as to how many deck passengers might safely be carried, but this provision was difficult to enforce, and carried the minuscule fine of a refund of passage fare, plus ten dollars. Additionally, the passenger was able to obtain redress only if he sued in court. It was hardly adequate compensation if the passenger failed to take advantage of the new (and problematic) escape measures.

7

Navigating the Inland Western Waterways

A pilot must have a memory; but there are two higher qualities which he must also have. He must have good and quick judgment and decision, and a cool, calm courage that no peril can shake.[1]

The techniques and intricacies of navigating the ever-changing Mississippi, Ohio and Missouri, as well as the innumerable tributaries, presented challenges that the pilot had to learn from the bottom up. Early crews proceeded by trial and error, for there were few guidelines, none of which pertained to actual steamboats. Never had so heavy a vessel, nor one as cumbersome or dependent on mechanical power, been dreamed of in the early years of western settlement, and many a vessel was sunk before rivermen became adept at avoiding the various natural and man-made obstacles confronting them.

River navigation was the sole responsibility of the pilot. In the early days, when boats did not operate during the hours of darkness, only one pilot was necessary. With the advent of round-the-clock navigation, two pilots worked the same boat, taking alternate turns at the wheel. They plotted the course and, ideally, steered the boat safely to her destination. This task included setting speed, ordering the engineer to supply power as needed, reading the river, the banks and the weather, and being familiar with every type of obstruction encountered along the route. Mark Twain and George Merrick each gave intricate descriptions of piloting the river, and their discussions are used in this section to relay the various situations a western river pilot encountered.

Using the jackstaff as a navigational tool (as opposed to it being solely a pole upon which to fly a flag) in steering by daylight, and the stars at night, the pilot "straightened the boat up" after the captain backed her away from the wharf. (That and directing the boat into port were the only actual navigational duties ascribed to the captain.) Maneuvering the large, heavy, multispoked wheel with both hands, the pilot "shaved" the docked boats still in port, maintaining a position as close to the shoreline as possible. The rationale for this was that the easy (calmest) water lay along the banks, while the current in the river presented a more formidable challenge.

Once in the river, a boat progressing upstream hugged the bank to take advantage of the slack current. Navigating against the flow of water was easier than going with it, especially in bad weather or low water, for the still water exerted less pressure on the vessel, vastly improving maneuverability. Progressing downstream, the boat stayed well toward the middle, as the rapid current exerted a powerful force. While it was relatively easy to get the bow around a crooked bend, it was considerably more difficult to prevent the stern from swinging from side to side. Once out of control, the back of the boat was vulnerable and might strike a rock or other solid obstruction and sustain serious damage.

Putting Out the Lead

When a boat came to an area of shoal (low) water, the pilot sensed the change in depth by the way the boat dragged and the strained motion of the paddle-wheels. This necessitated a prompt reduction in speed. (When in doubt, the rule of thumb was to "ring the stopping bell and set her back."[2]) Labboard (port; left) and stabboard (starboard; right) leads were then lowered into the water. The leadsmen called out the depth, and their cries were repeated by "word-passers" on the hurricane deck. By calling "Mark three! Quarter-less-three..." the depth of the river was ascertained and the boat put in her "marks" (points used by pilots to navigate).

On the upper rivers where water depth was shallower, a twelve-foot pole was also used in determining depth. Rather than what the words implied, the expression, "No bottom!" meant the depth was greater than the sounding pole, or deeper than two fathoms.

"Leading" was not only used for immediate problems such as determining how to get the boat over or around bars, but also to enable the pilot to keep and maintain an accurate picture of the river bottom. The numbers he received at various points along the river were mentally compared to those taken on the last trip. The variable numbers indicated the evolution or dissolving of reefs, prompting the pilot to adapt his marks for future voyages. Sounding, therefore, enabled the navigator to create a picture of the river bottom as accurate as the one he possessed of the banks and bluffs.

Along the Upper Mississippi, Cassville, Brownsville, Trempealeau, Rolling-stone, Beef Slough, Prescott, Grey Cloud and Pig's Eye were the "nightmare" sections of river. In dangerous situations, the engines were turned off and the vessel allowed to drift with the current. When the boat touched bottom, the order was given to put the engine "hard down" (give it all the power possible), and the paddle-wheels churned over the obstruction.

Depth generally increased from the headwaters to the mouth of a river, and from a tributary to its outlet. The Ohio River at Pittsburg measured eighteen inches at its lowest point and increased to three feet at the mouth; the Mississippi measured two feet at St. Paul, four feet at St. Louis and five feet at Cairo. The only times these ratios reversed was during periods of rapidly rising or lowering waters.[3]

The current depended upon the slope of the river and the stage of water. In the deepest pools during very low water, the current might be one mile an hour or less; three miles per hour at a moderate stage; and up to four or five miles per hour at the peak of flood season.[4] Areas by obstructions and rapids had much higher rates.

Sounding the Bad Places

Sounding was done in areas of the river where the water was especially low. The boat was tied up above the shoal crossing and the off pilot (the one not on duty) took a crew out on the river in a yawl, or "sounding-boat." They searched for the deepest water, being observed from the pilothouse with a spyglass. On rare occasions, if the pilot at the wheel was not satisfied with the positioning, he signaled the crew with the boat's whistle, directing them into position.

Sounding was done with a pole 10–12 feet long, which was used to take measurements at the shallowest places. When these areas were identified, a crewman in the yawl set the buoy, typically a board 4–5 feet long, with one end turned up. (Mark Train described it as a reverse schoolhouse bench with one of the supports removed.) The buoy was anchored in the shoal water by a rope attached to a stone. At night, a paper lantern with a candle was fastened atop it, permitting the signal to be seen for a mile or more.

Once the work was completed, a crewman in the yawl raised an oar, indicating the steamboat could come forward. The boat either called on all her steam and went grinding over the buoy and the sand or followed the yawl as it led them through the deepest areas. If a mistake was made and the boat floundered, it might take several days of sparring before the craft was freed.

Reading the River

Varying weather conditions and the available amount of light affected how the river and its banks were perceived. Learning the shape of the river was one of the earliest lessons of a cub pilot. Starlight threw heavy shadows over the shoreline, making every tree branch appear solid, while pitch blackness evened out the coast, hiding true obstructions. Grey mist erased all shape, and various types of moonlight changed the configuration in different ways, making knowledge of all phases essential.

Being adept at navigation meant possessing the ability to read the river, including the color and texture of the water's surface, to determine what lay beneath. Knowing the composition of the river bottom enabled the pilot to predict what type of obstruction might have formed there. Equally important, he had to have a feel for each particular boat he worked, hearing changes in how the engine strained and determining distances from the echo of the boat's whistle off the surrounding hills. Resistance of water against the hull varied by the water's depth, and he had to be alert for how that force affected maneuverability.

Passing over a reef required "mounting," or going over. The technique demanded significant power from the engines, for a boat resisted the maneuver. When the tiller shifted and jerked, that was another indication the water was too shoal. Occasionally, one wheel was stopped while the other was utilized for maneuverability. If the boat "smelled" the bar beneath the reef, it headed into it, and maneuvering away sometimes sent it careening out of control and plunging ahead. In that case, it might take a mile or more for the pilot to bring it under control.

Piloting by observing the bank was of equal, or greater importance than relying on the leads. The leads might "lie," but to the seasoned pilot, the bank accurately reflected the stage of the river. By observing how far up the water table was, the pilot could ascertain whether the river was rising or falling, and by watching how far the water reached to the tree line, he could mark the boat's progress. Driftwood usually indicated a rising river, for the current swept dead limbs into the water, but wood also continued to float ("strand") as the river fell, so a pilot had to be aware of the difference. Lines of sand or fine sediment along the bank also indicated a falling river.

Learning these marks (observation points along the river) was no guarantee of continued success, however, for many changed from year to year, so that a pilot was continually learning and relearning the river.

Close (narrow) chutes were not run on a falling river or when going upstream, and few were run going downstream. A rising river covered bars that were impassable in low depths, and typically low places were filled to a level that permitted easy passage. Marks told the pilot where to steer. When a pilot determined to venture through a crack, or chute, he had to be certain of his decision, for once begun there was no going back. These passageways were too narrow to turn around in and too crooked to back out of. He needed to be equally confident of the stage of the river, for shoal water was always found at the head of chutes. If he were caught on a falling river and grounded, the boat would likely be stuck there for six months, or until the river rose.

On a rising river, the incidence of floating and half-submerged debris made navigation difficult and dangerous, and required a constant shifting of position within the channel. If a log was too large to be avoided, the engines were stopped and one wheel would "walk" over the log, careening the boat and creating a terrible noise. If a large log were struck, it might lodge and become wedged across the bow, backing the water up behind it. Careful maneuvering was required to get around it.

Rising water also brought out another sort of obstruction, one not typically found in catalogues on the subject: rafts, heavily laden with people and cargo. Despite the law requiring such vessels to carry warning lights, many did not, earning them a place as an additional river hazard.

On very dark nights, light was the enemy of the pilot. He ordered all illumination blocked, which included everything from the red tip of a cigar to the boat's furnaces, which were curtained with huge tarpaulins made of heavy canvas, called "shroud," or "mufflers." No fire was allowed in the pilothouse, and even the skylights were closely blinded. This prevented reflections from appearing on the water, which might be mistaken for floating objects. The darkness also kept the pilot's pupils expanded, so he might absorb images through the gloom.

In point of fact, illuminating the river by artificial light would have done little good as far as navigation went, for the steamboat was not steered by landmarks close at hand, but rather by those a mile distant. The absolute blackness permitted a better view when separating the dark of sky and river from the graded shades of the shoreline and surrounding terrain.

Crooked Rivers and Shoving It into Gear

Steamboat engines utilized a lever and poppet-valve system that was heavy and awkward to manipulate. When the order came from the pilothouse to reverse engines, the camrod, connected to the rock shaft (weighing fifty pounds), was manually lifted off the bottom hook and the lever thrown over, thereby raising two heavy valve levers. The rod was then lifted about three feet and dropped onto the upper hook.

This procedure, called "shipping up," had to be performed by the engineering staff repeatedly in a short time while the pilot worked the steamboat around bars, obstructions, and twisted sections of river. It did not take long to exhaust the crew, and there was often little opportunity for them to catch their breath before the bells began ringing again. On side-wheeled boats where the engines worked independently, the engineer typically handled the starboard engine and his cub worked the port engine under his supervision.

If the procedure was not properly performed, the engine "centred," thus disabling the vessel. During periods of difficult navigation, "centring" could cause the boat to hang up on a sand bar or run into the rocks. In contrast, when making a landing, the need to reverse engines might happen only one or two times before the vessel was safely secured in her berth.[5]

Working the river boats might have seemed glamorous on paper, but was hardly so in reality. Cub engineers who tired easily or were guilty of repeated mistakes did not survive in the post, for as George Merrick observed, a man's reputation belonged to the entire river community. Those who failed to perform adequately were quickly ostracized and sent packing.[6]

8

The Trades and the Trade-Offs

If any one fact made steamboating unique, it was that the ownership and operation of a boat was open to nearly anyone. Very little upfront capital was required. All a man had to do was scrape together enough money to buy a boat—through a partnership or a loan—hang out a shingle, and open for trade. Should the budding entrepreneur wish to buy a new vessel, a smaller variety could be had for as little as $100 or less per measured ton. Larger vessels cost in the neighborhood of $40,000.

An even better idea was to purchase a used steamboat. Considering wear and tear and the significant depreciation, a serviceable craft might be bought for several thousand dollars. The huge number of boats on western waters is adequate proof that many took advantage of the opportunity and pursued the 19th century American dream: to get rich quick.

Perhaps an even more significant aspect of steamboating that drew the fiercely independent western frontiersman into the business was the appeal of freedom. Rivers were everyone's property: no laws prohibited the common man from taking his boat wherever he pleased. Unlike stagecoach lines that required way stations and roads, the construction of which were largely dependent upon state and local governments, a captain had the right to follow the water and put in at any port he desired. He adhered to no schedule, following the freight and passenger trade at will. The risks were his alone, and so, too, the profits. He succeeded or not as his own skill—and the caprices of Mother Nature—saw fit.

In the first decades of western steamboating, this casual style of operation dominated the field, but as more boats entered into competition, definitive routes naturally fell into place. There developed what became known as "the Trades," or established starting and ending points between which operators habitually ran. The Fulton concern actually established the first "trade," concentrating on the short, relatively easy passage from New Orleans to Natchez, over which freight and passengers were carried on the *New Orleans*.

In the parlance of the times, a "local trade" meant a trip between river cities within 20 to 30 miles of one another, while a "distance trade" meant a journey of at least 1,000 miles. Larger cities became hubs for smaller trades. On the Mississippi, St. Louis was the cutoff point. It was uncommon for a boat sailing the New Orleans–St. Louis trade to sail beyond that point, up the Mississippi or the Missouri River. The same held true in reverse: boats working the upper rivers seldom moved into the trades below St. Louis.

There were several reasons for this. Once familiar with the towns and settlements along the route, the captain made numerous business contacts. Presuming he provided adequate service and charged fair rates, he could count on these merchants for repeat business. Being familiar with local needs and customs, there was no reason for the steamboat owner to expand his territory unless times were particularly hard or a lucrative cargo fell his way that he could not refuse. Logically, pilots were seldom familiar with sections of the river they did not habitually navigate. Conditions varied greatly between stretches of river, and tributaries presented even greater challenges. Therefore, most pilots were hesitant to navigate on unfamiliar waters. If a captain did decide to change trades from one major section of the river to another, or if he moved operations to another trade altogether, the odds were great that he would also hire a different pilot who was accustomed to the territory. After the new regulations stemming from the Steamboat Act of 1852 came into general use, such a change of pilot was mandated by law, unless the pilot held a license for multiple trades.

On the Ohio River, Louisville and Cincinnati became major trade centers for through traffic, as well as serving the lesser trades catering to the tributaries near these cities. New Orleans was situated to become the center of steamboat operations on the Red, Yazoo and Arkansas rivers, as well as the river traffic coming and going from the Gulf of Mexico.

Trades were typically called by their points of travel: thus, in the long distance trades, Louisville to New Orleans, Pittsburg to St. Louis, and St. Louis to Fort Benton were major trades covering over 1,000 miles. Trades were further marked by accessibility. The points beyond which only smaller vessels could navigate became other divisions. Those included:

Nashville, on the Cumberland
Florence, on the Tennessee
Peoria, on the Illinois
Natchitoches (and later Shreveport) on the Red River
Little Rock, on the Arkansas
Leavenworth, St. Joseph, and Omaha on the Missouri

Trades were, by nature, seasonal, incremental and subject to change depending on the vagaries of commerce. They also gained or lost importance depending on circumstance. Steamboat men were quick to adapt, and at a moment's notice might alter a schedule or entirely cancel a trip. Such blatant disregard did not sit well with passengers or merchants, but their complaints usually came to naught. Once cargo was loaded or passage bought, receiving a refund was difficult and usually impractical.

As in the Upper Mississippi, the Lower Mississippi trades were dependent upon weather and water levels. Many of the larger-class vessels were compelled to put up during the extreme heat of summer when the depth of the rivers fell, or during the winter, for the same reason. Spring and fall brought annual rises, and with them eager passengers and an abundance of cargo. At times like these, work was plentiful for all, and prices adjusted accordingly.

An image of the *Argosy* taken in Middleport, Ohio, in 1867. A covered wood flat is attached to her side, waiting to be loaded. She worked the Wheeling–Pittsburg line.

Tramps, Packets and Lines

Western steamboats fell into three classes: tramps, packets and lines. The "tramps," or less descriptively, "transients," were the smaller-class boat, usually owned and operated by one or two individuals. Dependent on commerce, they went where the work took them, easily transferring from one small trade to another. Their fees tended to be lower than larger-class boats and their schedules less dependable. This freedom of movement for all river traffic differed sharply from conditions affecting oceangoing ships. Ships sailing the Atlantic and the Pacific were subjected to customs' restrictions and the necessity of clearance papers. They were required to register their arrivals and departures, submit bills of lading and pay regulatory fees, whereas steamboat captains were checked only by the operative limits of their boat, and ranged the rivers at will.[91]

Typically, packets were individually owned or the title was held by small number of partners. They differed from transients in that their routes were more formally organized, keeping to one trade for the duration of the season, and owners made a more concerted effort to keep to established times of departure and arrival. Packets were generally of a higher quality, offered more amenities to passengers and had a better safety record. In consequence, rates charged were somewhat higher than with tramps.

The Steamboat Line consisted of a number of steamboats operating as a single unit along a set trade route, with the underlying object being to limit competition. In the antebellum years, lines were informal arrangements, each boat being independently owned and operated. No central board dictated terms, and each owner kept the monies he earned. The structure of a line dictated only that a boat arrive and depart on time; thereafter, it was left to the owners to obtain freight and passengers as he was able. Because there were no guarantees of success and no profit sharing, the proposition was always touch-and-go, and captains were free to abandon their places in line with few, if any, sanctions. They would then be replaced as quickly as possible by another willing to operate on that segment of the river.

The principal disadvantage of line service as pertained to the individual captains was the necessity of leaving port on time. That often meant departing before a full cargo was obtained, lessening the profitability of the trip. The concept did, in fact, limit competition for those wishing their merchandise delivered within a set time, and provided more passengers. The chief drawback to lines was that they tended to operate only during the peak seasons, leaving the field open to transients and packets the remainder of the time.

Permanent routes developed along the major trades, with lines providing service from cities such as Pittsburg and Cincinnati on a weekly, twice weekly, or weekly departure. Longer voyages enabled a shipper or passenger to make direct connections with the same line, as opposed to seeking and researching the history of a new boat and pilot at every stop, which might include as many as six transfers along a 500-mile journey. For the most part, however, this "through service" required the traveler or merchant to buy transportation separately for each leg of the journey.

This was particularly irksome considering the reputation of western steamboats and their propensity to blow up. The wary traveler was well advised to research his boat of choice and solicit opinions from the locals as to the safety record of individual vessels, as Charles Dickens remarked in *American Notes*:

> Our next point was Cincinnati; and as this was a steamboat journey, and western steamboats usually blow up one or two a week in the season, it was advisable to collect opinions in reference to the comparative safety of the vessels bound that way, then lying in the river.

Once established on a line, however, that made the choice somewhat easier. Being "in combination," however, did not completely eliminate rate wars among member boats. Captains, desperate for profit, did whatever it took to survive, even if that meant cutting one another's throats, or steaming with reckless abandon. The *Commercial Herald* stated on July 3, 1843:

> STRIFS [sic] AMONG THE STEAM BOATS.— There was considerable strife witnessed here on Saturday last between the steamers WAYNE and WESTERN. The Wayne advertised to carry passengers from Milwaukie to Buffalo, cabin, for $5 — deck, $1. The Western left 12 hours behind the Wayne and carried for $10 in the cabin and $7 deck.— What the meaning of two boats chasing each other up and cutting under, when in combination, we can't divine. The Western was only 12 hours behind the Wayne and she was making her best time to overhaul Capt. Cotton.— Capt. C. informs us that hereafter, he shall land at the Pier. All right.

Through freight had an added complication in that it could not book its own continuance. Sometimes captains would take on the task as a favor to the shipper, but more often, local merchants would act as intermediaries. This required time and care, while conveying unwanted legal responsibilities. There were also charges that these merchants favored one line over another (especially if they happened to own stock in that line), and did not obtain the best or most expedient transport. Those problems were solved by the creation of a new occupation: steamboat agent. Acting as a paid consignee, the agent would accept merchandise, pay transfer fees, store the merchandise if necessary and drive a hard bargain for the best fees on outgoing boats. For this, he received a commission.

The role of steamboat agent rapidly evolved into a jack-of-all-trades position. Not only did he act as a middleman for shippers, but he served the same capacity for captains, securing cargo, collecting freight bills, and having needed foodstuffs and other supplies ready as soon as the boat reached port.

From a personnel perspective, crewmen were more willing to sign onto one leg of a line because this provided them with a far more stable life. Working the same short route week after week enabled the deckhand to get home with regularity, keeping his family together for the greater part of the season.

Before 1850, there was no pressing need for line service, as transients and packets provided frequent, if not regular transportation. Although New Orleans was the principal destination of goods from the Ohio Valley, and the port through which goods came from the eastern seaboard as well as Europe, there was no line service between that city and Louisville until 1851. The 1850s, however, wrought (at times unwillingly), great changes to river transportation, despite some resistance. The rapidly expanding Iron Horse (railroads) proved schedules could, and should, be met, compelling rivermen to take another look at their casual attitudes.

Steamboat lines were generally worked out over the winter months (low water season), when captains and/or owners had the opportunity of gathering in one of the larger cities. A name was typically adopted for prestige and ready identification. Lines primarily operated along routes of 500 miles or less, but as they evolved, more audacious routes were structured. These lines sought to attract the finest boats, and the success of the venture depended more on the reputation of the captain and the finery offered than by any rigid adherence to schedule. These enterprises usually obtained the best freight contracts and the high-end travelers, but again, immediate success did not equate to longevity. Lines seldom lasted more than a season or two before they were broken up, either by too many captains dropping out or by a failure to capture a significant proportion of the business.

The *Milwaukie Sentinel,* July 1, 1843, reports a typical dissolution, with an interesting editorial comment:

> SAILORS' RIGHTS — We learn from the Maumee River Times that the steamboats Com. Perry and James Allen have released themselves from the steamboat monopoly, and determined to run independent. We hope the community will sustain all such boats and thus aid in breaking up one of the most vicious monopolies that impose upon the traveling and business community. — *North Western.*

The only truly successful way to run a line was by having a single ownership group carry title on the participating boats. That enabled them to operate with stricter guidelines and, as captains were employees rather than individual owners, eliminated the fear of losing a link in the chain. This system did not come into effect until after the Civil War, and by that time, steamboating had suffered considerable losses to the railroads.

Joint-Stock, Partnerships and Cooperation

The unique character of steamboating lay in the fact that it was an enterprise of small business owners. Approximately two-thirds of all steamboats throughout the era were owned by four men or fewer, and nearly one-quarter were owned by a single individual. The greater number of boats on the water meant lower rates for all, and heavy competition led to a greater difficulty in obtaining full loads of freight. By 1839, certain individuals saw this as a drawback, and hoped to limit competition by forming business associations. The earliest attempt at creating private associations came in the form of the Joint-Stock Company.

The Joint-Stock Company was formally organized with directors, trustees for holding property and various other petty officials. Contracts were drawn, and complex tables for the calculations of profit and loss were created. Westerners were not overly fond of written contracts, however, and rigid rules went against the national grain. It did not take long for the Joint-Stock Company to be replaced by a less formal Association of Part Owners, or Partnerships.

Partnerships worked one of two ways: one partner oversaw the operation of the boat while the other dealt exclusively with business matters, or the two worked in conjunction with one another, handling all aspects of operation together. Each partner was liable for the success or failure of the venture equally, and if one assumed a debt, the other was legally bound to honor it.

The Association of Part Owners was a less exclusive enterprise. The ownership of a boat was easily transferred if one partner opted out, and partners could sell their shares without the consent of the rest. Part owners could not bind their associates in all matters relating to business (unlike the partnership), although part owners could act as the agent for the boat in matters of ordinary transactions, like repairs and hiring, but not in obtaining insurance.

As the concept of part owners developed, it became a common practice to solicit one or more owners from major port cities, and recruit others who resided in smaller towns. This way, the enterprise had local men in strategic places who could exert their influence on friends and acquaintances to acquire freight contracts. In turn, they were able to divert their boat to local shops for supplies, repairs and outfitting.

With association men dispersed throughout the route, this also permitted a more direct supervision of the operation as a whole. These men were also familiar with the wealthier members of the community who had money to invest. It did not take much convincing to enlist their money in the venture, for "owning" a steamboat brought tremendous prestige to anyone involved.

Not everyone could afford to own a boat, or even buy a full stake in an association, but they still wanted to invest in such an enterprise. That desire led to the development of the sale of fractional shares of "steamboat stock," ranging from halves to hundredths. These shares were bought and sold at will without affecting the overall running of the steamboat. Often, "certificates of enrollment" were issued, listing the owners with the amount of stock purchased.

A typical advertisement for stock sale ran in the Huron *Reflector,* June 4, 1839:

For Sale,
$600 STOCK in the Steamboat

> "GREAT WESTERN," on
> Lake Erie. Apply Soon.
> GEO. A. KNIGHT
> New Haven, May 20, 1839.

One other type of steamboat ownership was a "Cooperation." This type of structure involved rivermen who pooled their money to build or purchase a boat. Profits were divided according to how many shares held, and the same was true of the "plum" jobs available, typically that of captain and clerk. Giving important (and well-paying) positions on the merit of stock ownership often caused trouble, however. Worse, with so many rivermen involved, each usually felt he knew what was best, and arguments were known to turn acrimonious.

Suing a Steamboat

Under maritime law, the boat was liable for debts incurred in the course of operation. That meant the vessel itself could be claimed for payment. In 1825, the United States Supreme Court ruled that the admiralty jurisdiction was limited to tidewaters (coasts). This effectively deprived creditors from attaching liens to the steamboats. When merchants went against the owners, it proved difficult to determine who was actually liable. Under the convoluted associations and cooperations, owners readily escaped legal action by selling their shares or moving out of the area, so that it was difficult to trace them.

Around the time of the Supreme Court ruling, most western states, beginning with Indiana in 1824, began enacting the maritime concept of allowing creditors to attach claims against the boats. By 1840, Alabama, Kentucky, Missouri, Ohio and the Wisconsin Territories had followed suit.

In particular, the Missouri law personified the steamboat, giving it the privilege of entering into contracts and assuming debt. This meant that as an "individual," the boat was responsible to creditors and could be sued by name.

In 1851, the admiralty jurisdiction was declared to extend into the navigable rivers, but state laws continued to take precedence over court cases until 1867, when they were declared unconstitutional on the ground that exclusive jurisdiction was vested in the federal government.[2]

Not Quite "Faster Than a Speeding Bullet"

Merchants and shippers were often irked by delays in steamboat departures, but in truth, there was seldom any pressing need for cargo to reach its final destination on a certain date. Those most inconvenienced by casual schedules were, of course, the passengers.

Published cards circulated the waterfronts, advertising schedules filled the newspapers and chalkboards proliferated on the wharves, promising "for certain" departure dates and times for "way passengers" (those not going the whole route) and "through passengers." Even after the vessel left port, nothing was etched in stone, for steamboats were run for profit. Once a fare had been paid, all obligations were superfluous to whatever opportunities might present themselves along the route.

Besides the frequent stops for wood, every plantation along the bank had its own landing (however unrecognizable), and captains would often put in to see if there was cargo to ship. The same held true for the hundreds of small settlements and wayside towns that sprang up like mushrooms. These towns would put out a flag to signify they wished the next passing boat to stop, whereupon a delay of fifteen or thirty minutes might be occasioned by the embarkation of a single traveler, a bundle of goods, or a letter to be carried. The time might stretch to an hour if any hard bargaining was done over fees.

Steamboat Landings (Average)

St. Louis to New Orleans (1,169 miles)	1,327 landings
Nashville to Smithland, the mouth of the Cumberland (192 miles)	121 landings
Shreveport to the mouth of the Red River (306 miles)	530 landings
Baton Rouge to New Orleans (135 miles)	1,025 landings

"Terminal" Facilities

Unlike a sailing ship that required extensive wharves and rowboats to bring people and cargo to and from the vessel, steamboats pulled directly into a port of call. All that was required then were a few boards serving as a gangplank, or later, a more stable stage reaching from the lower deck to the shore. These gangplanks had no railings, and the boards often slipped out of place, casting the unwary deckhand or roustabout into the muddy water below. They also tended to become slippery with mud or water, and

A typical scene of crew and passengers aboard a steamboat. Note the plank used for going ashore or boarding. All persons as well as freight had to use this narrow plank and balancing was often difficult. During rain or other bad weather, the few steps could prove dangerous. If a boat put in at a small landing and a crewman slipped and fell off, his life was at risk, for he might easily drown before being rescued.

learning how to balance on such flimsy supports proved essential to crew and passengers alike.

Along the route, landings were often no more than an embankment, presenting steep climbs up slopes or down ravines. These were succinctly described as "one of the greatest hindrances to river traffic." If no one were present to accept a consignment at one of these smaller ports of call, it became accepted practice to leave the cargo with a nearby resident. If there were no agents or townsmen willing to accept responsibility, the captain would order a few toots of the whistle as fair warning that cargo had been delivered. This signal legally discharged him of his duty, and the boat steamed away. If freight left in this manner were stolen or damaged by weather, it was no longer his concern.

As the era progressed, larger cities improved landings by grading the waterfront and even straightening the slope to ease transfers. Eventually, some were paved, thus somewhat eliminating the ever-present threat of mud that affected everything from shoes to cargo. Wharves ranged from 200 yards to over a mile in length. Some provided temporary shelters for cargo awaiting shipment, while incidentally allowing the passengers to escape inclement weather.

An advertisement running in the Milwaukie *Sentinel,* April 5, 1843, gives a good idea of the development of wharves and piers throughout the river system:

FORWARDING and Commission Merchants, Dealers in Provisions, Grain, Flour, Pork, Plaster, Water lime, Copper, Lead, Shot, Glass, Lumber &c., &c.

At the opening of navigation we shall be receiving Pittsburg Window Glass and Nails; also Plaster, from Michigan Grand Rapids — a full supply of each article.

N.B. We are slowly but steadily progressing with the Pier, and hope to be in readiness at the opening of navigation to do business, and at which time we shall have a good supply of seasoned hard wood on hand, for Steamboats and Propellers. All goods left on our Pier will be delivered to *consigners* without delay, not asking whether such goods are subject to lightarage, or not. And we shall expect that public patronage will enable us to do business at prices to suit the times.

We will also state for the benefit of the public that we have made such arrangements, with a few responsible Lines upon the Lakes and Erie canal, as will enable us to contract for the delivery of Produce at New York, Boston, or Montreal at the very lowest prices. *(Transient or Wild Lines, we do not pretend to compete with.)**

At peak times of the year, steamboats crowded the wharves of New Orleans, St. Louis and Cincinnati. Docked side by side, sometimes nearly touching each other and positioned two or three rows deep, they quickly unloaded their wares. At the same time, the captain or agent negotiated for new cargo. These piers were busy places, with hundreds, if not thousands of people waiting for passage, collecting goods, or merely taking in the sights. Boys hired by local hotels screamed favorable room rates while eagerly grasping baggage; others hawked the latest newspapers and sold fried cakes, or fruit and nuts. In New Orleans especially, street performers juggled balls, trained dogs did tricks, bands played and the omnipresent pickpocket worked his way through crowds.

Bales of cotton were piled on top of one another, forming high, temporary barriers that impeded walkways; barrels of whisky and molasses, crates of heavy machinery, travel trunks and carpetbags were shoved into every available corner, obliterating any semblance

**Italics added.*

of free passage. Sermons were preached to outbound travelers, alms were begged, Indians loitered about, trappers with coonskin caps sharpened knives, and hopeful travelers of all descriptions searched for berths.

Few laws were made to regulate the transfer of cargo, even in the most populous cities. It came under a general rule of thumb that deliveries were to be made on weekdays during regular business hours. The captain or steamboat agent was expected to make "reasonable effort" to report the delivery to the local consignee, and that constituted legal transfer. If the cargo were to be forwarded, it was the responsibility of the consignee to make arrangements, although some captains, primarily those of tramps and packets, did handle these duties as a courtesy, always remembering that good will led to repeat business.

City officials charged wharfage fees as a means of building and maintaining the paved surfaces and maintenance facilities; charges were based on the size of the boat and length of stay. In the early years these "taxes" were fairly low. However, as it became customary for every little stopover to charge a fee, they quickly added up. Upkeep on wharf improvements was irregular and often nonexistent. When rising river waters flooded the wharves, surging tides left behind tons of silt and mud; potholes, the bane of urban development, rapidly opened gaping holes along the crosswalks, and summer dust covered objects like a shroud. Straw, gravel and crushed shells were sometimes put down to protect the landing sites, but they soon became part of the overall nuisance.

Floodwaters also ruined whatever might have been stored on shore, or washed the goods away, leaving financial loss in their wake. Storing merchandise farther away from the banks would have eliminated that problem, but the expense of lugging heavy cargo the additional distance past the water line seems to have ended discussion of that solution.

Large towns came to employ a "wharf-boat," usually a steamboat past its prime, onto which freight could be unloaded and left until it was convenient to collect it, or where passengers spent the night in its cabins before disembarking, or to wait for a boat to depart.

Tow boats were common sights along the waterfront, used to transfer cargo from incoming vessels to shore, or to maneuver disabled steamers into repair areas. Similar to their river-navigating cousins, they were not immune from disaster, as demonstrated from this account, taken from the *Republican Compiler,* June 5, 1843. The account is also a glaring representation of the dangers to the engine crew, who risked their lives on a daily basis for minimum wage.

> The New Orleans Bulletin of the 24th inst. contains the following particulars of a disastrous occurrence near that city on the previous day. The steam tow boat Phoenix, Captain F. Annable, in moving the ship Flavice to the stave yard near Huntsville, in the act of casting off from the stop, three of her boilers exploded, and displacing the 5 remaining ones, carrying with them nearly the whole of the boiler deck, chimneys, &c. No cause can be assigned to the explosion. The boilers were only three years old, and were thoroughly overhauled and repaired last fall. Capt. Annable on nearing the point where he was to leave the ship, had as usual ordered the fire doors opened and the steam cooled down; and when the ship was cast off, only a moderate head of steam was up.
>
> The following is a correct list of the killed, wounded and missing:
>
> | Robb Ross | fireman, | killed; |
> | John Flynn, | do* | do |
> | John Robertson, | do | badly scalded; |

* *ditto*

Geo. Manning	do	do
Thomas —,	deck hand	slightly;
Geo. Stanwood,	do	do
John Passair,	do	missing;
Chas. Davis,	do	do
John H. Clark,	pilot,	badly scalded;
James Skinner,	2d engineer,	do
Lawrence Forest,	deck hand,	missing;

The first engineer, S.C. Fish, and the rest of the officers, escaped unhurt.

Eventually, wharfage fees became onerous, to the point that captains refused to land cargo at cities that charged what they considered unfair rates. Several cases were brought to court and resulted in a reduction of charges. At New Orleans and Vicksburg, where fees were assessed on the size of the boat, the federal courts eventually ruled in favor of the steamboat operators, declaring that such fees were actually "tonnage taxes," and unconstitutional on the grounds they interfered with interstate commerce. This victory was mitigated by other court rulings declaring wharfage fees legal on the basis that they represented "charges for service," rather than being outright taxes.[3]

There were also canal fees with which to contend. The Louisville and Portland canals charged 50 cents per measured ton for a boat to pass. For the larger class of boat working eighteen round trips per year (or thirty-six times through a canal), the total assessment of fees approached $4,320. On the Cincinnati–St. Louis trade, a boat passing through thirty times would have to pay $5,400 per annum.

Like wharfage costs, the price was not prohibitive, but it hurt small owners who could hardly bear the brunt of any additional subtractions in their ledger books. These fees sometimes dictated the trades a transient worked, compelling some captains to avoid canals altogether unless the waters were high. In that case, they could steam around the canal and thus avoid paying any more out of pocket than necessary.

9

Economic Conditions During the Steamboat Era

In order to fully comprehend the wild machinations of the steamboat era, roughly spanning 1807 to 1861, a background in the political, social and economic conditions occurring over those decades is essential. How the United States fared reflected on those who worked, traveled on, or invested in the technology of steam-powered vessels.

Politically, the country was beginning to lose its close connection to the Founding Fathers. Men who were children or not yet born during the Revolution came of age; Andrew Jackson, who became the seventh president, was the first to hold that office without actually having fought in the war. Bitter fights over the constitutionality of the Bank of the United States dominated economics; immigration swelled, with proponents and opponents on both sides demanding to be heard. Unresolved issues of human bondage were coming to a head; abolitionists fought slaveholders, using pulpits, printing presses and occasionally punches. Medicine was caught in a transition period between science and superstition; free will and self-determination became watchwords. A man's word was his bond, unless he turned out to be a confidence man, swindler or quack.

Over a span of fifty-four years, the new country faced international crises, a second war with England, a war with Mexico, and the Texas Independence Movement. Boundaries swelled to rising cries of Manifest Destiny; the struggle swelled against the admission to the Union of slave states; gold was discovered and specie lost. The stock market rose to unimagined heights and crashed just as quickly. Steamboats opened up the North and West, exacerbating the Indian Question; land sold for $1.25 an acre, and there was plenty of it for the taking. Speculation ran rampant: some investors became wealthy while others jumped out windows.

The rapid development of the steamboat along the inland waterways was not an isolated movement, separated from the East by the Alleghenies and the West by the Rockies. Although those rugged pioneers of the Ohio, Mississippi and Missouri liked to think of themselves as independent, they were as tied to New York and San Francisco as a package with the bow set squarely at Washington. Rivermen depended on trade as their source of income; states needed money for canals and roads; corporations took out unsecured loans from wildcat banks. Forts had to be established, manned and supplied with food and munitions; courts were required to interpret laws, and, ultimately, the federal government had to pass acts regulating everything from currency to river safety.

No one was exempt from the world around him. From the clustered, rapidly expanding cities of New Orleans, Pittsburg, Cincinnati, St. Louis and Chicago, to the most isolated fur trader at the Falls of St. Anthony and the lead miner at Galena, the common thread of country bound them together.

Turning Back the Clock: The Panic and Depression of 1819

Robert Fulton launched the *Clermont* in 1807, during a period of national upheaval. England continued to impress American seamen; international relations were strained as the new nation struggled with its foreign policy. The good news brought westward by the development of steam transportation was both inspiring and frightening. A new era was dawning. No one knew what changes it would bring; whether the wind blew good or ill.

It did not take long for verbal barbs between the Mother Country and the new United States to lead to bloodshed and armed conflict in the War of 1812. The capitol in Washington was burned; pride, and perhaps freedom, hung in the balance as both sides struggled for ascendancy. Money was scarce, factories fell on hard times and families went hungry. In New Orleans, Henry Shreve and the *Enterprise* performed good service for Andy Jackson. Jean Lafitte and his pirate band served the commanding general in the capacity of artillerists. The city held: heroes of flesh and blood and wood and steam proved the deciding factor.

With peace came prosperity. The United States reasserted its sovereignty and established itself as a power with which to be reckoned. New territories opened up and the promise of cheap land drew farmers by the thousands. Demand for produce and manufactured goods, both domestic and foreign, grew in leaps and bounds as people reveled in the good times. With the breaking of the western Fulton-Livingston monopoly, steamboat manufacturing entered the first real phase of construction. Captains began testing the limits of the inland waterways, while explorers challenged the unknown distant territories.

But all this came at a price, and that price happened to come on borrowed money. Speculators bought land with currency and sold it at a profit. Instead of paying down their debt, however, they borrowed more to buy additional land. This pattern was repeated time and time again. Banks printed their own paper money to meet the demand. Inflation reared its ugly head, but no one stopped to pay attention — not when times were flush.

Cyclic factors were already in play, however. After the War of 1812, the desire for luxuries was finally satisfied and the demand for manufactured goods fell off sharply. The stock markets began declining. Prices on the principal articles of domestic growth fell. Laborers lost their jobs or saw their wages slashed. This only accentuated the habit of borrowing, and more and more individuals and families fell into serious debt.

Banks responded by curtailing the amount of money they printed, in part because they had to make good this paper currency as specie, as gold and silver suddenly become difficult to find. This prevented families from taking advantage of lower prices for domestic goods and produce, and manufacturers from hiring more workers at reduced cost.[1]

The struggle continued. On June 19, 1819, the Nashville *Gazette* reported:

The Farmers' and Mechanics' Bank of this place yesterday suspended specie payments. We are authorized to state *positively*, that the Bank of this State, and the Nashville Bank, and their Branches, are resolved to continue their payments in specie as usual.

But it would not last. With capital in short supply, the Bank of the United States, as well as state banks, called in their loans, with the result that land speculators went bankrupt. This, in turn, drove local banking institutions to fail, adding to the growing crisis. In Europe, a similar crisis developed, severely cutting sales of manufactured products. Continental merchants turned to America, dumping cargo after cargo on its shores in an effort to sell here what they could not at home. Prices were slashed, undercutting the sale of locally manufactured goods. This took another toll on jobs, as more U.S. factories closed their doors. Bank after bank failed. The rosy picture after the War of 1812 faded into an economic panic.

In order to alleviate the problem, a call for the return of specie in place of paper money was heard loud and clear. Men believed their trouble stemmed from too much money that could only be curtailed by using the old standard of gold and silver to transact business. A bank may fold, they reasoned, making its currency worthless, but a man with gold in his pocket still had real value. Others scoffed at the idea, deriding these proponents as wanting nothing more than to turn the clock back. They argued that the influx of cheap foreign goods had destroyed prosperity, and a new cry was heard across the landscape: B*uy American!*

President James Monroe cited factors leading to the Panic in his Third Annual Message to Congress, December 7, 1819: cheap foreign goods, low or nonexistent wages, the shortage of currency, and the unusual drought that prevailed in the middle and western states. His conclusion offered scant comfort:

> It is deemed of great importance to give encouragement to our domestic manufactures. In what manner the evils which have been adverted to may be remedied, and how far it may be practicable in other respects to afford to them further encouragement, paying due regard to the other great interests of the nation, is submitted to the wisdom of Congress.

Good intentions aside, the country fell into a depression, from which it did not fully recover for six years.[2]

The Panic and Depression of 1832

As the country slowly worked its way back to prosperity, states began investing in internal improvements such as canals and roads, and many of them took out loans to pay for the work. It seemed that nothing had been learned from the depression of the mid–1820s. Land speculation increased, and staggering sums were borrowed to support the same scheme of buying-selling-reinvesting. In the western states, steamboat construction grew into a major industry, and more and more boats saw service in the waterways, transporting freight and passengers. Wages grew, profits increased; it appeared the good times were here to stay.

In politics, Andrew Jackson had risen to the presidency on the call for a return to hard currency. Believing the Bank of the United States to be unconstitutional, he began plotting its downfall and eventual closure. In 1832, Nicholas Biddle, the head of the Bank, was urged by Henry Clay and others to ask the president to renew the Bank's charter four years early. Finding the opportunity too good to refuse, Jackson promptly rejected the offer, stating his intention that it not be renewed at all, thus effectively ending its tenure.

Not content to wait out the next four years, President Jackson removed approximately ten million dollars from the Bank of the United States and distributed it among state banks and privately owned financial institutions, immediately termed "pet banks." In a move to counter Jackson and restock reserves, Biddle immediately called in loans owed the Bank. Those businesses that did not have the ability to repay on short notice went into receivership. An economic downturn resulted, and dragged on through 1833 and 1834.

In his address to Congress, December 3, 1833, President Jackson attempted to defend his actions:

> The extent of its [the Bank of the United States] misconduct, although known to be great, was not at that time [the removal of funds] fully developed by proof. It was not until late in the month of August that I received from the Government directors an official report establishing beyond question that this great and powerful institution had been actively engaged in attempting to influence the elections of the public officers by means of its money, and that, in violation of the express provisions of its charter, it had by a formal resolution placed its funds at the disposition of its president to be employed in sustaining the political power of the bank.[3]

Not everyone approved of the policies or subscribed to the idea that President Jackson and his fights were good for the country, as revealed by the following bitingly sarcastic editorial of October 13, 1834.

> The people complained last winter that they were suffering unparalleled distress. The Globe told them that they were *mistaken*— that they were in a condition of the most perfect prosperity. *They,* of course, were no judges as to whether they *were* or were *not* distressed. Now the Globe assures them that *their pockets are full of gold*. Thrusting their hands into the said pockets, they cannot find a single particle of the metal, but still the organ *insists that it is there*. If they cannot find it, the fault is their own — not Gen. Jackson's. What is the evidence of a man's senses in comparison with the authority of the king's mouthpiece?[4]

Notwithstanding, by 1835 it appeared the United States economy had turned the corner. Business recovered, land speculation continued at a fever pace, and banks continued printing money, which they promptly lent out. Inflation soared, but so long as paper continued to be spent, the economy appeared to thrive. Hoping to take advantage of this sudden prosperity, foreign investors and governments resumed loans to American businessmen.

During this period of wealth, the question of federal involvement in internal improvements became of paramount importance. Private investors, states and local communities all wanted loans, whether to expand shipyards, improve the rivers, construct more locks and canals or build railroads. Between 1830 and 1837, steamboat tonnage on the western rivers rose from 63,053 to 253,661. In 1830, only 23 miles of railroad track had been laid. The following year, 94 miles had been completed, and by 1836, 1,273 miles stretched across the northeast. Finished in 1835, the Erie Canal immediately brought huge dividends, and people wanted to see more of the same in every state and territory in the Union.

Nor was it only manufacturing and transportation that thrived in this atmosphere which fostered the most extraordinary growth the world had ever witnessed. From 1833 to 1837, the cotton crops of the slave states of Tennessee, Alabama, Mississippi, Louisiana, Arkansas and Florida increased from 536,450 bales to 916,960 bales, while price fluctuations rose from 10 to 20 cents a pound.

Land speculation continued to dominate spending. The price of public land was fixed by law at $1.25 an acre. Lands were available to anyone without limit to acreage, and bore no residential requirements. This proved too tempting to resist, for here was a tangible commodity that increased dramatically in value. Anyone able to obtain a bank loan could purchase significant tracts, divide it into parcels and sell it piecemeal to small investors or settlers. People often bought farms no one had ever actually seen that were far removed from civilization. Families attempting to settle there were isolated, often surrounded by unfriendly Indians and left to their own devices as far as supplies were concerned. Those who settled in the Northwest were dependent on steamboats bringing cargo upriver and had to travel scores of miles to the nearest town to buy these goods. In wintertime, when the rivers froze, no boats navigated the waterways, and it was not unheard of for families to starve before spring thaws.

From 1820 to 1829, the annual sale of land averaged less than $1,300,000. It steadily rose over the next seven years:

Land Sales

1830	$1,517,175
1831	$3,200,000
1832	$2,600,000
1833	$3,900,000
1834	$4,800,000
1835	$14,757,600
1836	$24,877,179

In 1829 and 1830, Jackson attributed the modest increase in sales as proof of increasing prosperity. In 1831, his congratulations were hushed, but by 1835, even with the abnormal growth, he declared a state of economic prosperity. The Treasury of the United States paid off the national debt and accumulated a surplus, all from the sale of public land. By January 1, 1836, the total reached $25,000,000, and by June 1837, the amount totaled $41,000,000. This enormous advance was paid for in paper, which in turn formed the bulk of government deposits, kept by only a small portion of the approximately 600 state banks holding federal money.[5]

Land speculation was not contained to the western portion of the United States. Urban real estate values increased to dizzying heights, making it possible for investors to make a 75 percent profit on a $1,000 investment in Michigan, where the boom was especially high. In 1831, a farm near Brooklyn was offered at $20,000, with no takers. In 1835, the same property sold for $102,000.[6]

The Huron *Reflector,* July 18, 1831, reported on a significant rise in real estate:

> The Cincinnati Daily Advertiser mentions, that the Real Estate fever has recently broken out in that city- property is now thirty per cent higher than it was a short time since.

It was not until 1836 that Jackson finally realized the true significance of the enormous increase. Too much spurious paper money was in circulation, banks were overextended, and staggering inflation instilled a false sense of security in personal wealth, as laborers demanded higher and higher wages.

Confusion persisted over what was and what was not legal tender, and who would accept currency in various denominations. Regulations from the federal government did not make matters easier. The *Western Paper,* June 20, 1836, reported what was likely a common occurrence:

Frauds in the Land Office

A man came in, wishing to enter five half sections of land, and offered in payment 1,875 dollars of a New York Safety Fund Bank, and he was informed by the agent, that, by the orders of the Department of Washington, they were prohibited receiving any money of a less denomination than 5 dollars, or money of any bank other than the deposite [*sic*] banks. Here the man was in a predicament, for if he went to get his money exchanged for such as was receivable under these orders, some one might enter the very lands he selected to settle himself upon. What could he do? The agent very obligingly helped him out of his trouble by exchanging his own *private* money with him, for five percent premium, in which operation he pocketed the trifling sum of $83.75, and the man had to pay, in reality, $1,958.75, for five half sections, being $83.75 more than the law of the land requires. But mark the sequel. The same day a merchant wanting funds that would go to the East, paid this land agent 2 per cent for the same money, being $22.92 clear shave, out of two individuals, on the same money in one day.

The Specie Circular Goes Round and Round

Inflation rose out of control; eventually, this began to take its toll. Currency depreciated and the enormous extension of bank credit finally reached a saturation point. By 1836, President Jackson, then nearing the end of his eight-year term, issued the Specie Circular in an attempt to rein in the effects of too much privately printed paper money, and thus put an end to land speculation, which had fueled the unhealthy economy. Under this act, only gold and silver would be accepted as payment for federal land. Jackson hoped this would put an end to nonresidential ownership of vast tracts, a situation which had been an obstacle to the advancement of newly formed states. Further, he stated in his address to Congress that the Circular would open public lands to emigrants, who could then buy farmland from the government, rather than enriching speculators. Finally, he expected the Specie Circular to convey large sums of silver and gold to the interior, stabilizing the lenders.

It did not work out as expected. Speculators swarmed to banks on the expectation of exchanging paper money for coin, only to discover that the banks had overextended themselves by failing to keep adequate reserves of precious metals on hand. When they could not make good the exchanges, the financial institutions folded, thereby rendering worthless all the paper money issued in their name.

This caused further runs on those banks still open, causing them to fail. In Europe, foreign investors, fearful of accepting worthless paper in payment of loans, followed Jackson's lead and demanded only gold or silver. During 1837 alone, nearly 800 banks failed for lack of coin, virtually destroying the entire banking system of the United States.

A brief mention in the *Adams Sentinel* of November 13, 1837, is typical of the period:

The Franklin Repository says: "Upwards of twenty suits have been instituted in this borough, against the Fayetteville Savings Fund Institution, to recover payment on their notes."

Without gold and silver, and with paper money next to or actually valueless, employers had no legitimate money with which to pay workers. Massive layoffs resulted: laborers who had traded time, energy and sweat for promises were left destitute. During the Panic of 1837, nearly 10 percent of American workers were unemployed at any one time.

On February 14, 1837, several thousand people gathered in front of City Hall in New York under the call of what the *Commercial Advertiser* called "Jackson Jacobins." They shouted, "Bread! Meat! Fuel! The prices must come down!" Politically, they demanded the prohibition of bank notes under $100 and called for a return to the gold and silver standard. The meeting degenerated into a riot, with some of the participants, "chiefly foreigners and few in number," rioting in front of Eli Hart & Co.'s large flour warehouse. Tempers flared and the building was gutted, forcing the military to be called out. One of the owners later issued a card pointing out the grim truth that, "the destruction of the article can not have a tendency to reduce the price."[7]

Commercial failures began in New York by April 1, 1837. A week later, nearly 100 failures had occurred, including five foreign and exchange brokers, thirty dry-goods merchants, sixteen commission houses, twenty real estate speculators and eight stockbrokers. By April 11, the number rose to 128 failures.

The widespread devastation did have one beneficial effect: prices plummeted. Wages, rent and provisions tumbled, and goods and services of every sort lost staggering amounts of value. The problem was, of course, no one had any money to take advantage of the freefall.

In less than a week, the New York *Herald,* which had been keeping tabs on the failures, no longer bothered. There were simply too many to count. Workers wandered the streets, numb from shock, begging for work at $4 a month, and found few takers. Frontiersmen, who had participated in the dream of owning their own land, found themselves isolated. Due to the panic, most steamboats lost money in the years 1837–39, as merchants did not ship that which they could not sell. Wages rose and excessive competition dragged rates down to unprofitable levels.

Martin Van Buren, a New Yorker and Jackson supporter, was elected to the presidency in November 1836. He inherited the developing crisis, later complicated by bad crops in 1837 and a worsening reaction from Europe. English banks raised interest rates and reduced credit to United States businessmen, sending shock waves through the cotton market. Prices fell from twenty cents to ten cents a pound, and the agricultural industry fell into a period of unstable depression that would linger for six years.

Americans fully expected Van Buren to cure their ills with legislation and the loosening of credit, but he made no effort to do so, believing the economy would naturally turn itself around. In that, he erred, and his popularity, along with Andrew Jackson's legacy, sank.

The next two years saw a seesaw of economic gains and regressions. The price of cotton rose to sixteen cents a pound, fell in the summer of 1839, then dropped to a low of five cents a pound in 1840, adversely affecting steamboat operators who made much of their money in the transportation of cotton. With prices so low, many planters opted not to sell at all, or to delay shipments. Many northern banks had not yet resumed operations, and in the South, numerous plantations were subjected to execution sales.

England resumed its enormous exporting to America in 1839, and land sales also reflected a brief upward trend.

Customs Duties Received from England

1836	$23,000,000
1837	$11,000,000
1839	$23,000,000
1840	$13,000,000

Land Sales

1838	$3,700,000
1839	$7,000,000
1840	$3,000,000[8]

The government's failure to help the people in the Panic and Depression of 1837 and beyond resulted in nearly universal blame of the Democratic Party. In 1840, the voters elected William Henry Harrison, a Whig, to turn the country in a new direction. Harrison died three months after taking the oath, and his vice president, John Tyler, became president.

The panics and depressions of 1837–39, the results of which lingered until 1843, made one salient point perfectly clear to the citizens of the United States: they were just as vulnerable to economic disasters as were the older, more established countries of Europe. Adding to that harsh reality, they came to understand that recoveries were slow and painful. Such awareness, however, did not make them wiser or more prepared for the future.

An indication that hard times were felt all over the country in the transportation business was the following notice in the Milwaukie *Sentinel*, July 1, 1843:

CHEAP ENOUGH—The following is the fare from New York to Milwaukie Eougrand train of stars, steamboats, &c, viz;

New York to Albany,	25 cts
Albany to Buffalo,	$5.00
Buffalo to Milwaukie, by the Etson propellers,	$4.00
	$9.25

Really, this is cheap travelling-hard times taken into consideration. For $9.25 you can land at the pier in Milwaukie, bag and baggage.

A notice in the *Adams Sentinel*, September 3, 1845, should have stood as a warning, but did not.

NOTICE

It is hereby given that the following tracts of land sold for the taxes and charges then due in the month of December A.D. 1813, remain unredeemed in the office of the Clerk of the Board of County Commissioners of Iowa County, Wisconsin.

Now, therefore, unless the taxes, costs and charges added to the several tracts hereinafter described, together with all legal interests and charges that may subsequently accrue, be paid on or before the 21st day of December A.D. 1845, as described below in pursuance of the Statute in each case made and provided, the same will be forfeited for such taxes and charges. *Interest computed up to the 14th day of June, 1845.*

After which followed several pages of columns detailing sections, acres, "H'dths" (a division of land) and "Tax, Interest & Charges."

Political Satire Takes a Ride on a Steamboat

Even in dire economic conditions, the people of the mid–19th century were not above poking fun at the situation and their political enemies. This article, from the *Oshkosh Courier* (September 17, 1856), designed to look like an advertisement, provides an example:

> *Great Attraction!!*
> *FARE REDUCED:*
> *For the Saline Springs.*
> *The New Steamer*
> *REPUBLICAN*

With the staunch mud boat Know Nothing in tow, to carry the cattle, will start from the polls, at sundown, November 4th, 1856, bound up Salt River.

This boat has only run two seasons, but as her color is rather somber, and her arrangements mixed, a vast majority of the people will have nothing to do with her.

On this account the company will be very select and aristocratic.

Her frame is constructed of *Live Oak,* other lumber mostly *Beech*— or some other kind.

The timber of the mud-boat is not so easy to make out, without the use of a dark lantern.

They will reach the headwaters, and set about exploring, in search of the Fugitive Slave law, the Missouri Compromise and some other matters, "sent up" by them in advance, and to locate a political cemetery, on or about the 20th of November, weather permitting.

The steamer is rigged with a whistle of modern style, which will shriek nearly all the time.

She is offered as follows:

Captain— John Charles Fremont de Mapipean
Frst Mate—William L. Dayton, Sailor Lasher
Second Mate—Horace Greeley
Steward—Fred. Douglass
Clerk—Wm. H. Seward
First Engineer John P. Hale
Second Engineer—Joshua R. Giddings
Chaplain—B. Ward Schrecher
Pilot—Chancellor Kent
Firemen—Sumner, Burlingame, Booth, Jim Lane
and Gov. Reeder
Boot black Bennett
Bottle Washer—Billy Cramer
Waiters—Hecker, Hoffman, Roeser, and some other
  furrinners, who are thus promoted on account of
  their patriotism.
Chamber Main in Chief—Jessie.
Assistants—Dr. Lucy Stone Blackwell, Mrs. Harriet
  Beecher Stow and the rest of the strong minded.

The cargo will consist mainly of Beef Cattle, bodies of murdered Irishmen and Germans taken from graves in Cincinnati and Louisville, and other trophies of past Republicans and Know Knothing triumphs.

Freight or passage may be paid in California interest coupons, Mariposa gold or Kansas aid subscriptions, and must be applied for at Know Nothing Councils or Abolition newspaper sanctums, no matter which. N.B. The few furriners on board will be kept under hatches, that their presence may not injure the constitution of the other passengers, and be nearer their departed friends, who will be stowed below.

The boat will positively start at the appointed time, without fail.

By order of the committee of arrangements.

Madison Argus

The Panic of 1857

Just fourteen years after the Panic of 1843, the country suffered yet another recession, this time the result of a downturn in agricultural exports brought on by the end of the Crimean War (1854–56), as well as overspeculation in real estate and in the new and rapidly developing mode of railroad transportation.

As an example of how many citizens were involved in land speculation, the *Richland (Wisconsin) County Observer* of December 29, 1857, related this amusing but highly telling anecdote:

> It is said that a worthy minister in Indiana, who had become somewhat mixed up in land speculations, recently announced to his congregation at the opening of Divine Service, that his text would be found in "St. Paul's epistle to the Corinthians, section four, range three west"

Although of briefer duration than the panics of 1819, 1832 and 1837, the Panic of 1857 was brought to a climax much earlier and with added potency by the rapid spread of the telegraph. When a branch of the Ohio Life Insurance and Trust Company failed, news that once would have taken weeks to disseminate spread like wildfire and was known within hours around all corners of the wired country. This rapid mode of communication seemed to make the news that much worse, for there was not the time of weeks passing between fact and discovery to soften the blow. Reaching the major eastern cities within hours, the news triggered one of the first panic sellings in the stock market.[9]

James Buchanan became president on March 4, 1857. In his first Annual Message, he stated:

> We have possessed all the elements of material wealth in rich abundance, and yet, notwithstanding all these advantages, our country in its monetary interests is at the present moment in a deplorable condition. In the midst of unsurpassed plenty in all the productions of agriculture and in all the elements of national wealth, we find our manufactures suspended, our public works retarded, our private enterprises of different kinds abandoned, and thousands of useful laborers thrown out of employment and reduced to want.

The message continued:

> It is apparent that our existing misfortunes have proceeded solely from our extravagant and vicious system of paper currency and bank credits, exciting the people to wild speculations and gambling in stocks. These revulsions must continue to recur at successive intervals so long as the amount of paper currency and bank loans and discounts of the country shall be left to the discretion of 1,400 irresponsible banking institutions, which from the very law of their nature will consult the interest of their stockholders rather than the public welfare.

He went on to say that inflation had countered the positive effect of the country's protective tariff, making imports less costly at the expense of local growers and manufacturers. He promised the government would not suspend payments on its debts, but said, "no Government works already in progress shall be suspended, new works not already commenced will be postponed if this can be done without injury to the country. Those necessary for its defense shall proceed as though there had been no crisis in our monetary affairs."

Buchanan proposed that banks "shall at all times keep on hand at least one dollar of gold and silver for every three dollars of their circulation and deposits," and that a suspension of specie payments would result in a bank's civil death. To further increase money in circulation, he noted the Congress authorized the issue of $20,000,000 of Treasury notes, and should this prove inadequate, an additional $20,000,000 shall be added in June, 1858.

Ironically, and perhaps with foresight, Buchanan concluded:

> No statesman would advise that we should go on increasing the national debt to meet the ordinary expenses of Government. This would be a most ruinous policy. In case of war our credit must be our chief resource, at least for the first year, and this would be greatly impaired by having contracted a large debt in time of peace. It is our true policy to increase our revenue so as to equal our expenditures. It would be ruinous to continue to borrow.

Indeed. The bloody "War Between the States" was just four years across the horizon.

Wildcat Banks

If "Every man can be a captain" was the cry along the riverfronts, so, too, was the (unspoken) boast, "Every man can be a banker" among the townsmen of the western frontier. State banking charters were issued for the asking without proof of adequate gold and silver backing, and unscrupulous individuals set themselves up in business.

The only expenses to the wildcat banker were those of printing money, which he did with unbounded enthusiasm, and the renting of a warehouse in which to set up business. A sign out front, an open door and a courteous smile lured in the customers. Those depositing specie received paper currency when withdrawing funds; those seeking loans on land purchases were also given paper money, typically drawn on another bank in the "partnership," so that no bank gave out its own printed paper. This scheme protected the erstwhile banker, who did not have to make good on the currency of another institution.

Captain A. B. French, photographed when he built his first boat in Cincinnati. He used this picture of himself on his billing papers.

Loans to land speculators were the most common type, with the deeds of sale being held as security. Other loans to farmers, particularly out-of-state purchasers, were eagerly granted at an interest rate of five percent. If the farmer could not pay his mortgage, he was granted a pass for several months, although interest continued to accrue. When times were hard, it did not take long for the banker (who financed the original loan with spurious money) to foreclose, adding genuine value to his personal portfolio.

If an investor demanded gold or silver in exchange for his paper (as happened after the issuance of the Specie Circular) banks refused to exchange the currency of other institutions (which likely had already disappeared) or simply folded, with no one held accountable. This led to a run on banks and more closures.

George Merrick details how his boat was given a "Thompson's Bank Note Detector," a list of currency they were not permitted to accept as payment for freight or passage. Unfortunately, the list of undesirable bills changed rapidly, and they usually ended up with five hundred to one thousand dollars of worthless paper at the end of a trip. The clerks were usually able to "work off" some of these on the next trip. By the end of a season, he estimated the company had two thousand dollars of worthless paper, which the boat was debited with to balance the books.[103]

If the above could not actually be considered "counterfeiting," two notes of warning in the Hagerstown *Mail,* May 7, 1830, do:

COUNTERFEITS.—A counterfeit note of the Bank of the United States; of the denomination of $100 was offered and detected yesterday at one of our banks. It is of the letter G, payable to C.J. Nicholas of the Richmond Branch, dated August 1st, 1825, and signed Thos. Wilson, Cashier, and N. Biddle, President. It is said by an astute officer of the Bank to be well calculated to deceive.

* * *

Counterfeit Notes. Counterfeit $5 notes on the Farmer's Branch Bank at Easton, Md. are in circulation. They bear the date 1827, and are under number 2100.— The public had better be on the look out (*Elkton Press*)

Fully twenty years later, the schemes continued, this one reported on April 22, 1850, by the *Adams Sentinel*:

Counterfeit $10 notes on the Farmers' and Drovers' Bank are in circulation. They are of a plate similar to the genuine one, but the paper has a more light and flimsy appearance, and the engraving generally bad.

No rest for the weary.

10

Making Money on the Rivers

The early and mid–1800s were a time of nationalism, growing pains, wild speculations, depressions and unbridled individualism. Nowhere did these concepts get greater exposure than along the waterways, and all of them involved steamboats.

When Robert Fulton brought his patent west in the form and shape of the *New Orleans*, men promptly saw opportunities for making their fortunes. Not in the dry, eastern businessman's ways of stocks and bonds, but by using both their brains and their brawn, simultaneously satisfying their needs for adventure and those of monetary gain. In a new country lacking royalty and nobility (a frequent complaint of British tourists), any man might catapult himself to the heights of social standing. With hard coin in his pockets, the steamboat master was the equal — or the better — of his elitist eastern brethren.

Money, as the saying goes, all spends the same, and there was money to be made on the rivers. At least such was the case in the early years, when everyone from keelboatmen to merchants entered the transportation business in one form or another. Boatyards sprang up at Pittsburg, Cincinnati and Louisville. (The only significant steamboat construction outside the Ohio Valley was at St. Louis, but only after 1840. Several boats were made in New Orleans, but virtually none after 1825.) Mechanics flocked from the northern states, local blacksmiths learned how to make engines, carpenters crafted hulls and new boats were quickly put into commission.

Individuals without the upfront money to commission a boat bought one secondhand, took out a mortgage on the boat itself, or raised funds by selling stock. It was not uncommon during the flush times for a farmer to mortgage his land, buying shares with the money received. Everyone wanted a part of the new enterprise, and share not only the spoils, but the instant prestige associated with steamboats.

An account from the Deed Book at the Alexandria, Virginia, Courthouse, June 14, 1837, describes a typical transaction:

"On November 30, 1836, John W Leathers of Campbell County stood indebted to his father, John Leathers, of the same county in the sum of $5000 for money advanced to John W, so that he could build and construct the steamboat Sun flower which was about to leave 'this port for the Southern trade.' John W was 'desirous to secure and endemnify' [*sic*] his father in the money, so he mortgaged to him the undivided half of the steamboat 'with all and singular her apparattus [*sic*], Engine, furniture and other belongings to said boat....'"[104]

Steamboating was clearly a western fascination, for although firms were set up through eastern banks or stockbrokers, ownership remained local. After the 1830s, few easterners' names appeared on certificates of enrollment.

Perhaps the greatest advantage to steamboat ownership came from the fact that after the purchase of a boat, there were few overhead costs associated with starting the venture. The captain had merely to hire a crew, set out a shingle and wait for the customers to come to him. The average cost of a 211-ton steamboat in the years 1830–50 was $9,073. In a typical season of eight and a half months, the owner stood to make $26,596.

	Operating Expenses (in percentages)		
	1834	1849	1855
Wages	36	38.5	40.5
Fuel	30	32.2	22.3
Stores	18	25.6	29.6
Other	16	3.7	7.6

Annual operating expenses could be 125 to 200 percent greater than the original cost of the steamboat.[2]

A Bigger and Better Boat

Profit was the impetus, if not the soul, of the business, and the object was to make as much money as fast as possible. A boatman starting out with a small vessel running the lesser-traveled tributaries — an investment of $5,000 — could double or triple his money in three years. The owner then took his net gain and bought a bigger boat. With several more years of profitability, the same owner could pool his resources with several others and buy a $25,000 steamboat, capable of carrying 150–200 tons of freight.

Continued success provided the partnership with a small fortune, even considering the limited life span of three to five years for the typical steamboat. The grand scheme depended on luck as much as skill, however, for the owner or owners were just as likely to be ruined if their investment struck a rock and sank, or the boilers exploded.

Smaller boats capable of operating in the least amount of water, and thus working the trades longer than their fancier and heavier rivals, made the best investments. The sternwheel boat of average tonnage, capable of carrying the most cargo (albeit at a slower pace, for speed was a primary factor in expense), represented the best chance for profitability.

Steamboat Repairs: More Than an Oil Change Every 3,000 Miles

Steamboating on the western waters was a hazardous occupation. Hidden obstructions could damage the hull, and floating logs often destroyed the buckets, or worse, the paddlewheels themselves. Common repairs, such as replacing buckets or fixing guardrails, were done aboard ship. Very early in steamboat development, it became apparent that blacksmithing equipment was required for minor repairs, and a forge soon became standard equipment on all boats. Usually, the engineer or ship's carpenter (a newly created position aboard larger vessels), were responsible for handling these duties.

If the hull was pierced, immediate action was required to save the boat from sinking. Blankets, pillows or tarpaulins were shoved into the hole, and if a seal was made, the boat would continue on its journey. If this could not be safely accomplished, due to the size or the placement of the breach, every effort was made to bring the vessel onto a bar where shallow water afforded the opportunity to make repairs, or the craft might be taken to shore. If possible, the hull was rammed onto a flat part of the bank where the damage could be better assessed from land. Carpenters would then try to repair the hull just well enough to get the boat to a city with adequate docking and repair facilities. Unfortunately, the larger mechanical parts required for steamboat repair were often available only at the major boatyards along the Ohio River.

Boats could be towed, but this could cost upward of $1,000, to say nothing of lost time and discomfort to passengers. For this reason, a captain would do everything in his power to effect the necessary repairs. Occasionally, long, deep holes were dug in the sandy bank to provide room for repair crews to reach breaches otherwise inaccessible. If the boat happened to be carrying blocks and tackle, chains were attached to nearby trees and the boat lifted by makeshift shears with poles set under the framework, actually raising the heavy front portion of the hull out of the water. Significant danger was associated with this type of extensive repair. Excessive movement or improper fastening of the chains could result in the superstructure breaking away. In that case, unfortunate repairmen could be crushed. Additionally, if the river rose unexpectedly, an exposed hull easily filled with water, ruining the interior mechanisms. In that case, it was likely the steamboat would be a total loss.

Rising rivers also hampered rescue operations after a disaster. The Alton *Telegraph*, March 23, 1836, reports:

> Steam-Boat Accident. The steam-boat Diana, owned and freighted by Messrs. Pratte, Chouteau, &co. bound for Council Bluffs, Missouri river, on the 16th inst., when about 300 miles above this place, struck a snag and commenced filling rapidly, notwithstanding the great exertions made with the pumps. The boat was directed to the main shore at the most convenient spot for landing the freight, all of which was got out in three or four hours, though much of it in a damaged condition, there having been two feet in the hold. The damage would not have been so great had not the goods been exposed during the night to a heavy fall of rain. A short time after the accident, the river commenced rising, and before the goods could be removed the bank was washed away and many articles of freight with it. The freight was generally insured.

So-called marine railways were in place by the early 1850s, whereby damaged steamboats were drawn broadside out of the river and up the slope of the river upon wheeled cradles running on tracks. These were few and far between, however, and saw limited service.

Ordinary upkeep consisted of overhauls to the thin wooden hull — always at risk from minor damage by river obstructions — patching, painting and caulking. Non-essential maintenance was done as infrequently as possible to avoid the cost and delays afforded by such work. Whenever possible, work was done in the off-season to ready the vessel for the spring rush.

For the average-priced steamboat of the 1840s, which sold for nearly $20,000, laying up for repairs cost $1,200 for each of the first two years after launching, and up to $900 for the remaining 18 months of its life expectancy. Most of this came from repairs to the fragile hull. Captains without ready cash were forced to take out loans at high interest rates. Failing that, the boat might have to be sold for salvage, forcing the owner to take a considerable loss, and possibly bankrupting him.[3]

"Perils" of the River and Acts of God

One expense not typically brought to mind when considering the operating costs of a steamboat was insurance: the necessity of paying up front for what the owner hoped he would never need. Insuring western steamboats was primarily a local concern, employing the basic tenets of maritime policies with adaptations for river conditions. Unfortunately, this was not as easy at it appeared, with the predictable result of bitter argument and contentious litigation.

The early catch-all phrase used by insurance companies when determining liability was summed up in four words: *perils of the river*. Lawyers for the policy issuers interpreted this to cover "natural accidents"—those preventable by "prudence." That put the onus on the steamboat operators to prove accidents were unavoidable, and thus reimbursable.

"Natural accidents" were defined as accidents caused by snags, rocks, obstructions, storms and collisions. It soon became apparent that one other factor had to be included, and the phrase "perils of the river" was amended to read: *perils of the river and fire*.[4]

In the 1830s, court decisions principally favored the insurance companies. Justices declared that damage not owing to external causes (boiler explosions and accidents to the machinery) were ipso facto the result of negligence, and that negligence on the part of the mariners was historically established as not to result from a "peril of navigation."

In 1838, the Bangor (Maine) *Daily Whig & Courier* ran a notice on October 25, commenting on the loss of a steamboat and the hardship it caused the insurance agency:

> *Another Steamboat Lost:* The steamboat *Irene*, ascending the Mississippi on Thursday night last, when about five miles above [unreadable] on the Upper Mississippi, ran upon a snag which split her hull. The [boat] sank immediately in about eight feet water. The hull, it is supposed, will be a total loss; the engine and most of the cargo will be saved, but all articles liable to injury from the water will be more or less damaged. Her cargo consisted chiefly of merchandise, and was quite large, only about $10,000 of which, we are told, was insured. The boat was insured in the Missouri and Union Insurance offices at $15,000.
> She is the third heavy loss sustained by the owners and Insurance offices in less than two weeks. The losses to the Insurance offices on the Rolla, Governor Dodge and freight amount, exclusive of the risks on the cargoes, to about $46,500.

In 1842, a case stemming from the *Moselle* disaster (see chapter 9) reversed previous legal precedent by declaring that losses due to explosions, even when negligence may have been involved, were commensurable under the language of current insurance policies. The insurance industry responded immediately to this temporary setback. Policies were quickly rewritten, specifically excluding boiler explosions and similar accidents. Hazards specifically exempted in most steamboat policies eventually included all forms of boiler explosions and breakage of the engine and machinery; explosions from gunpowder; and accidents caused by overloading.[5]

Underwriters formed insurance groups at major cities, creating rules and conditions meant to protect their singular interests. These stipulations gave them a great deal of power in the ultimate management of the steamboat. Insurance rules:

(1) Defined and forbade overloading
(2) Regulated the stowage of combustibles
(3) Discouraged the practice of jettison
(4) Required that boats be well-equipped, manned and piloted
(5) Prohibited issuance insurance to any master who refused to comply with these rules
(6) Charged higher premiums for stern-wheeled boats (considered less manageable and less safe), or rejected them entirely.[6]

Insurance companies also developed the practice of refusing to insure boats at full value with a view toward discouraging careless handling or intentional destruction of boats in order to collect insurance money. Through the 1830s and 1840s, limits were generally fixed at three-quarters of value. By 1849, this had lowered to two-thirds value. By the late 1860s, the percent insured fell to only one half a boat's value. Captains often got around this threat by buying a number of policies from different insurance companies, up to the full value of the boat, while part owners sometimes privately insured their interests so that in the case of disaster, they personally would not sustain a devastating financial loss.

Before issuing a policy, insurers considered the age of the boat and the dangers of the trades it normally worked. A side occupation soon arose in the form of the insurance investigator. It became his job to inspect vessels before policies were issued, and this in turn created another bone of contention between insurance agents and steamboat owners: depreciation.

Although a steamboat might cost $20,000 to $40,000 coming out of the boatyard, its value rapidly declined, due to rough handling and the omnipresent river obstructions that even the most experienced pilot could not always avoid. Annual depreciation charges ran as high as 25 percent of the boat's original cost, less salvage value. By 1849, the average life of a steamboat was calculated to be three and a half years, and depreciation was fixed yearly at 24 cents per capital invested.

Stiff depreciation values often meant that in case of total loss, the owner recovered only a fraction of his original investment. Hopefully, he had taken this into consideration, however, as the attrition rate on steam-powered vessels was staggering. Nearly one-quarter of all steamboats launched between 1811 and 1849 were sunk or destroyed, with the average age being just 2.86 years. Clearly, owning a steamboat was not for the fainthearted.

Insurance policies could be taken out by the trip or the season, but typically six-month coverage was purchased.

Insurance Rates

By the month	1–2 percent
By the year	8–18 percent
Inferior boats paid	25–30 percent

Fees charged were generally one-half to 1.25 percent of the value of the freight. Rates on the Upper Mississippi were higher because of dangerous navigation. By 1850, the standard was 1 percent of value per every 1,000 miles traveled.

In an era of risk-taking and working "on a paddle-wheel and a prayer," nearly 50 percent of captains opted to forgo insurance. As late as the 1850s, insurance was considered a luxury.

To Underwrite or Not to Underwrite: Working on a Paddle-wheel and a Prayer

Carrying insurance may have been a luxury to some steamboat captains, and bad seasons or overwhelming disasters such as the Great Fire at the St. Louis wharf (see chapter 22), could break an insurance company, but overall, the business was generally lucrative for those involved in the risk-taking trade. The annual statement for the Merchants' Mutual Insurance Company of 1850 (carried in the Milwaukee *Sentinel Gazette,* March 13, 1850), provides a vivid picture.

The following is the first annual statement of this Company for the year ending February 4, 1850, made and published in conformity with its charter:

Total am't Marine premiums rec'd	$39,921.42	
" Fire " "	6931.28	
	$46,852.70	
Dodnet [?] return premiums	207.19	
Total premiums		$46,645.51
Earned Premiums ascertained Feb 4, 1850	$43,139.21	
Paid during the year as follows:		
Marine losses, Re-insurances and Commissions	$19,149.51	
Fire losses	1,384.20	
Expenses, Salaries, Stationary, [*sic*] &c	4,382.53	
		$24,936.25
Net earned premiums	18,228	
Assets		
Cash on hand	$15,553.73	
Guarantee capital, Judgement Notes	$86,000.03	
Bills receivable	2,497.41	
Due from Agents and individuals	5,316.17	
Salvages	1,427.92	
	$107,989.25	
Deduct for unsettled balance	1,038.39	
	$104,930.56	

Upon the earned premiums in the above statement, the Trustees have declared a dividend of thirty (30) percent, and parties interested will please call at the office and receive their scrip, on or after the 1st of March next. The trustees have also declared a cash dividend of 7 percent for interest on the guarantee capital for the year 1849.

L.W. WEEKS, President.

William J. Whaling, Secretary.

Interestingly, boats and freight were insured separately, making merchants responsible for purchasing their own coverage. Insurance rates on cargo were adjusted by the trade, and were gauged by the reputation of the boat. Throughout the era, nothing was more important to success (or more promoted by the owner) than the reputation of the captain and pilot. Insurance fees were considerably higher on older boats, making those with a first-class rating preferable. However, with that taken into consideration, rates for freight on the better boats were also higher, compelling the shipper to make a choice between saving money and hoping his merchandise made it safely to its destination, or opting to pay up front with some assurance that he would be compensated for loss in case of accident.

Referencing common law, the doctrine held that common carriers (which included steamboats) were responsible for the cargo they accepted, and must compensate the shipper for all losses "except those due to Acts of God, or public enemies." At no time was this doctrine enforced, speculation being that little was to be gained by suing captains or owners who possessed little personal wealth.[7]

With an eye toward protecting themselves, however, it became standard practice by the mid-1820s for steamboat owners to insert a clause into the bill of lading exempting the steamboat from liability to cargo caused by "unavoidable dangers of the river and fire."

A typical advert ran in the Sandusky *Clarion*, April 14, 1824:

NOTICE

Pursuant to a resolution of the Board of Directors of the Lake Erie Steam-boat Company, notice is hereby given, that all goods, wares and merchandise, furniture, plate, jewels and specie, which may be shipped or transported on board of the Steam-Boat Superior, shall be at the risk of the respective owners or shippers thereof; and the said company, or the stockholders, will not pay nor hold themselves, either individually or as a company, liable or responsible for any loss or damage which may happen in the shipment, transportation or delivery of any of the articles aforesaid; and that the captain of the said Steam-Boat is to receive no freight, excepting on the conditions in this resolution mentioned.

J.I OSTRANDER, SEC'RY.

Albany, February 10, 1824.
*** *Notice by Davis & Center.*—Goods or Packages forwarded by the above Boat, or destined to go by her, directed to the care of Davis and Center, will receive particular attention from them.

It is important to note that the advent of western railroads dramatically altered the policy of requiring merchants to carry their own insurance. Railroads also fell under the doctrine of common-law carriers, requiring them to insure their cargo. Here the practice was enforced, and thus gave the railroads a considerable advantage over steamboats, as it eliminated the additional expense incurred by the shipper.

Making Ends Meet

The two principal avenues of revenue for steamboat owners were freight and passenger service. Contrary to popular belief, income from the transportation of cargo far exceeded that of filling the staterooms with well-paying ladies and gentlemen. figures from the times reflect the fact that gross revenue derived from carrying cargo was greater by one-third to three-fifths when compared to the passenger business. After adjustment for expenses required for the latter but not the former (food, lighting, linen and service), the cargo business takes on even more significance.

As with rates on the Upper Mississippi, the prices a Lower Mississippi captain charged for freight depended on a number of variables: the season, the level of the water, the availability of cargo, competition and the overall health of the nation's economy. The season of the year and the rise of the rivers were dependent upon Mother Nature; cargo awaiting shipment and the number of boats offering service were somewhat negotiable. During periods of stiff competition, captains often lowered their demands, occasionally agreeing to terms at or below their operating costs. This certainly favored the merchant, but wreaked havoc with a boat owner's bottom line.

RATES CHARGED FOR 100 POUNDS OF FREIGHT (UPPER MISSISSIPPI)

Before 1820: Keelboat Service
    New Orleans to Louisville $5.00
    St. Louis to Pittsburg $3.00
    Cincinnati to Pittsburg $1.50

Downstream rates were one-third to one-quarter lower than upstream rates.

1820:	Steamboat Service (during a severe depression)	
	New Orleans to Louisville	2 cents per pound
1823:	New Orleans to Pittsburg	$1.00
1828:	New Orleans to Louisville	37 ?–50 cents
Early 1830s:	New Orleans to St. Louis, Louisville and Cincinnati	62½ cents
1839:	New Orleans to Louisville	33 cents – $1.50
1842–44:	New Orleans to Louisville	62½ cents

The downward trend in costs between 1828 and 1830 was the result of a greater number of large boats capable of carrying more freight, as well as competition, more efficient operation of steamboats and the general reduction of time required to make the journey. The rise of costs in the early 1830s reflected poor economic conditions around the country. The Panic of 1832 drove many operations out of the business. With competition lessened, prices tended to rise. From 1835 on, the country entered a period of prosperity, and again more operators entered the business. The prices of the period 1842–1844 reflect growing inflation.

A captain never left port without a full cargo (unless compelled to do so by prior contract agreement; see the comments on steamboat lines), and to get one he would go to great lengths. Failing to procure the necessary tonnage, he might delay departure for hours, days or even a week or more, to the ire of those whose merchandise had already been loaded. Rather than incur the expense of having his shipment unloaded and finding other transport (with the equal likelihood of experiencing the same delays a second time), shopkeepers were compelled to wait out the delays. If they became protracted (as in times of fierce competition), a captain might actually discharge his crew and put up until the situation improved.

Shipments down the Ohio and Lower Mississippi, consisting primarily of raw and processed agricultural products, far exceeded typical return cargos of sugar and molasses, eventually leading to downriver rates being slightly higher that upriver ones. By the end of the 1840s, the price had exaggerated until there was a 50 percent difference in cost. Shippers moving merchandise from a major city had their choice of boats and paid lower fees than those in small or out-of-the-way locales. Plantation owners who shipped from their own wharves and shopkeepers doing business out of little river towns paid substantially more to ship and receive goods.

Freight was classified into two categories: light and heavy. Light freight was defined as dry goods and merchandise of European and eastern American manufacture, possessing an intrinsic value greater than their weight. Heavy freight was defined as products of rolling mills, foundries, glass works, hardware and bulky machinery or farming implements.

Rates were also determined on the particular containers in which goods were shipped: barrels, hogsheads, sacks, crates and bales. Instead of charging by the pound, they were assigned distinctive rates.

Freight Containers

Barrels	Flour, brined pork, whisky, vinegar
Hogsheads	Tobacco, molasses
Sacks	Corn, oats, coffee
Crates	Fresh fruit and vegetables
Bales	Cotton

Livestock and horses were also shipped by boat, being stabled on the main (lowest) deck, where deck passengers were compelled to share the limited space.

Upstream Rates (Upper Mississippi), 1857

30 miles or fewer*	6 cents per mile
(* no charge less than twenty-five cents)	
30–60 miles	5 cents per mile
Over 60 miles	4 cents per mile[8]

A boat leaving Galena on Friday evening usually arrived at St. Paul, unloaded her cargo, uploaded the new consignment and was ready to depart by noon on Tuesday.

Downstream Rates

30 miles or fewer*	5 cents per mile
(* no charge less than twenty-five cents)	
30 – 60 miles	4 cents per mile
Over 60 miles	3 cents per mile[9]

Downstream rates were lower because less fuel was consumed when the boat worked with, rather than against, the current. If a boat made four trips a month (without significant downtime or accident), the ledger sheet presented a very favorable picture:

One downstream trip (passengers and freight)	$1,240.00
One upstream trip (passengers and freight)	$4,450.00
Total income for one round trip	$5,690.00

If a steamboat made four trips a month, the figures are:

Income	$22,760.00
Less operating expenses	$11,500.00
Net profit for month	$11,260.00

Considering that an Upper Mississippi steamer typically was able to work only five months of the year, from the middle of April to the middle of October, the ledger ran:

Income (5 months)	$113,800.00
Expenses (5 months)	*$57,500.00*
Net profit (year)	$56,300.00[10]

A steamboat owner might well say of his boat "When she was good, she was very good; and when she was bad, she was very bad."

Commerce in the West

Fewer than twenty years after steamboats reached the inland waterways, their presence and necessity were well established. In 1820, the *Republican Compiler* of May 31 offered the following statistics:

> Since the introduction of steam boats, considerable attention has been paid at Cincinnati to exportation; & from October, 1818, to March, 1819, it amounted to $1,334,080, and consisted of flour, pork, bacon, lard, tobacco &c.; while the amount of imports, for the same period, amounted to only $500,000. In 1817, the imports amounted to $1,442,266, and in 1818, to $1,619,050! They seem to be convinced that the only way to relieve the western states, from their "present embarrassments," is to *export more and import less,* which will soon effect a rapid change in their affairs.
>
> About 60 steam boats, from 25 to 700 tons, and many of them, finished in a style of

The *Brilliant*, photographed on the Upper Mississippi in 1864. Note the name painted on the wheel guard, very typical of the period.

elegance and taste, are now in successful operation, and most of them have been built within two or three years.

A year later, in 1821, the *Torchlight & Public Advertiser* of September 4 bragged that the new mode of transportation was so rapidly advancing "as will outstrip any moderate calculation." Preceding a table of tonnage was the statistic that as of January 1, 1821, there were "not less than *seventy-two employed*" on the Mississippi and Ohio rivers.

Steamboat	*Tonnage*
Feliciana	408
Tennessee	416
Manhattan	427
Columbus	450
United States	647

The editorial concludes with a fine verbal picture of the early 1820s:

The river is occasionally animated by the smoking boats passing each other; in many stretches, several boats are sometimes visible; and along the levee at New Orleans, twenty steamboats or more are sometimes to be seen together. The people of the Mississippi ought to erect a statue to the genius of Fulton, in the most conspicuous part of New Orleans.

By the mid–1820s, the western frontier had much to brag about, as evidenced by the *Adams Sentinel* in an editorial published September 14, 1825:

Cincinnati, Ohio.—We have after all but an imperfect idea of the vast resources that are constantly developing in this country, and the magic effects of that spirit of enterprise which in a few short years has erected flourishing cities in the midst of pathless forests, and established immense marts of trade on the margins of inaccessible rivers, whose surface but a short time since ruffled only by the wild fowl and the otter, or the occasional glidings of the Indian canoe, is now ploughed by rafts of commerce, and majestic steam boats laden with the produce of one of the most fertile regions of the world. And yet we are by no means sensible of the great activity in business which pervades the towns and cities of the West. Who will not be struck by a statement of the fact that about two months since there were building on the stocks in Cincinnati, Ohio, *eleven steam boats* of from one to four hundred tons burthen, besides three smaller ones that had just been launched, all destined for the navigation of the Ohio and Mississippi rivers!

The year previous, eleven steam boats were built at the same place and put in operation. In addition to this there is more building going on in that city than at any former period.

The Alton *Telegraph* provided statistics for the late 1830s:

At the close of a statement of the number of trips, with the names of the boats, which ascended the Missouri river from the 6th of March, 1838, to the first of January, 1839 showing an aggregate of 18 different boats, and 93 voyages [was] published in the *Missouri Republican*.

The paper goes on to provide the following data:

On the Upper Mississippi.

From April 1, 1838, to March 31, 1839, there were made up this river 270 trips by 55 different boats—180 by 10 regular traders, of which the Quincy performed 38; 84 by 34 boats from the Ohio river, and 6 by 2 New Orleans boats.

On the Illinois

During the same period, the number of trips made up this river amounted to 187, by 47 different boats—136 by regular traders, of which the *Chariton* performed 22; 40 by 26 boats from the Ohio river, and 5 by 4 New Orleans boats.

In addition to the above, 240 trips, terminating at Alton, have been made within the time specified, by boats from various points; of which 57 were from New Orleans by 34 different traders.—The total number of voyages, therefore, made to and above this city, during the year ending on the 31st ult. amounts to 697; and, as it is notorious that, owing to the extreme lowness of the water, the navigation on the Upper Mississippi and Illinois was almost wholly suspended, for the span of about six weeks out of the twelve months designated, it is reasonable to assume that, under more favorable circumstances, many more trips than those enumerated would have been made.

On September 13, 1843, the Milwaukie *Democrat* glowingly recounted the following information:

With reference to the increase of commerce, and the facilities of commerce, in the Mississippi region, a Cincinnati paper says: "The Merchant's Magazine states our whole tonnage in 1817 at 6300 tons, and as our older readers know, a trip to New Orleans or New York, at the time occupied weeks. We had only a few Steam-boats, and the best of those required a large portion of the season to complete their voyage between this and the Crescent City. What a change since then! In 1831, the tonnage on our waters amounted to 39,000 tons; in 1842, to over 100,000 tons; there being at the first period 250 steamers, and at the last mentioned, 450 steamers, averaging 200 tons, floating upon them! And the time of travel reduced between Cincinnati and New York, and Cincinnati and New Orleans, to six days!"

Another paper says—last year the navigation of the Mississippi included 450 steam boats, averaging each 200 tons, and making an aggregate tonnage of 90,000: the cost above $7,000,000; and navigated by nearly 10,000 persons. Besides these steamers there are about 4,000 flat boats, which cost each about $150, manned by five hands each, or 20,000 persons, at an expense of $1,380,000. The estimated annual expense of the steam navigation, including 15 per cent for insurance, and 27 per cent for wear and tear, $13,618,000.

Of the increase in commerce upon the lakes we have not the requisite data at hand to enable us to give a full or correct estimate. In 1836, the lake trade amounted to $16,460,000. In 1841, according to official returns at Washington, it amounted to $65,460,000, and this year, from the progress in past years—*ten millions of an increase per annum*—a fair estimate would be $83,000,000!

In 1847, the Cincinnati *Gazette* printed a table reporting a detailed statement of the tonnage of several principal cities of the West, "giving a pretty good idea of the vast commerce which has sprung up in a region which fifty years ago was a perfect wilderness." (Note the word "Pittsburgh," spelled with an "h" at the end.) The table covers the period, from January to June.

City	Boats	Tons
Pittsburgh	1,405	161,933
St. Louis	1,150	224,564
Cincinnati	1,782	347,044
Total	4,334	734,441

The St. Louis numbers exclude keelboats and flatboats and the Cincinnati numbers include flatboats but exclude hulls, canal boats and keels.

The newspaper further remarks that these totals come to within a fraction of equaling "as much as the whole of the tonnage employed in the immense foreign trade of Great Britain and Ireland, which extended all the ports of the United Kingdom during the first quarter of the year 1845, and also in the same period of 1846." It goes on to say:

> And yet, the *Washington Union,* the organ of this Democratic Administration, whose warmest supporters are found at the West, frowns upon a convention assembled for the purpose of putting into operation measures for the safety and preservation of this incalculable rich commerce. The Atlantic seaboard has not stronger claims upon the protection of the General Government than has this seaboard of what Mr. Calhoun has pronounced a great inland sea.[11]

The construction of steamboats on the Ohio provided most of the vessels for use on the Mississippi.

Year	Number of boats built
1847	*
1848	150
1849	124
1850	73
1851	126
1852	175
1853	115
1854	*
1855	102
1856	150

The aggregate of vessels built was 1,260. Of these, 1,088 were steamboats, making an average of 108 boats constructed per year. The entire number of steamboats built in the United States in the ten years was 2,158, making more than half that number constructed in the Ohio Valley. (Of the remaining 181 vessels, two were ships built at Cincinnati, two were brigs, one was a schooner, and the remainder were sloops and canal boats.)[12]

Gross Revenue: The Deck Passenger

While much has been made in lore and screenplay about the glories of passenger steamboat travel and the implied profits of such service, the hard truth is that deck passengers—those who paid a minimal fee for transportation—were a greater source of revenue for the steamboat owner. Less welcome than tolerated, in terms of gross revenue, the unheralded deckers' fares were clear profit. They were not only self-handling, but paid a rate three to four times the amount they would have brought, baggage included, on a straight pound-for-pound ratio.

Cabin passengers, on the other hand, who were so courted and lavishly dined and entertained, actually paid less for their conveniences than what was expended. As far as balancing the books, this actually made them a liability, for they came in under the red ink column, but steamboat owners had a peculiar sense of business. More concerned with profits at the end of the season rather than balancing out individual aspects of the business, they were content to absorb this type of loss in exchange for the prestige and sense of pride that running a "brag boat" brought.

---

*The numbers from 1847 and 1854, combined, are estimated at 254.

Steamboats and the Business of Communication

Stagecoach operators were authorized agents of the federal government, carrying sacks of mail from way station to way station, but just as important, drivers and passengers brought the latest news by word of mouth as surely as they carried newspapers and periodicals from distant cities. Dropped in saloons or barber shops, the papers were read by many, their pages stained by dozens of tongue-licked fingers. In this way, current events, society "doin's" and gossip made the rounds of the known and sometimes even the uncharted regions of the expanding country.

One of the steamboats' earliest gifts to the West was the swift and sure transportation of news. Newspapers went up and down the rivers with a facility never before imagined, and the contents of what they carried were quickly transferred to local papers and discussed around drinks at nearby watering holes. Of equal importance were going rates, timetables, weather and river conditions, much appreciated by crews working toward those areas of interest. Pilots would often congregate to share the development of new snags, speed of currents or water levels with their fellows in a spirit of goodwill and cooperation.

From the beginning, steamboat captains agreed to carry letters from civilians and businessmen without charge, dropping them off at the town closest the recipient's home or place of work. Pigeonholes were often filled with such letters and small packages, and this courtesy was greatly appreciated by everyone but the United States Postal Service. In 1815, Congress attempted to put a stop to the practice of delivering mail without due postage by enacting legislation dictating that all letters given crewmembers were to be handed to the captain, who was then ordered to turn over such letters to the nearest post office for a small remuneration. The legislation was generally ignored.

Two years prior, in 1813, the postmaster general had been granted the authority to contract steamboat operators for the official delivery of mail. The system worked well in the East, where steamboat captains received annual payments of $310 per mile for this service. Peculiarly, the postmaster general was far stingier with his payment plans to western captains. For transport of mail on the Ohio, the going rate was $33.00 per mile; from Louisville to New Orleans, $16.53, and from St. Louis to the same destination, the very low fee of $4.99 per mile. Further complicating matters, the postmaster general deviated from the successful eastern system by refusing to offer yearly contracts, dictating that western captains accept commissions on a trip-by-trip basis.

Needless to say, the offers were not highly regarded, and those undertaking the job occasionally did so with something less than enthusiasm, as indicated by this article from the Cleveland *Herald*, reprinted in the Milwaukee *Sentinel*, June 2, 1840:

> *Steamboat Mails*—If boats are paid for carrying the mails they should do the fair thing. It is the business of the officers of the boats we presume to see that the mails are properly delivered at the post-office in this city. We are told that the mail brought by the Erie yesterday was tumbled into a warehouse, and left there. Towards evening it was reported to the Post Master that a mail had been left at one of the warehouses, and he promptly had it brought up. The combination had better not raise the anti-accommodation steam too high. The public may possibly help a blow up if they do.

By 1845, only nine steamboat mail routes were in operation. Although much discussion was had on the subject, for mail delivery held the alluring prospect of profit, nothing substantial was done until 1852 and 1854, when routes between New Orleans and Cairo, and Louisville/Nashville to Memphis benefited from new laws and rates of payment. As was so often the case, the advent of railroads siphoned off this business, particularly after the Civil War.

Steamboats Carry News of the Cable

The steamboat's role in carrying news was also reduced by the successful stringing of the trans–Atlantic telegraph cable. But they did not go down quietly, playing a significant role in transmitting this news item to eager Americans.

In 1857, an attempt was made by the U.S.S. *Niagara* and the H.M.S. *Agamemnon* to lay a cable across the Atlantic. The cable broke, however, and a second try was made the following year. The boats involved suffered the same fate. It required four tries before they succeeded, but when they did, the achievement was momentous. The first message transmitted announced:

> Europe and America are united by telegraph. Glory to God in the highest, on earth peace, and good-will to men.

Captain Daniel Smith Harris of the *Grey Eagle* determined to be the first to bring the news to St. Paul a 265-mile voyage. A race developed between Harris and Captain David Whitten of the *Itasca,* which had only 200 miles to cover. Carrying copies of the Dubuque and Galena newspapers, Harris set out on August 17, 1858. Hard-firing the engines with pitch, butter and grease, Harris flew upriver, stopping only when absolutely necessary to discharge cargo and bribing way-passengers to stay aboard by the offer of free meals and the honor of being the first to bring such wondrous news to St. Paul.

Harris broke all speed records and, unbelievably, made up the additional sixty-five miles, bringing his boat to within a nose of Whitten's at Dayton Bluff, just one mile outside of St. Paul. Whistles shrieking, cannon blasting, the *Itasca* beat the *Grey Eagle* by seconds. But while Captain Whitten lowered the stage, deckhands from Harris' boat tossed newspapers affixed to a piece of wood into the waiting arms of an agent, and in so doing won the technical, if not the physical race, to St. Paul.

Thus, Queen Victoria's triumphant message to President James Buchanan, was read to the cheering throng:

> The Queen desires to congratulate the President upon the successful completion of this great international work, in which the Queen has taken the greatest interest. The Queen is convinced that the President

Captain Daniel Smith Harris, an early Upper Mississippi river captain of great fame. He commanded the *Grey Eagle*, and participated in the race with the *Itaska* to bring news of the transatlantic telegraph to St. Paul in 1858.

will join her in fervently hoping that the Electric Cable which now connects Great Britain with the United States will prove an additional link between the nations whose friendship is founded upon their common interest and reciprocal esteem. The Queen has much pleasure in thus communicating with the President, and renewing to him her wishes for the prosperity of the United States.[13]

The *Grey Eagle* ran from Dunleith to St. Paul in 24 hours, 40 minutes, including landings and wooding stops, making a little over eleven miles an hour upstream, shattering the record set in 1853 by the *Die Vernon* by over three hours.[14]

Captain Daniel Smith Harris became a legend, but, tragically, the *Grey Eagle* was lost when she struck the Rock Island Bridge on May 9, 1861. She sank in less than five minutes with seven lives lost. Captain Harris was in the pilot-house with a specialty "rapids pilot" when a sudden gust of wind blew the vessel against the abutment. Brokenhearted, Captain Harris sold out his interest in the Packet Company and returned home to Galena. He retired and never again commanded a steamboat.

11

Man Overboard! Steamboat Disasters on Western Waterways

In the 1800s, the word "steamboat" was analogous to the word "disaster." Or more graphically, "Explosion!" equated to "death." New technology on the scale of mass transportation always came with a price: the cost of research and development in this case was borne by the early pioneers who experimented on the basis of trial and error at their own expense. What worked became an improvement. What did not was discarded or revised. That was the accepted standard, and it worked well. From the earliest years of low-pressure to high-pressure engines, copper boilers and auditory steam gauges, bells and whistles, within half a century American ingenuity took the concept of a small flatboat with an engine and converted it into the mythical floating palace. Lives lost along the way were considered a regrettable price to pay for success. In the 1800s, no natural disaster or man-made calamity claimed as many lives as the fires, explosions and sinkings of steam-powered boats. Loss of life on the rivers was surpassed only by war and Indian massacre, but these numbers primarily involved soldiers, professional men who accepted the possibility of death as part of their occupation. Steamboat accidents were indiscriminate. They took women and children to watery graves with blind indifference.

No official statistics were taken before the Steamboat Inspection Act of 1852, but graphic newspaper reports and eyewitness accounts provide rough estimates. They tell a chilling story of scaldings, burnings and floating corpses that claimed headlines from New Orleans to Boston. As frightening as these stories were, they failed to keep people off the boats. The prospect of speed, the allure of glamor and the insatiable desire to see new sights drew people from all walks of life and from around the globe, all willing to accept the common homily that death was for someone else and not for me.

The odds were in their favor, but just barely. Nearly 30 percent of steamboats built before 1849 were lost due to accident.[1] The Milwaukee *Sentinel,* February 25, 1840, provides some data:

> *Steamboat Accidents and loss of life.*—According to a table in the American Almanac, for 1840, it appears that the number of lives lost from 1816 to 1838, inclusive, is 1676. The number of wounded is 443. 99 of the accidents were by explosions and collapse; 28 by fire, 25 were wrecked from gales, collisions &c; 52 from snags and sawyers, and 21 from different and unknown causes.

Steamboat Accidents between 1810 and 1851

Cause	Number	Average Loss in $	Property Loss in $
Collision	45	8,635	379,933
Fire	166	10,948	1,817,428
Explosion	209	13,302	2,780,118
Snags & other obstructions	576	6,391	2,681,297

The loss of life from these disasters was estimated at 2,300, and a large number of people were scalded, wounded or crippled permanently. The pecuniary loss was estimated as much as $86,000,000, and this is probably a low estimation.[2]

In 1841 alone, the Ohio *Repository* of November 4, estimated that since January, the steamboats and cargo sunk between St. Louis and the mouth of the Ohio were valued at $620,500.

Onto the Rocks

While rocks were primarily an Upper Mississippi problem, the Hagerstown (Maryland) *Mail* of May 8, 1835, gives an account of a disaster occurring on the eastern seaboard that demonstrated the dangers common to all inland vessels.

> The steamboat *Chief Justice Marshall*, Captain Waterman, which left New York on Monday afternoon for New London, in consequence of the gale, came to anchor at 2 P.M. on Tuesday off Saybrook. She parted both cables, and both boilers rolled overboard. The captain attempted to take her into New Haven, but did not succeed, in consequence of her having become water-logged. She was thrown on the rocks near the light, about one mile from New Haven, and is stated to be a total loss. The pilot took to the small boat, and was thrown over by the waves, and drowned. The crew and passengers were saved.

Snagged by a Sawyer

Snaggings accounted for more accidents than all other causes combined, but were responsible for fewer fatalities. This was due to the fact that snags seldom penetrated the hull, doing the most damage to overhanging guards or paddlewheels. These damages were often expensive to repair, but unless a passenger or crewman were struck by flying debris or fell overboard, most of the injured survived this type of mishap.

If the hull were breached, an immediate action had to be taken to prevent the boat from sinking. Pillows, blankets or other handy items were shoved in the hole. If these did not work, the pilot attempted to pull the crippled boat to shore or settle her on a sandbar. If the accident occurred in deep water, the likelihood of the boat going under became a real possibility. In that instance, the lone yawl or sounding boat might be used to transport passengers to safety. Passengers who jumped overboard, whether by necessity or choice, often drowned, for few men and virtually no women could swim. Those who tried were often drawn below water by the undertow. Because of the nature of the disaster, however, there was seldom opportunity for this means of escape. An example of just such a tragedy comes from the *Republican Compiler,* August 11, 1835:

> *Shocking Disaster.*—The Cincinnati Gazette mentions that the steamboat Rob Roy, on her way up from New Orleans, on the 19th ult., about 15 miles from New Madrid, struck

a snag, which caused her to twist in such a manner as to break one of her branch pipes, and scalded 10 or 12 persons, of whom 4 have died, and two more are not expected to recover; at the same time a number of persons leaped overboard of whom 3 were drowned.

In case of sinking, fire or explosion, the situation for deck passengers was always more dire than those residing in the upper cabin deck. Because of severe overcrowding on the lower, or main, deck, deckers were closest to the water line. When a boat sank for whatever reason, this area went under first. If they were unable to escape up the few stairs leading to the higher levels, deck passengers were quickly submerged. The same fate awaited the engine room staff and those off duty, for the crew's quarters adjoined, and were often intermixed with, the deck passengers'.

During a disaster, cabin passengers were herded to the highest point on the boat. If only the lower portions were covered with water, they were saved from drowning. However, in the instances where the boat became totally submerged, there was no safety for anyone. Nor was there much time to react, for a damaged vessel went down in two to five minutes. The Southport (Wisconsin) *Telegraph* of November 9, 1841, gives two accounts of the sinking of the *Eliza*. The disaster was one of the worst ever caused by snagging.

Fearful Disaster—Twenty or Thirty Lives Lost!

We have this morning another account of the horrible loss of life upon the Western waters.

The steamboat Eliza, Capt. Littleton, of St. Louis, struck a snag in the Mississippi, four miles above the mouth of the Ohio, and sunk in THREE MINUTES, leaving but two feet of her hurricane deck above water. The Cincinnati Gazette of the 20th says that nearly all on the lower deck and in the cabins perished, and adds:

Reports speak of some forty or fifty that were drowned, but our informant thinks the number about 20—certainly not more than 25 persons. Among the dead are the captain's wife and two children, and some eighteen or twenty deck passengers!

A melancholy record, this, and the more melancholy too, because this and similar accidents might be easily prevented! Is life of so little value that western people will make no effort to render travelling on our waters safe? Are the lives and property of our citizens of so little moment in the estimation of the Government, that it will do nothing to protect them? Let us hope not. Let us hope that the fearful accident will rouse people of the west to something like united action, as regards the improvement of our western waters, and convincing our rulers of the necessity of a speedy and efficient movement on the subject.

The Tribune's Baltimore correspondent says the Eliza at St. Louis, has above 100 passengers. When she struck, the greatest alarm instantly prevailed, the wheels became waterlogged, many plunged to swim ashore and were drowned, and the deck passengers, with cattle, horses &c. were carried down with the sinking boat. The captain's wife leaped overboard with two children: her husband leaped to save her but only rescued one of the children. The particulars of the disaster were not accurately known, but it was believed not more than 40 perished.

The following reports by the Hagerstown *Mail,* May 8, 1835, list disasters caused by snags. The first does not fail to mention the other important fact in casualty reporting: the state of insurance.

Steamboats Lost

A Florence (Alabama) paper states that while the steamboat *Nashville,* belonging to that place, was on a passage from N. Orleans, and when near the Iron Bank, Mississippi, she struck a snag, which so much injured her, that it was found impracticable to keep up; she was consequently run on shore. Considerable part of her cargo has been saved, but we understand that the boat will be a total loss. She was insured as well as most of the cargo.

* * *

On the 11th Instant steamboat *Express,* from New Orleans to Pittsburg, struck a snag 5 miles below Rodney, and sank in 5 minutes in 40 feet of water. Boat and cargo irrecoverably lost, and a black man drowned — none others ascertained.

Being a "crack" boat did not necessarily mean a safer boat, as a note from the Milwaukee *Daily Sentinel and Gazette,* June 4, 1850, relates:

> The Missouri Sunk — The steamer Missouri, one of the largest and finest boats on the Mississippi struck a snag last week and sunk in deep water. The boat is a total loss. Loss $17,000, insurance $8,000.

Sawyers presented another danger. If the sharp, lance-like tip of a submerged tree struck the boat in a vulnerable area, great damage was done. Occasionally one broke through the staterooms, went up through the hurricane roof and sheared the Texas deck in two, toppling the structure over the side. (See chapter 4 for a contemporary account of just such an accident.)

Most steamboat accidents took place in darkness and on upstream trips. Striking snags, sawyers and other types of obstructions were generally considered "acts of God," and no blame was accorded the pilot.

Sideswipes and Collisions

Two boats passing close enough to touch each other was referred to as sideswiping. Usually, little or no damage resulted, and aside from the excitement afforded passengers, little thought was given to the mishap. A typical notice of such an event, from the Madison *Express,* May 12, 1841 reprinted in the Galena *Gazette,* reads:

> The steamboats Monsoon and Rosalie lately accidentally came in collision, and the former boat was so much injured as to be obliged to go on the dock to be repaired at St. Louis. No lives were lost.

A dispassionate account of a collision reads as follows:

> Captain Menaugh, of the *Polander,* had just left the wharf at Cincinnati, about eight o'clock P.M., the night being dark and foggy, when she encountered the *Hornet,* which was coming into port. Both vessels were considerably injured, and the Captain of the *Hornet* was crushed to death. One of the crew of the same vessel was severely wounded. No further particulars have been published.[3]

The following, from the Hagerstown *Mail,* dated June 1, 1832, gives just those "particulars." (Note the different spelling of the captain's name; probably neither reporter knew for certain and guessed at the phonetic spelling. One of them also reversed the name of the boat the unfortunate captain commanded.)

> On Wednesday evening the 18th inst. between seven and eight o'clock, the steamboat Hornet, one of the regular packets between Maysville and Cincinnati, on her downward

trip, when about two and a half miles above the latter place, came in contact with the Steam boat Polander, which carried away her wheel house and guard, the main anchor at the same time taking the starboard side of the cabin and remaining connected with her. The captain of the Hornet, G.W. Menau, was standing on the guard at the moment, and it is supposed was crushed to death and was swept into the river. His cap, one slipper, pocket book, two teeth, and a lock of bloody hair, was found on the guard. A youth named *Bakewell*, a student of Augusta College, who was standing near the captain, had his collar and breast bone broken, and received several severe contusions in other parts of his body; but it is supposed he will recover. The collision between the boats, we learn, was purely accidental, and no blame is attached to neither [sic] commander.

Despite the credited width and appearance of most rivers, the actual space for safe navigation was far narrower. This was particularly true of island chutes, shoal waters, rapids, and low water, where uprooted trees, shrubs, and silt from the banks tended to congregate. When two boats steaming in opposite directions met on the river, certain rules were established for how to proceed. Crafts descending the river were to keep to the channel and sweep the bends; those ascending were to hug the bars. The downriver boat was to stop its engines and drift, placing responsibility for avoiding a collision on the ascending boat. At night, both vessels were to stop their engines, ring their bells and show lights. The ascending boat would then take the side of the river its captain preferred. These rules were as often disregarded as followed. The only practice which became standardized was that one tap on the bell signified a boat would pass on the right; two taps, to the left.[4]

In the case of head-on collisions, the ascending boat typically sustained the most damage. Danger in these instances came from breaching the hull, or in the rupture of steam lines or the snapping of tiller lines. Without power or the ability to steer, the injured boat fell to the mercy of the current. While most struck boats did not sink, the danger of scalding from escaping steam caused serious wounds and occasionally death to the engineering crew and nearby deck passengers.

The *Adams Sentinel,* December 2, 1833, paints a rather succinct description of a collision:

> We are informed that the Navarino, on her way to Louisville from New Orleans, was run into by the Constitution, and had her wheel-house, wheel, guards, &c. carried away.

The most tragic collision occurring before the Civil War came in the late fall of 1837. The *Monmouth,* a small, 135-ton boat, was ascending the Mississippi to the trans–Mississippi West with 500–600 Creek Indians who were being removed to new lands. The boat was overcrowded and conditions were unsafe. The boat was struck by the *Warren,* descending the river with a boat in tow. The *Monmouth* sank, with the loss of life put as high as 400. The owner of the *Monmouth* blamed darkness and rain and the fact the captain of the *Warren* had failed to put out warning lights. The defense countered that, according to accepted standards, the *Monmouth* was running in an area of the river she had no right to be, and therefore the captain of the *Warren* should not have been expected to anticipate her sudden appearance.

The *Adams Sentinel* of November 13, 1837, reported:

> STEAM-BOAT ACCIDENT!
> 300 INDIANS DROWNED!!

The steamer Monmouth, with 600 Indians on board, on their way to the west, was run into by another steamer in the Mississippi, on the night of the 27th Oct. and immediately sunk. About 300 of the Indians were drowned, together with several of the hands belonging to the boat.

Another short notice ran in the *Miners' Free Press* (Mineral Point, Iowa County, Wisconsin Territory) on December 22, 1837:

> It is stated in the New Orleans Picayune of the 8th inst. that there were 602 Creek Indians on board the steamer Monmouth, recently sunk in the Mississippi, and that 234 were killed and drowned, instead of the 311 as mentioned in our paper some days ago.

Although such a monumental accident normally would have attracted headline news and long editorial discussions, the tragedy occurred far from normal trade routes and reporters had no access to the scene. More tellingly, as the vast majority of dead were Indians, the incident passed off with little publicity and less ire.

Interestingly, this was not the first collision involving the *Warren*. On January 9, 1837, nearly a year before the October tragedy, the *Adams Sentinel* ran a report, dated "New Orleans, December 19, 1836":

> *Steamboat accident and loss of life:*—The steamer WARREN left here recently for Vicksburg, was run into by the ALGONQUIN at an early hour yesterday morning. The concussion was so great as to move the whole furnace some inches, breaking the connecting pipes, and scalding one man (a deck hand) instantly to death; also four very badly, and two slightly; of the former, two have since died. The passengers on the Warren represent the morning as being clear and starlight; but notwithstanding, say they are far from attaching any blame to the officers of the other boat.

Fire: The Familiar Enemy

Remarkably, accident due to fire was considered an acceptable danger associated with steamboat travel, and before 1852, little was done to lessen this omnipresent danger.

Causes of steamboat fires were apparent to everyone. The very nature of the vessel rendered it susceptible to ignition and the rapid transport of killing flames. While the hull and main deck were built as a solid foundation, the massive upper structure was designed to be light and insubstantial. Floors, walls, siding and framing were made with soft, resinous woods, and exteriors were sun-beaten, overdried by river breezes and covered in layers of oil- and turpentine-based paints. Open spaces along stairwells and promenades were perfect conduits for the spread of fire.

Wood used for fuel was also a major contributor. The basic steamboat design placed open furnaces at the front of the boat, completely exposed to the elements. This was done to catch the draft, augmented by the chimneys and forward motion of the vessel, which, in turn, aided the combustion in the furnaces. That machinery, as well, contributed to the danger, for the metal remained red-hot. Furnace doors were often left open and bits of burning wood fell through to the hearth, where they occasionally spilled out. When additional power was needed and more wood and pine tar was added to create steam, the process generated terrific heat and a profusion of sparks. An ember landing in the wrong place could easily burn long enough to cause damage, if not create flames.

Additionally, freight was stored in every imaginable corner in highly flammable materials: burlap sacks, twine bindings, cotton cloths. Straw was used as packing and cushioning; oil, whisky, liquor and turpentine barrels leaked; gunpowder casks swelled at the seams. Bales of cotton, although tightly packed, would burn and hold heat, while sparks, lifted by river breezes, were borne upward, easily capable of igniting small fires that could soon become infernos. Once started, a blaze was nearly impossible to contain.

Preventative measures were scant. The hurricane deck and the roof of the pilothouse were sanded to lessen the chance of catching fire from sparks falling from the twin chimneys, and on more expensive boats, captains occasionally laid down a thin covering of sheet-iron, but that was the exception. In recognition that fire was likely to occur, barrels of water were stationed around the boat to use in case of emergency.

The need to insulate the metalwork of chimneys and furnaces from contact with wood was only slowly acted upon, and fires from overheated chimneys were so common—and disregarded—that over the course of a boat's life, she might be exposed to fifty such blazes.

Less obvious but readily apparent sources of ignition came from lighting. Candles and whale oil lamps were used for illumination. When brought to the cabins by passengers, the lurch of the boat or a sudden stop or start could tip the open flame and catch the bedclothes.

When a fire started near the furnace or chimneys, the draft coming in from the river quickly spread the flames. The partially open design of the lower deck and the open galleries of the boiler deck aided the spread, and overpacked areas made it difficult to get to a fire, much less contain it. Fortunately, the thick hull was not easily burned. By battening down the hatches, air was cut off, and fires likely went out from lack of oxygen. The situation was diametrically opposed if a fire started above the main deck. Once begun, flames could sweep through the entire boat in a matter of minutes.

Possibly the most crucial and least glamorized equipment on a steamboat were the tiller ropes. Connected to the rudder, these hemp ropes ran much of the boat's length before finally reaching the pilothouse. The pilot's ability to turn the great wheel and steer by changing the direction of the rudder depended entirely upon this tenuous connection. Unfortunately, hemp burned well and a fire of any magnitude quickly burned through the ropes, thus rendering the boat incapable of being maneuvered. Only in later years was hemp replaced by more durable and more fireproof material.

The year 1832 was particularly bad for steamboat fires: three boats were destroyed, with up to 90 lives lost. The worst disaster (predating the Steamboat Act of 1852), occurred in May 1837, when the *Ben Sherrod*, en route to Louisville, burned with the loss of 120–175 persons.

The *Ben Sherrod*, a 400-ton boat heavily loaded with cargo and more than 200 passengers, was alleged to have been in a race with another steamboat. Eager to come out the winner, the captain plied his firemen with whisky to stimulate them in their work of shoving pine knots and resin into the superheated furnace. Apparently no one noticed when sparks or embers caught in a nearby pile of cordwood. Contemporary reports state the fire was well out of hand before the deck passengers were alerted to their danger, by which time it was too late for most to get to safety.

Fire burned through the tiller ropes before the captain realized the grave danger, and his order to steer for shore came too late. The engineer abandoned his post (an extraordinary failure and not typical of that class of officer), and the boat careened out of control. Deck and cabin passengers jumped into the river to save themselves; those who were not awakened in time were killed in the ensuing inferno as barrels of brandy and gunpowder blew up. Explosions of the boilers followed.

Steamboats in the vicinity, attracted by the eerily lightened night sky and the sounds of igniting gunpowder, came to the rescue within half an hour of the disaster. One survivor who was picked up by the *Columbus* later wrote:

> The screams of men, women and children pierced the air for miles around while in the bright light that went up from the waters, the hanging forms of the poor wretches as they clung convulsively to the burning sides of the boat struck the deepest anguish into the hearts of the spectators.[5]

Worse, the captain of the *Alton*, another boat arriving on the scene, refused to aid the rescue attempt, abandoning the scene with such haste he ran his boat "over many, and drowning others by the waves created on his passage."[6]

An aside to the above (which contains no other information), comes from the Alton *Observer,* June 8, 1837:

> The amount of specie on board the Ben. Sherrod, steamboat, lost on the Mississippi, is ascertained to be $230,000, which had been drawn from the New Orleans banks. It is supposed none of it will be recovered.

A public meeting was held in Natchez, and a committee of twelve was appointed to investigate the disaster. They "condemned the ignorance, negligence, and rashness of the officers as the invariable cause of steamboat accidents, denounced the practice of racing, adopted resolutions censuring strongly the conduct of the captains of the *Sherrod* and *Alton,* and urged legislative regulations to make transportation and travel safe.[7]

Subsequently, the Steamboat Act of 1838 briefly addressed fire hazards, but it would not be until 1852 that a stronger, more thorough act was legislated.

The disaster of the steamer *Belle of the West* was covered by the Covington *Journal* in its Saturday, April 27, 1850, edition. In this case, the culprit was an ordinary box of matches. The writer explains that the boat left Covington on Monday afternoon (April 22) for St. Louis, having on board 200–300 passengers, among whom were a large company of emigrants bound for California. At midnight, about two miles below Warsaw (62 miles from Covington), the boat was discovered to be on fire.

> At first the officers endeavored to suppress the flames without alarming the passengers, but when their efforts were found to be of no avail, the pilot was directed to run the boat ashore and their passengers were roused from their sleep and warned of their imminent peril. The scene which ensued beggars description. Passengers were seen hurrying wildly through the smoke in their night clothes — many of them delirious with excitement, and not a few leaping helplessly into the water, to escape the threatening danger of the flames.

The Cincinnati *Dispatch*, April 25, 1850, gave other details:

> As near as we can discern, the lives lost will number over eighty. None of the crew were lost. A portion of the goods were brought to this city in a damaged state. Over thirty of the dead and dying are at Florence. The boat was fully insured in this city, and we learn from good authority that a protest was issued yesterday.
>
> The watchman is of the opinion that the fire originated in the hold from a large box of matches taken on board here.... The captain, after he discovered the fire, immediately gave the alarm, and hastened to a forward hatch where a keg of powder had been stored, and threw (it) overboard, saving in all probability, a dreadful explosion.

The two versions are noteworthy in their discussion of how the fire was handled: the author of the *Journal* article notes, "At first the officers endeavored to suppress the flames without alarming the passengers," while the reporter for the *Dispatch* relays the opposite: "The captain, after he discovered the fire, immediately gave the alarm...." It is likely any riverboat gambler worth his salt would take odds on the former being true.

Even more telling, the article in the *Dispatch* conveys information in an order relevant to readers in a western river city:

(1) Lives lost will number over eighty
(2) None of the crew were lost
(3) Good transported to Cincinnati were in a damaged state
(4) The dead and dying were brought to Florence
(5) The *Belle of the West* was fully insured
(6) A protest (involving insurance) was issued two days after the accident

Three of the points had to do with money and one quickly dispatched any fears for the crew. These were not uncommon priorities.

Two other articles on the disaster bear inspection. The first, dated "Cincinnati, April 23, 1850," states:

> It is confidently asserted that not less than one hundred persons were burnt to death and drowned.... The officers saved their lives by immediately jumping overboard, and swimming ashore. The Belle of the West was owned in this city, and was insured for $8,000. She is said to be totally lost.

And the second, dated "Madison, April 28, 1850," gives the report of an eyewitness, who states the boat appears to have been burned rather than being destroyed from the collapse of her boiler. In addition:

> The fire was discovered at about 12 o'clock, in the hold, when she was immediately run ashore, where she was made fast, and stage planks run out. Up to this moment the flames had not burst forth. The after-hatch was then raised for the purpose of getting water into the hold, but such was the pressure of the flames, that all efforts to quell them were of no avail. The total number of passengers is estimated at 400, among whom were two companies of California emigrants, and about thirty families removing westward.
>
> It is ascertained from the register that over sixty souls perished, and probably as many more have been lost whose names were not enrolled. Such was the progress of the fire that, before the passengers could get out of the state rooms, all communication between the after cabin and forward part of the boat were cut off, and all either were compelled to jump overboard or perish in the flames. At the time the deck fell in, a lady and gentleman, with a child in his arms, were standing between the chimnies [*sic*].
>
> A large number of horses and cattle were nearly all burnt to death.

An overlooked cause of fire, but one equally devastating, came in the form of intentional burning. A note in the Covington *Journal* of Saturday, May 25, 1850, reports:

> On Thursday night last, the steamer Enterprise lying in Licking was destroyed by fire. The fire is supposed to have [been] the work of an incendiary. The boat was insured to its full value.

A second notice is a stark reminder that life on the western frontier was arduous:

> The Pine Bluff steamer caught fire and sank in the Ohio River near the mouth of the Licking River in December, 1866. Loaded with ore, it plunged to the depths in the early

hours of a Sunday morning. No one was injured, but damage was set at $20,000. Arson was suspected.[8]

Not all passengers aboard vessels destroyed by fire or other means were fortunate enough to have their fates known. Occasionally, a steamboat traversing the oceans just disappeared off the face of the earth. One telling account in the *Madison Express,* August 4, 1841, provides a fascinating insight into one discovery.

> Considerable interest was elicited here this morning (3d inst.) by the announcement of the fact that Capt. Sawyer of the brig Augusta [has] fallen in with the hull of a large Steamship, burned to the waters edge. The Augusta is 24 days from Trinidad, and reports to have fallen in with the wreck in question on Thursday last, in lat. 43 30, long. 73 26. One guard was under water and the other much charred and burned, and although he bore down close upon it, he was unable to ascertain any thing more than it was the remains of a large steam vessel, and that she had been destroyed by fire. These two facts he is positive of — The general belief is that this is the remains of the long missing and unfortunate steamer President, and as it entirely changes the previously entertained belief of the manner of her loss, hopes are expressed that at least a portion of her crew and passengers may be safe. Anxiety is now intense to learn if her destruction was by fire. Some censure Capt. S. for not boarding the wreck and gathering all the particulars.—*Philadelphia correspondent of the N.Y. Herald.*

A Tradition of Explosions

Boiler explosions continued to capture America's imagination. Monstrous stories passed by word of mouth, detailing volcano-like boilers bursting and spewing forth steam, boiling water and chards of metal. The awful consequences were also fodder for the tabloids, periodicals and monthlies, attracting the same audiences who read about the atrocities of Jean Lafitte and John Murrell.

Heroes in the case of steamboat accidents were likely the captain, the pilot or a self-sacrificing passenger; the villain almost certainly the engineer.

Steamboat explosions were a fact of life. Incomplete records kept before 1852 indicate that one-half of all boating deaths were caused by explosions. The situation never got much better. Between 1860 and 1889, explosions accounted for 40 percent of steamboat deaths.

In 1858, the explosion of the *Pennsylvania* hit a young cub pilot particularly close to home:

> It was six o'clock on a hot summer's morning. The Pennsylvania was creeping along, north of Ship Island, about sixty miles below Memphis on a half-head of steam, towing a wood-flat which was fast being emptied.... There were a good many cabin passengers aboard, and three or four hundred deck passengers.... The wood being nearly out of the flat, now, Ealer [the pilot] rang to "come ahead" full steam, and the next moment four of the eight boilers exploded with a thunderous crash, and the whole forward third of the boat was hoisted toward the sky! The main part of the mass, with the chimneys, dropped upon the boat again, a mountain of riddled and chaotic rubbish — and then, after a little, fire broke out....
>
> By this time the fire was beginning to threaten. Shrieks and groans filled the air. A great many persons had been scalded, a great many crippled.... Both mates were badly scalded, but they stood to their posts, nevertheless. They drew the wood boat aft, and they and the captain fought back the frantic herd of frightened immigrants till the wounded could be brought there and placed in safety first....
>
> By this time the fire was making fierce headway, and several persons who were impris-

oned under the ruins were begging piteously for help. All efforts to conquer the fire proved fruitless; so the buckets were presently thrown aside and the officers fell-to with axes and tried to cut the prisoners out.

One of the clerks fell into the water, but swam back to help save the wounded. He performed his service, but suffered grievously, eventually being taken to a makeshift hospital in Memphis, where he died six days later. The clerk happened to be Henry Clemens, and the sad duty of writing about his death, which he witnessed at the bedside, fell to his brother, Samuel.[9]

Ironically, the time of the boiler explosion of the *Pennsylvania* was an anomaly: the majority of explosions actually occurred during landings or departures.

One of the most graphic descriptions of a boiler explosion found in contemporary newspapers came from the Cleveland *Herald* of October 6, reprinted in the Milwaukie *Commercial Herald,* October 13, 1843. Note the last sentence in the horrific account.

> Explosion of the Clipper.— The explosion of the steamer Clipper at Bayou Saro, La., on the 19th ult. was one of the most appalling steamboat accidents we have recorded the present season. The boat was backing from the landing when all the boilers burst simultaneously. The explosion is described by the Chronicle, as terrific; and machinery, vast fragments of the boilers, huge beams of timber, furniture, and human beings in every degree of mutilation, were alike shot up perpendicularly to a great height in the air, and diverging like the jets of a fountain, fell to the earth in some instances full 300 years from the scene of destruction. The hapless victims were scalded, crushed, torn and mangled and scattered in every direction. Out of some fifty persons, but six or eight escaped unhurt, and some twenty were killed outright.
>
> The floors of the two large ware rooms were literally strewn with the dead and dying, and others pouring in as fast as it was possible to convey them — praying, groaning, howling and writhing in every possible contortion of physical agony. In the midst of this confusing din, up to their arm-pits in oil and cotton and bandages, we found our praiseworthy physicians — the good Samaritans doing good — quietly and silently, but with the energy and activity apparently of fifty pairs of hands — now washing a burn, now dressing a wound, and anon splintering a fractured limb. Indeed, our citizens generally, every man and mother's son, appeared only anxious as to how they might render most service to the poor sufferers — white and black without distinction.

An article from the *Adams Sentinel,* December 2, 1833, apprises its readers of the boiler explosion of the steamboat *Illinois.* Interestingly, it also includes some seldom-imparted news: the disposal of the dead.

> ANOTHER STEAMBOAT DISASTER AND LOSS OF LIVES. We heard from a gentleman who arrived here yesterday, who was on board at the time the occurrence happened, that the steam boat Illinois, on her way from St. Louis to Louisville, on Friday, 8th inst. in the Mississippi river, about five miles above the mouth of the Ohio, collapsed one of her boilers; and that from 36 to 40 persons were either lost or injured. The 2nd Engineer and the Steward were among those dead; the others were passengers. Nine were buried at the mouth of the Ohio, and four a few miles above; about 20 were left at the Smithland hospital. We understood one of the gentlemen to say, that nine persons were seen to sink. The boat has arrived at Louisville. *Cincinnati Daily Advertiser, Nov. 18.*

The Most Dangerous Period

In a sad twist of fate, the very first western steamboat disaster happened to Henry Shreve, aboard the maiden voyage of the *Washington.* Leaving June 3, 1816, from Wheeling, the innovative, high-pressure boat steamed 200 miles to Marietta, where the captain put in for some adjustments. The next morning, with the fires hot and steam up, all hands were assembled to weigh anchor when the boiler exploded. Many, including Shreve, were thrown overboard and nearly the entire crew was injured by scalding. A contemporary wrote of the scene, "It was terrible beyond conception." Another added, "Death and the most excruciating pain was spread around.... Six or eight were nearly skinned from head to feet, and others slightly scalded to the number of 17. In stripping off their clothes the skin peeled with them."[10] Eventually, sixteen died as a result of this terrible accident. The cause was attributed to an undue accumulation of steam pressure and the failure of the safety valve.

Steamboat explosions not only took a toll on life and limb, but the wild reporting on these disasters reached such a crux that editors in Cincinnati, Charleston and Philadelphia wrote dissertations on how disasters could actually discredit steamboat transportation, ultimately having a disastrous effect on future financing and development. Calls for legislation controlling safety matters were widespread.

Early Steamboat Explosions

Year	Boat	Locale	Casualties
1821	General Robertson	Cumberland	
1824	Etna	New Jersey/New York	13 or 14 dead
1825	Teche		20–30
1828	Car of Commerce		30 dead*
1830	Helen Macgregor†	Memphis	40
1835	Belle of the West		35–50
1838	General Brown		55
1838	Oronoko		200–300
1838	Pulaski	Baltimore/Charleston	100

The *Pulaski*, a 687-ton, low-pressure steamer, had been constructed in Baltimore in 1837 and ran the trade between that city and Charleston. The destruction of an eastern steamer sent cries of distress and outrage throughout the eastern seaboard, reminding everyone that disasters were not solely the province of the Ohio, Mississippi and Missouri rivers. (For particulars of the disaster, see chapter 13.)

The greatest antebellum tragedy on Western waters occurred in 1838, involving a "brag boat," *Moselle*. She was built between December 1, 1837, and March 31, 1838, in Cincinnati. At 150 tons she was not a large boat by the standards of the day, but was outfitted in grand style and developed a reputation for speed very early in her career. Under the command of part owner Captain Perrin, the vessel established a new speed record on her trip between St. Louis and Cincinnati. To accomplish this feat required the liberal use of resin and the dangerous practice of holding onto steam instead of allowing the excess to escape from the safety valve. The achievement earned Captain Perrin great accolades and within several days, his boat was filled with passengers attracted by the *Moselle's* reputation as a "crack" boat.

Late in the afternoon of April 25, 1838, the *Moselle* left Cincinnati for St. Louis, and steamed down the river a mile and a half to pick up passengers at Fulton (located

---

*The Wheeling Gazette of May 24, 1828, reported fifty-seven killed and wounded.
† Also spelled "McGregor" in some accounts.

opposite Brooklyn, Kentucky). After taking on a family and wooding up from a lumber raft, "the captain had madly held on to all the steam he could create, with the intention, not only of showing off to the best advantage the great speed of his boat, as it passed down the river the entire length of the city [Cincinnati], but that he might overtake and pass another boat which had left the wharf for Louisville, but a short time previous."[11]

The wheels had scarcely made their first revolutions when three of the four boilers exploded with a violent crash. Smoke rose several hundred feet in the air, and fragments of metal, wood and human limbs were scattered everywhere. The hurricane deck as far back as the gentlemen's quarters was blown away, and the boat immediately began to sink. A strong current pushed the *Moselle* from its initial position, thirty feet from shore, out into the river. Within fifteen minutes, only the chimneys and a small portion of the upper works were above water.

Passengers on the main deck, many of them Irish and German immigrants,[12] were immediately drowned, while cabin passengers "with a fatuity which seems unaccountable, jumped into the river." Many, including Captain Perrin, were flung onto shore, already dead or dying by the time their bodies landed. The absence of immediate help (the boat was above the ordinary business parts of the city), hindered rescue attempts and many bodies were carried downriver, never to be recovered. The lowest number aboard was put at 280; of those, 81 were known to be killed, 55 were missing and 13 badly wounded. Besides the captain, the list of dead included eight other crewmen: a clerk, a steward, a deckhand, a fireman, first mate M. Thomas, pilot J. Fleming (whose body was blown to the opposite side of the river), second engineer Halsey Williams, and first engineer, J. Madder. Four women, six children, one "colored" person, a merchant and a barkeep also perished.

The list of badly wounded included Captain Perrin's brother as well as six crewmen: the first clerk, a deckhand, fireman, ship's carpenter, and the second and third cooks. The tally of missing contains a reference to a minimum of 17 children and nine women.[13]

The *Moselle* disaster was the worst of its kind in the entire world, and shocked the nation. A public funeral was held, and thousands flocked to Cincinnati to pay their respects. Business was suspended for the day as long lines of caskets were taken to the cemeteries.[14]

Between 1816 and 1848, there were 185 explosions on western waters, compared to 45 for the remainder of the United States. The loss of life compared 1,443 to 384. Of this time period, the period 1836–40 was the worst for both sections of the country, with 451 dead in western explosions and 168 in eastern disasters.

Mid-Era Steamboat Explosions
(Greater than 35 Deaths)

Year	Boat	Casualties
1841	*Creole*	36
1842	*Edna*	50
1844	*Lucy Walker*	50–60
1847	*A.N. Johnston*	50
1848	*Clarksville*	40+
1848	*Kenney*	50
1848	*Edward Bates*	53
1850	*South America*	40
1851	*Brilliant*	over 60
1852	*Glencoe*	60
1854	*Georgia*	40[15]

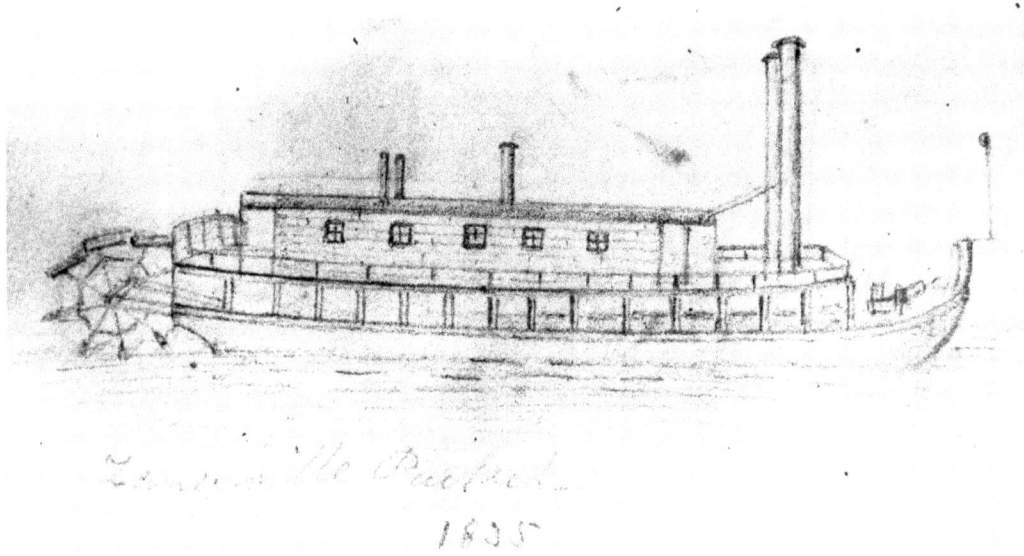

A sketch of the *Zanesville Packet*, on the Muskingum River, date unknown. This boat was built in 1844, if it is actually the *Zanesville Packet*. More likely it is the *Zanesville*, built in 1838. The handwritten note identifies it as the *Zanesville Packet*, 1835, but there was frequently confusion over similar names. Steamboat captains and owners had no qualms about calling boats by the same or similar names as others working the western rivers, or giving the same name to a boat after a previous vessel had been destroyed or retired.

In a point of irony, steamboat owners did not appear to harbor any superstitions and frequently called their replacement boats by the same name as the one destroyed, merely adding "the Second" or "the Third" to the title.

Boilers and Other Hazards

Snags, sawyers, collisions, fires and explosions all possessed the potential to be dramatically and devastating to lives and property, but they were not the only dangers presented to the 19th century river traveler. Steam pipes broke, cylinder heads burst, packing blew out of cylinders and flues collapsed. Usually, the only persons suffering injury were the engine room personnel and the owner, who had to pay for the dreaded repairs. In a somewhat tongue-and-cheek style, George Merrick noted:

> His [the pilot's] answer to the question, whether the water is below the safety point, comes as he feels the deck lifting beneath his feet, and he sails away to leeward amid the debris of a wrecked steamboat.[16]

Of these accidents, the worst involved the boilers. Internal flues were subjected to the crushing pressure of steam, and when they gave way, the ends were torn loose from the boiler heads, creating rents through which steam and boiling water escaped. The problems were obvious: under certain circumstances, the pressure of steam became greater than the boiler could sustain, creating an explosion. How to correct this remained a problem, over which engineers and scientists labored for many long years.

The Hagerstown *Mail,* January 12, 1838, reports a "typical" incident involving boilers:

> *Steamboat Disaster.*—The Steam Boat Black Hawk, on her way from Natchitoches to New Orleans burst her boilers on the night of the 27th ult., killing instantly the Pilot and Engineer, and severely and mortally wounding several others—a number of the passengers are supposed to be lost.

A more complete accounting ran in the *Adams Sentinel,* January 15, 1838:

> THE BLACK HAWK. The steamer Black Hawk, Captain Taylor, on her passage from Natchez to Natchitoches, burst her boilers on Wednesday night the 27th inst., a short distance above the mouth of Red River. She had a full freight, a large number of passengers and horses, together with $90,000 in specie, belonging to the U. States. The pilot and engineer were instantly killed — several more were supposed to be lost — number not known. Four or five were severely and several mortally wounded. Most of the passengers were saved by the timely arrival of a flat boat, which conveyed them to shore. The principal part of the cargo was saved in the same way. Seven horses were lost. $75,000 of the specie saved. The hull, partly under water, remains near where the accident occurred.

In such reports it was common to read the precise number of horses lost and the exact accounting of specie recovered, but only estimates of the loss of human life.

One factor usually receiving scant attention in the long list of steamboat disasters is weather. The fragile superstructure and the high pilot house made the western steamboat particularly susceptible to high winds. Usually, no more serious damage was done than blowing the vessel off course or toppling the upper apartments, but on occasion the damage was much more severe. One account from the *Maysville* (Kentucky) *Monitor* of December 29, 1836 (reprinted in the *Adams Sentinel,* January 9, 1837) is typical:

> *Reported Steamboat Disaster.*—It was reported here last week by boats from above, that the entire cabin of the Steamboat Mariner, on her way up, was blown off during a gale that prevailed on Monday or Tuesday night, and that 20 or 30 passengers had been aroused from their quiet slumbers to find their death pillows beneath the ruthless wave. The appearance, a few days after this report, of large pieces of a wreck floating down with the ice was calculated to confirm the most melancholy intelligence.

An example of perhaps the least common type of steamboat notice came from the *Adams Sentinel,* December 2, 1833:

> The steamboat Caspian, which we published an account of the destruction of, from the St. Francisville Phoenix, we rejoice to have it in our power to say, arrived without accident at New Orleans, and is reported in the Louisiana Advertiser of the 4th ult.

The Probables of Steamboat Accidents and Remedies

Although the theory was discredited by 1836, many rivermen held that a buildup of gas inside the boiler caused the machinery to explode. Researchers at the Franklin Institute proved this to be incorrect, but it continued to hold sway well into the 1840s.

The most widely held theory was that boiler explosions resulted from low water in the boilers. Low water resulted from stoppage of water, failure of the supply pump to maintain adequate levels, or from a dramatic shifting of the boat. Every pilot knew that it was important to keep the main deck on an even keel, and cargo was loaded with an eye toward maintaining as near a perfect balance of weight as possible. Any list from one side to the other caused boiler water to be too high on one side and too low on the other.

Those parts thus exposed overheated, and when brought in contact with water after proper levels were restored, caused an explosion. Sudden stops or starts had the same effect of shifting water levels. When the boat was righted and the water stabilized, explosions occurred. This idea basically placed the blame for explosions on engineers and their assistants by arguing that carelessness in not maintaining proper water levels led to disaster. It was felt that by manipulation of steam gauges, opening the throttle or safety valves, danger could be avoided.

Evans' Safety Guard to the Rescue

The Zanesville *Courier* on June 7, 1849, printed an article urging legislation that would mandate use of Evans' Safety Guard, a remedy for boiler explosions that was already utilized on some boats and had been available for a decade. Evans' Safety Guard was mentioned frequently in earlier news reports and steamboat advertisements as an inducement to draw passengers, as well as explain the cause of preventable accidents.

> Explosions of Steam Boilers.— The Honorable Edmund Burke, late Commissioner of Patents, in his interesting report on this subject attributes the bursting of steam boilers mainly to the use of cast iron boiler heads, and the absence of such guards as inventive genius has furnished. It is more than twenty years since the Government of France passed a law prohibiting the use of cast iron boiler heads, after some destructive explosions from this cause; yet our government has neglected to follow the example.
>
> The report strongly recommends the addition of a safety guard invented by Mr. Evans of Pittsburgh, son of the inventor of the locomotive engine.— It has been in use nine years, and has been tested by 125 boats on our Southern and Western rivers, and out of the whole number, only three accidents have occurred, and those were caused by the lever of the safety guard being obstructed from opening.
>
> From the report it appears that within the past twenty years there have been killed by explosions in this country, 926 persons, and 298 wounded, many of them cripples for life, while nearly a million of dollars in property has been destroyed. The Commissioner urges the passage of an act by congress compelling all high pressure steamboats to use Evans' Safety Guard, until something more perfect may be discovered.

This is a typical announcement from the Huron *Reflector,* December 10, 1839:

> THE IRON BOAT.—The *Valley Forge*, the new Iron Boat built at Pittsburgh, is to leave that port for New Orleans on her first trip, on the 5th inst. She is snag proof, fire proof, and explosion proof, being protected with Evans' Safety Guard. The safety Guard invented by Evans, we notice, is coming into very general use on the river boats. An explosion of the boilers is considered impossible where the Guard is used.

The *Republican Compiler,* October 16, 1843, gives an account in which the Evans' Safety Guard is specifically mentioned — as not being used:

> STEAMBOAT ACCIDENT.— *Three Lives Lost.*— We learn from the Pittsburgh papers that the steamboat Muskinggum Valley collapsed one of her flues of her starboard boiler on Wednesday last, about forty miles below that city. Three men were scalded, two of whom jumped overboard in their fright, and were drowned. The other lingered a few hours in great agony, and expired. The names of the sufferers are William Butler, of New York or Baltimore; Daniel Kaughman (colored) of Zanesville, Ohio; Gibson Fazier (colored) of Zanesville, Ohio. The boat was not provided with Evans' Safety Guard.

An advert run in the *Hawk Eye,* March 7, 1844, provides reassurance for perspective travelers:

REGULAR WEEKLY PACKET

For Nauvoo, Fort Madison, Burlington, Oquaka, Bloomington, Davenport and Rock Island. The well known and light draught steamboat

SARAH ANN,

E. H. Gleim, master, will run as a regular packet between the above ports, leaving St. Louis every Thursday, 11 o'clock M. The accommodations of the Sarah Ann are inferior to no boat on the Upper Mississippi. She is provided with Evans' Safety Guard, to prevent explosion of boilers, as well as is attached a fire engine and hose in case of fire. For freight or passage apply on board.

The theory of low water in boilers causing explosions was, in fact, true, but water levels were only one cause of explosion. Unfortunately, as the belief spread that a well-filled boiler could not explode, attention was directed away from other, equally serious problems.

Other Causes of Boiler Explosions

(1) A cast-iron boiler head weakened by blow holes or cracks
(2) A badly made plate
(3) A defective supply pump
(4) A clogged connection pipe
(5) A corroded safety valve
(6) An accumulation of mud in the boiler
(7) A rag or broom left inside the boiler after cleaning
(8) A poorly fastened rivet
(9) A progressive wear and tear of corrosion

A disaster ascribed to uneven water levels and the practice of putting used equipment in ostensibly new boats ran in the *Experiment* (Ohio), March 29, 1843. The italics have been added.

Explosion and Loss of Life.— As the Cutter, a medium sized steamer, well known for her fast running, was leaving the wharf at Pittsburg on Friday afternoon, and had reached the middle of the river, her starboard flue collapsed with a loud explosion. Several were blown or jumped overboard, some of whom were picked up by skiffs from shore, the second engineer was horribly mutilated and died instantly, the first engineer and a colored fireman were dangerously, and a lady, a cabin passenger, very badly scalded.— Three others were slightly scalded. The Coroner's jury returned the usual verdict that said "inquest thinks there can be no blame attached to any officers of said boat." The Chronicle however, says:

"The Mate of the Cutter (we learn) states that all the flues were examined and put in order in Cincinnati, previous to her last trip, with the exception of the one which collapsed. *The Captain ascribes the explosion to the usual practice of passengers — that of suddenly rushing from one side of the boat to the other, when leaving port. This, he says, together with the force of the wind, caused the boat to careen to starboard, and the water filling the starboard boiler occasioned the accident.*

"The Cutter was built in Cincinnati, and has made but three trips; her boilers were old, having been for several years used on the steamer Richmond; they were originally made in Wheeling. There was no Safety Guard on the vessel.

"*The practice of putting old boilers on new boats cannot be too severely reprehended; such trifling with the lives of the travelling public should not be suffered, and steamboat owners, to save a few dollars, be allowed to send their fellow-men to another world.* We wish to place as charitable a construction upon the conduct of others as it will bear, but we cannot refrain from stating that the person who placed the boiler which is injured upon the Cutter, or the government inspector who certified to its sufficiency, must be considered as responsible for the accident." (Cleveland Herald.)

As with the disaster involving the *Moselle*, captains also employed the grandiose technique of "shaving off" with a full head of steam as the boat left port. Although two-thirds of all explosions up to 1852 occurred during the brief interval while the boat left the wharf, the practice of shaving off was integral to the business.

Diving Bells and Submarines: The Salvage Business

The idea of salvaging freight off a sunken boat was not a new concept, but in the era of sails, unless the ship foundered on shore or in very shallow water, there was scant chance of retrieving anything from the wreck. Inland waterways had significantly less depth, but if a boat were lost in deep water as was often the case, the hulk lay there until low water season. By that time, the machinery had corroded, the hull had broken up and whatever cargo had been aboard was likely ruined.

Frances Trollope, upon entering the Mississippi from the Gulf of Mexico in 1827, observed:

> Only one object rears itself above the eddying waters; this is the mast of a vessel long since wrecked in attempting to cross the bar, and it still stands, a dismal witness of the destruction that has been, and a boding prophet of that which is to come.[17]

Wrecks were found everywhere, up and down the inland rivers and tributaries. Many became navigational obstructions of considerable importance. Without the ability to disinter them from their watery graves, the old boats remained as a constant reminder of the common fate of Man. They also harbored ghosts, if Mark Twain is to be believed:

> More than one grave watchman has sworn to me that on drizzly, dismal nights, he has glanced fearfully down that forgotten river as he passed the head of the island, and seen the faint glow of the spectre steamer's lights drifting through the distant gloom, and heard the muffled cough of her 'scape-pipes and the plaintive cry of her leadsmen.[18]

The captain held no interest in the cargo, so his only hope of regaining some value from a sunken ship was to salvage the mechanical equipment. Although engines and boilers were subject to serious wear and tear and possessed a limited life expectancy, that did not stop crews from pursuit, retrieval and resale. While boilers had limited value, that did not prevent their occasional reincarnation in a tramp steamer. It was more common to reuse engines, and by 1849, at least one half of new steamboats were outfitted with old engines and machinery. Occasionally, engines saw service a third or fourth time before being relegated to the scrap heap.

The diving bell was put into salvage operation as early as 1838 at St. Louis, but its further development was attributed to James B. Eads. Going from clerk to inventor, he adapted the device and went into business for himself in 1842. Constructing what he called "bell boats," bearing the name *Submarine,* he became very successful retrieving sunken hulks.

Insurance companies quickly realized the value of his work and often put out contracts on all covered boats. Like treasure hunters, salvage companies received a portion of what they recovered, ranging from 20–75 percent.

As the salvage industry developed into big business, the newspapers were filled with lawsuits pertaining to costs charged, contracts made, and the independent operation of salvage companies. A detailed article in the Milwaukee *Sentinel and Gazette,* February 14, 1849, stands as a prime example. It states the case of O. Gager vs. the Steamboat *A.D. Patchin* and *Harry Whittaker.*

> Interesting to Steamboat Owners.
>
> This was a suit for salvage on account of services rendered by the steamer *Albany,* of which the libelland was owner and master to the steamer *A.D. Patchin,* while hard aground on a ledge of rocks, upon which she had carelessly run at Racine Point, in the State of Wisconsin....
>
> A suit of salvage *in rem* as well as *n personam* may be maintained in the Admiralty for services, which, if rendered voluntarily would be salvage services, notwithstanding they were rendered in pursuance of a pervious [*sic*] agreement specifying the amount of the remuneration therefore, and although the contract was made with the owner of the vessel sued; especially if the owner was also the master, and the fact of his proprietory interest was unknown to the other party.
>
> Although all the suitors for salvage ought to join in the suit against the property sued, the non-joinder of the crew of the salvage vessel in a suit by the master and owner, is no impediment of such suit, especially if the services of the crew required no extraordinary execution, and were attached with no extraordinary personal danger.
>
> Although the amount agreed to be paid for salvage is not binding on a court of Admiralty upon the party at whose instance or for whose benefit the service was performed, and shall be reduced by the Court if exorbitant, and especially when assented to by such party under the pressure of alarm or distress, yet where the agreement has been deliberately entered into and does not appear to be oppressive, it will be enforced according to its terms.
>
> Eli Cock for the Libellant. George Underwood for the Claimant.
>
> The decree of the Court was in favor of the libellant (Gager) for $1,800 (the amount of the per diem contract for use of the Albany) deducting $300 already paid, with interest from the 7th day of June last, when the service consummated.

In a case of a salvage vessel (luckily) being in the vicinity of a disaster, the following article from the *Sheboygan* (Wisconsin) *Mercury,* April 22, 1848, provides details of a suit for property recovered without any formal agreement.

> The said libel stating in substance that the said libellants on the 21st day of Nov. 1847, about four o'clock A.M., while on board said propeller Delaware, the said propeller then lying at Sheboygan, discovered a vessel on fire about 17 miles north-east of Sheboygan — that thereupon the said propeller Delaware was got underway and at arriving at the vessel in distress, she was found to be the propeller Phoenix, a vessel of about 305 tons burden, bound from Buffalo to Chicago. Said propeller Phoenix was found to be burned to the main deck fore and aft, and had been abandoned by the master, crew and passengers. The said propeller Phoenix was made fast to the propeller Delaware and towed to Sheboygan afore-aid, a distance of about 17 miles, and was hauled onto the beach, and said libellants prayed that by reason of the risk and hazard they run and the service they performed in saving said propeller Phoenix and cargo therein, they are entitled to competent salvage for such service, together with all charges and expenses attending the same, and that process might issue according to the course of said court against said propeller Phoenix and cargo, and therefore an order was made by the Hon. Andrew G Miller, judge of said court, for the arrest of said propeller Phoenix, her tackle, apparel, and furniture, and the cargo,

goods, wares and merchandise therein, and that all persons pretending to have any right, title or interest in said vessel or cargo, be cited to appear and answer the allegations in said libel contained.

The U.S. Marshal was directed to attach the wreck according to a warrant and hold the vessel until the claim of the *Delaware* could be settled.

A determination of salvage from the Newport *Daily News,* December 3, 1847, gives a more succinct verdict:

> We learn from the St. Augustine Herald of the 18th inst. that on Saturday, the 13th, the claim of Captain Donnell and others for salvage upon the goods and machinery saved from the steamer Narragansett, wrecked near Mosquito Inlet, was decided in the U.S. district court. The court awarded 30 per cent. The evidence acquitted the captain of all blame.

But perhaps the most interesting — and unique — record of salvage comes from the *Daily Sanduskian,* January 30, 1850:

> Shocking Steam-Boat Accident. Cincinnati, Jan. 29. The steamer St. Josephs from New Orleans, blew up and burnt to the water's edge near Napoleon, at the mouth of the Arkansas river, on the 23rd inst. She had a large cargo on board and a large lot of emigrant passengers, 15 of whom were killed and 38 scalded, many of whom have since died.
> Capt. Baker of the St. J., hailed the steamer South America to land the wreck, and take the survivors off, as the yawl of the St. J. was engaged in picking up those blown into the river. After the boat was towed ashore, Capt. Baker, with the assistance of part of his crew and some passengers, succeeded in saving the coin chest containing some 13 or $1400. He took it and counted it and handed it to the clerk of the South America for safe keeping. The next day he asked him for it to pay off the crew when the clerk refused to give him more than $300, claiming the rest as salvage.
> Capt. Baker and his clerk stopped at Memphis to compel by law the clerk of the South America to give up the money. There was much excitement at Memphis in relation to the matter. Serious threats were made to mob the boat if the money was not given up.

Unfortunately, the newspaper did not relate the end of the story

By 1850, the development of powerful pumps made the recovery of boats, rather than just freight or engines, more practical. After constructing a bulkhead around the damaged hull and closing the hatches, the salvage crews pumped out water and raised the boat. If the struggles over salvage had been more graceful, it might have been a happy ending to a sad event. But as in most cases, the issue of money took precedence over fairness and common sense.

A Descent to the Bottom of the Lake

In most cases, the word "descent," when referenced in the same paragraph with any mention of water, usually connoted a negative story: one filled with explosions, natural disasters and loss of life. Such was not the case, however, in 1852, when the *Democratic Banner* of November 26 ran a story entitled "Diving Dress." It reprinted an article full of fascinating details on the newest — and seemingly the most outlandish — method of salvage operation.

> The mode adopted to reach the property in the steamer Atlantic, now lying at the bottom of Lake Erie, is a curious and interesting one. Mr. Mailefert assisted by his skillful diver, Mr. Green, it is known, has been for some time occupied in the endeavor to

recover the safe of the Express Company which contains much valuable property. The Buffalo Commercial published the following description of the attire worn by Mr. Green, on the occasion of his descent to the bottom of the lake. As an evidence of the perfection of science, it is full of interest.

The marine armour consists of a perfectly air-tight India-rubber dress, topped by a copper helmet, with a clear, thick plate of glass in front. The pipes which supply and exhaust the air lead from the top of the helmet. The pumping requires much labor; four, and sometimes six men being employed upon it at the same time, and compelled to work hard at that. A great pressure of air is experienced by the diver on his lungs, equal to 75lbs to an inch, and very few individuals could bear it for any length of time. When first going into the dress, the sensation of oppression is very overwhelming, but passes away in a great measure after entering the water. When a depth of ten feet is reached in the descent, the dress becomes entirely emptied of air, and collapsed to the body, causing a pressure all over the diver equal to a ten pound weight, excepting as to the head, which is protected by the copper helmet. The difficulty in breathing now becomes great, and a painful sensation is experienced by the diver, the jaws become distended, and the head as if splitting. This continues until after descending another ten or twelve feet, when the pain is relieved, the diver feels comfortable, and experiences no further inconvenience. When about sixty feet from the surface, hundreds of the legitimate inhabitants of the water surround the diver, nibbling at the strange visitor as tho' he were "food for the fishes." After reaching seventy-five feet, all is perfectly dark — a black impenetrable darkness — an electric flame plays around the inside of the helmet, caused by the friction of the pump. At about one hundred and sixty feet the water is very cold, being in the present season within four degrees of freezing.

This diving suit was adopted for use in retrieving valuables from Mississippi wrecks, as well. This method would have been especially important in retrieving cargo after the Specie Circular was issued, for boats like the one mentioned in The *Daily Globe* (September 23, 1854) carried a king's ransom:

> New York, September 23 — The steamship St. Louis sailed to-day with forty-eight passengers and $466,000 in specie. Whether such efforts affected the ghosts that lingered around the scenes of tragedy remains unanswered.

12

The Steamboat Race Is On!

Proving who was the biggest, the best and the fastest has always played a huge role in the development of speed. Early footraces led to the Olympics; horse racing began as soon as people learned to ride, and breeding champions became an obsession that has lasted throughout the centuries. Men raced carriages, stagecoaches and dogcarts; owners of canoes and flat-bottomed boats challenged one another across rapids and down meandering rivers. Mighty-masted schooners put their sails to the wind and attempted to beat their competitors to Europe, China or San Francisco.

It is not unnatural, therefore, that as soon as steam technology was used to power boats, contests to determine bragging rights on the rivers became part of American lore. Appropriately enough, the honor of participating in the first steamboat race in history belongs to Robert Fulton. In 1810, a rival company to the Fulton-Livingston Line was established at Albany, New York. The partnership constructed two boats to challenge the monopoly: the *Hope*, under Captain Bunker, was launched March 19, 1811, and the *Perseverance*, under Captain Sherman, somewhat later.

As soon as these rivals were put into operation, a race for dominance became inevitable. On July 27, 1811, the *Hope* and the *Clermont*, under Captain Bartholomew, left Albany at nine o'clock in the morning. The *Hope* took the early lead and kept that advantage until they reached an area of the river within two miles of the Hudson. Lighter in draft than the *Hope*, the *Clermont* attempted to pass in the shallow water of the channel. A collision resulted, injuring no one but putting an end to the contest. Captain Bartholomew promptly challenged the doughty Bunker to complete the race, putting a $2,000 wager on the outcome, but Bunker declined. The race settled no legal disputes between the two companies, and contentious litigation dragged on over many years.[1]

Money, Egos, and Racing

Less than a year after Henry Shreve took the *Washington* on her maiden voyage from Wheeling in 1816, it is reputed that he became involved in the first western steamboat race — and, subsequently, involved in the second boating explosion of his career.

The *Constitution* (previously named the *Oliver Evans*) blew up near St. Francisville on the Lower Mississippi, and "destroyed all the most respectable passengers, to wit, eleven."[2] (The reference is to cabin passengers, or those who paid full fare, the implication being that deck passengers were "not respectable," a commonly held prejudice.) In all, thirteen people died. The cause of the accident was said to have been two boats "running in opposition" (racing). If reports are true, the other boat involved was the *Washington*. This was the first recorded explosion that involved racing, and set the tone for the popular belief that racing was the cause of most explosions.

A reference to the accident relates the story of a gentleman who witnessed the explosion. The gentleman:

> confirms the account respecting the Steam Boat Constitution, and states that she is a very small boat with an iron boiler, the bursting of which was owing to the ridiculous vanity of the captain in ordering the steam to be raised for the purpose of propelling his boat faster than a larger one which was coming down the river at the time.[3]

It would be a misconception to believe that profit was the only deciding factor in the development of steamboats. While making money was instrumental in maintaining a boat or a line, the ego of the owner and/or captain was more often a determining factor. As demonstrated with the desire to attract cabin passengers, owners spent more than they brought in for the privilege of bragging rights. Toward that end, racing became instrumental in the establishment of reputations. "Crack boats," known for their speed, were the envy of all, and winning a race occasionally assumed more importance than life and limb. Legislation to the contrary, measures were taken to get around regulations, sacrificing safety for the unwritten codes of rivalry and competition.

Captains sought engineers with proven track records, men who were not above pushing their engines to, and sometimes beyond, the limits of endurance. These so-called hot engineers hired on, fully aware of what was expected. Well versed in the use of resin, lard and pine pitch, engineers stoked the fires until they were searing hot, creating steam in abundance. Another trick of the trade was to take advantage of the reserve power of high-pressure engines by holding, or weighing down the safety valves, creating steam pressures higher than was originally intended by the manufacturer.

Steamboat racing was not the sole province of officers, however. The public at large was enamored of the idea of speed, with hundreds of newspaper articles and periodicals printing and reprinting stories of prowess and accomplishment. Speed represented the western view of life: success at any cost, and damn the consequences. "Brag boats" drew crowds, and crowds attracted money. Wagers were made and accepted, brass bands accompanied the screech of whistles as boats steamed up, and onlookers flocked to the wharves, each seemingly with an individual stake in the outcome. Boat whistles were tuned so that they might be distinguished from those of other boats, and people familiar with the river could identify the arriving vessels by sound long before they came into sight. Just as fans today have favorite baseball teams, racing was a great western sport, with a "winner take all" mentality.

There were three distinct types of western racing: competition between two boats, the application of speed to set records between destinations, and challenging the clock.

Brag Boats on the Upper Mississippi

Records were meant to be broken, and the captain of a "brag boat" always had an eye toward seeing his name inscribed in the annals of the best and fastest.

Destination times were dependent on the ever-improving technology of steamboats, the condition of the river and the chosen course. While the most famous trials involved boats working the Lower Mississippi where routes were straighter and the rivers relatively free from permanent obstructions, captains plying the northern waterways were not exempt from setting and breaking the accepted norms of travel.

Steamboats on the Upper Mississippi were primarily concerned with freight. Early prototypes were capable of achieving feats of speed, but captains in these trades were more concerned with the safe, steady transport of heavy cargoes. Which is not to say a handful of more adventurous skippers had their eyes on records. The following table shows the advances technology and men rising to the challenge were able to achieve in a matter of decades.

Date	Boat	Destination	Time
July 1836	Frontier	St. Louis–Galena	3d, 6hrs
1837	Smelter	Dubuque to Cincinnati	5
		Cincinnati to Dubuque	5
1840	Omega	St. Louis–Galena	3
1841	Indian Queen	St. Louis–Galena	3d, 12hrs*
1841	Little Ben	St. Louis–Galena	2d, 12hrs
1841	Iowa	Round trip	
		St. Louis–Galena–St. Louis	6d, 15hrs
April 1844	J.M. White	New Orleans to St. Louis	4, 18hrs
	Lewis F. Linn†	St. Louis to Galena	2d, 9h, 20min
1844	J.M. White	New Orleans to St. Louis	3d, 23hrs, 9 min§
1845	Monona	Round trip	
		St. Louis–Galena–St. Louis	4d, 12hrs
March 1845	War Eagle	St. Louis–Galena	2d, 8hrs
April 1845	War Eagle	St. Louis–Galena	2d, 1hr
May 1845	St. Croix	St. Louis–Galena	45h, 45min
May 1845	War Eagle	St. Louis–Galena	43h, 45min[4]

*with a keel in tow
†This was a tandem effort; the cargo and passengers were transferred from the J.M. White to the Lewis F. Finn at St. Louis and brought to Galena, for a combined record of traveling from New Orleans to Galena in under one week.
§This record stood until 1870.

This last record, the *War Eagle* under Captain Daniel Smith Harris, stood for many years. At the conclusion of the trip, passengers complained that only one dinner had been served, inasmuch as the boat left St. Louis after mealtime on Tuesday and reached Galena before noon on Thursday. The taciturn officer replied that "if they were traveling for dinners, they should have to take a slower boat."[5]

The *War Eagle* was likely the fastest steamboat on the Upper Mississippi before 1850, and under Captain Harris she garnered much respect and acclaim.

The War Eagle

Built: 1845 in Cincinnati
Tons: 155
Length: 152 feet
Beam: 24 feet
Hold: 4 feet, 6 inches

At mid-century, Upper Mississippi steamboat captains turned their attention toward speed and beauty. After the admission of Iowa and Wisconsin as states, and the creation of the Territory of Minnesota, passenger traffic became more important, making speed and luxuries the order of the day.

Captain Harris was well aware of his reputation, and took great pleasure in flaunting his acumen. He was known to run alongside competing boats and encourage his passengers to jeer the slower vessels. He would occasionally hang behind until the next port of call appeared, then pull ahead with a burst of steam. Arriving first, he then had his pick of the best cargo and passengers.

With his new boat, the *Dr. Franklin No. 2*, Harris engaged in a race with the *Dr. Franklin* (aka the "Old Doctor") in May 1851. Captain M. W. Lodwich of the "Old Doctor" raced neck-and-neck with *Dr. Franklin No. 2*, and threatened to pass. Rather than accept such an inglorious fate, Harris steered the stern of his boat across the path of Lodwich's boat, forcing him to reverse direction in order to avoid a collision. A second incident nearly caused another crash, and only the skillful piloting of the "Old Doctor" prevented it.

As reported in the *Minnesota Democrat* (St. Paul) of May 27, 1851, Captain Harris sprang from the pilothouse to the hurricane deck with a rifle, forcing the *Dr. Franklin* to the side of the river and threatening to shoot the pilot if he made any further efforts to pass the *Dr. Franklin No. 2*. For this act of unsportsmanlike conduct, he was bitterly reprimanded by the passengers of the *Dr. Franklin* in a statement published in the St. Paul newspaper.

The Dr. Franklin No. 2

Built: 1848 in Wheeling
Length: 173 feet
Beam: 26 feet, 6 inches
Hold: 4 feet, 4 inches
Tons: 189

This incident began a three-year rivalry, with local townsmen and passengers all taking sides. In order to compete with one another, the warring captains often reduced fares to an unprofitable level, with some travelers paying no more than fifty cents for passage from Galena to St. Paul. The usual fare was eight dollars, so both captains lost money, and eventually the Minnesota Packet Company, which owned the *Dr. Franklin*, offered Harris a position. He accepted a directorship, thus ending the "war" and saving himself from the potential disaster of being driven from the Upper Mississippi trade.[6]

The Great Race of 1852

The feud between Captain Daniel Smith Harris and the Minnesota Packet Company had come to an end, but things were not all well on the Upper Mississippi. A new rival appeared on the scene. The Keokuk Packet Company decided to extend operations above Galena as far north as the Falls of St. Anthony to take advantage of passengers booking berths for the "Fashionable Tour." Both lines vied for the trade and there was only one way to settle which was best: a race between the brag boats of each line. For the Minnesota Packet Company, that meant Captain Harris with his new boat, the *West Newton*. For the Keokuk Packet Company, Captain Rufus Ford commanded the *Die Vernon*.

The *Die Vernon* left St. Louis on Monday, June 13, 1853, and reached Galena on Wednesday. The passenger fare for the complete trip to St. Paul cost $25. Aside from the thrill of being an observer, passengers were promised Longworth's Sparkling Cabinet and Still catawba wine to toast their anticipated victory. Captain Ford also had on board an army of waiters to treat his paying guests to every luxury available.

The Course

Galena to Lake Pepin	190 miles
Through Lake Pepin	22 miles
Beyond Lake Pepin to St. Paul	55 miles

Captain Harris was waiting for his foe, already having taken aboard all the tar and resin in Galena. Additionally, he had used his considerable influence to secure the cooperation of every wood boat for 150 miles. The wooders would sell only to him, thus leaving Captain Ford the unenviable task of delaying his trip long enough to chop wood along the way.

The *Die Vernon* backed out first into the Fever River, with the *West Newton* "blowing off steam and making more noise than a stalled freight train." Harris sought to pass his enemy at the outset and set the pace for the remainder of the race. Ford, however, was not so easily outmatched. He ordered the safety valves held down to create an abundance of steam and actually led all the way to Dubuque.

The Die Vernon

Built: 1850 in St. Louis
Cost: $50,000
Tonnage: 455 tons
Length: 255 feet
Beam: 31 feet, 2 inches
Berths: over 100

Captain Ford put in at Dubuque but Captain Harris steamed through, taking the lead. Leaving the town, Ford plied his firemen (stokers) with "whisky toddies to assist them in making steam," but he did not catch the *West Newton* until the following morning near La Crosse.

Accounts from passengers aboard the *Die Vernon* state that Harris was compelled to land to keep from being passed under; Captain Harris' friends deny the allegation. But the *Die Vernon* did pass, and then paused to take on fuel from a wood flat. It seems Captain Ford had anticipated Harris' trick of buying up all the wood, and had prearranged with Captain Louis Robert of the *Greek Slave* (an enemy of Harris) to have seasoned wood waiting.

The race continued, with Captain Ford and the *Die Vernon* coming out the winner. She made a record time of 84 hours, counting stops, making $9^{4}/_{10}$ miles an hour upstream. The *West Newton* averaged $9^{1}/_{10}$ miles per hour. Sourly returning to Galena, Harris made the trip in 21 hours and seven minutes, averaging $13^{7}/_{10}$ miles per hour.

An interesting side note to the race was that despite the fact the *Die Vernon* collected almost $4,000 in cabin passage fares, the lavish trip left a deficit of $1,100, proving again that ego often took precedence over profit.[7]

A Clean Sweep

The symbol of being the champion on the western waters was a big broom, proudly displayed on the pilothouse. The fastest boat was so identified and always carried her trophy until passed by a faster vessel. Protocol then dictated she take down the broom and keep it out of sight until she, in turn, could defeat the champion, and reclaim her crown.

The *Grey Eagle* was the undisputed champion of the upper river, and it lay between other boats to claim the right to display the broom. The *Northerner* of the St. Louis Line and the *Key City* of the Minnesota Packet Company were two that vied for championship rights. In a race between the two, which went neck-and-neck from Hudson (twenty miles beyond Lake St. Croix) to Prescott, the deciding point came when the *Key City* reached the foot of the lake. This gave her the right of way to turn the point and head upriver, compelling the *Northerner* to slow and wait before making the turn. Whistles and bells ringing, the crew hoisted the broom as the *Key City* won by the narrowest of margins.

When all was said and done—when the accounts were circulated in the newspapers, the letters written home by tourists delivered, and the fictionalized accounts of steamboats and river captains made their appearance in the monthlies—most races were business events rather than sporting contests. The boat that arrived at the wharf ahead of its nearest competitor won more than an inscribed plate or a silver cup. It came away with the most lucrative freight and best-paying passengers. And that, when day-to-day survival was at stake, proved the greatest reward. Racing was an expensive proposition in quantities of fuel burned and risks taken. Owners with an eye toward economy seldom engaged in such diversions.[8]

"I Yielded to the infection"

Races on the Lower Mississippi were as often impromptu as scheduled. No captain liked to be passed by another boat and frequently, with the urging of the passengers, a race would begin in the middle of the river. With the captain on the hurricane deck and the travelers hanging over the rails shouting encouragement, the fires would be stoked with resin, steam would belch and hats from the winners would be tossed into the air to celebrate a victory.

During such races, it became something of a standard feature that opposing pilots would steer as close to their rival as possible, locking guards in order to prevent their opponent from gaining the advantage. They were known to speed through the water meshed in this fashion for several miles before one broke free. Typically, little damage was done by these mid-river collisions, except to the guards.

Racing was performed for the establishment of reputations and, not incidentally, for the passengers. Long voyages tended to evoke boredom, and in order to break the tedium, passengers encouraged their officers to engage in a speed contest for temporary bragging rights. Even those who believed themselves above such contests often found the challenge too much to refuse.

I yielded to the infection, and was as anxious a spectator of the contest as anyone aboard. There were a few timid elderly gentlemen and ladies, who kept aloof; but with this exception, the captain of each boat had the moral strength of his cargo with him.[9]

There was a near-continual argument among steamboat people, townsmen and the popular press over the relative merits of racing, particularly as concerned safety. One side held that the liberal addition of resinous fuels and the practice of holding down the steam valves to increase steam (and thus speed) caused the most horrific accidents. This group lobbied for the regulation of such practices, if not their outright abandonment. Mark Twain, among others, preached the opposite side: that engineers were always more attentive during races, and this absolute dedication to their task (as opposed to duty aboard slow boats, where engineers drowsed, allowing chips to get into the "doctor" and shut off the water supply from the boilers) prevented explosions. He noted that after laws were passed restricting each boat to so many pounds of steam per square inch, the accidents were severely reduced.

Steam per square inch was important in that "hot steam" (that made with as little water as possible) generated more power to the engines. By opening the blowers (the forced draft) to increase oxygen and keeping the water levels low (below the second of three gauges), this hot steam was produced. During normal working conditions, the three gauges indicated appropriate water levels; when they read "full," the water supply was decreased. During racing, however, the third gauge registered empty, and it became vital to accurately determine how low the water actually was. When it became too low, cold river water was let in as fast as possible.

An engineer who miscalculated the appropriate levels subjected the boiler plates to red-hot temperatures. The sudden addition of water caused the mechanism to "jump." Steam pressure went up faster than the engines or safety valves could release it, and the boiler gave way.[10]

Twain went on to describe a typical race between two notoriously fast boats (either for bragging rights, or a challenge issued by the owners of a new boat to earn a quick reputation). The date of the race was set far in advance, and as the time approached, the vessels would be stripped of every encumbrance that added weight or resisted the free flow of wind and water. Spars and derricks were also sometimes removed, leaving the boat vulnerable in case it became grounded. Few passengers were taken, not only because they added weight but also because they would never "trim boat," or keep to the sides in order to maintain an even keel. In their excitement, passengers were likely to run from side to side, depending on which offered the best view, whereas staying in one place would maintain balance. He noted that no "way passengers" (those departing at small stops) were allowed aboard and that coal and wood flats were contracted beforehand to alleviate the time and trouble of stopping at wooding stations. Not surprisingly, he wrote that the boat with the best pilot won, for steering was a fine art: "One must not keep a rudder dragging across a boat's stern if he wants to get up the river fast."

Routes and Racing Times

Year	Route	Miles	Run Time
1843	Louisville–Cincinnati	141	12h, 23min
1843	Louisville–St. Louis	750	2d, 1hr
1844	New Orleans–Natchez	268	19h, 45min

Year	Route	Miles	Run Time
1844	New Orleans–Cairo	1,024	3d, 6 hrs, 44min
1844	New Orleans — Louisville	1,440	5d, 42min
1844	New Orleans — St. Louis	1,218	3d, 23hrs, 9min
1851	Cincinnati-Pittsburg	490	1d, 16hrs
1852	New Orleans–Donaldsonville	78	5h, 42min
1853	St. Louis–Alton	30	1hr, 35min

Racing Against Time

The *Carrollton Mirror Extra*, April 3, 1852, describes the saga of the steamboat *Redstone*, which happened to suffer the ultimate fate after racing to beat the clock.

> About half past two o'clock the Madison and Cincinnati packet, *Redstone*, passed this place. When about three miles above, it seems that she landed on the Kentucky side, and as she was backing out, her boilers exploded with a tremendous noise, tearing the boat to atoms and causing her to sink in less than three minutes in 20 feet water.
>
> Several gentlemen here whose attention had been attracted to the boat's racing and the great quantity of steam she was working, saw the explosion. Her chimneys were blown half way across the river. It is said that all on board have perished.
>
> LATER
>
> We have returned from the scene which is entirely indescribable. Comparatively but few if any of her passengers were saved. The only officers saved are the Captain and first Clerk. The former will in all probability die. There are from 80–100 passengers, 60 to 75 of whom must be lost. The force of the explosion may be judged from the fact that two bodies and a part of the boiler was blown more than a 1000 [*sic*] yards from the wreck....
>
> Every thing of her material and freight was destroyed. The *Redstone* had been recently place(d) in the Madison trade in opposition to the regular packets. At the time of her explosion she was racing against time.

Additional reports included the fact that torn clothing and other items littered nearby trees. The *Redstone*'s first clerk, O.M. Soper, was blown into the middle of the river but was unhurt. Two girls who had gone to their cabin to read the Bible were rescued in part because the ladies' cabin was the first place rescue workers searched. Estimates place the dead at 35.[11]

Racing thus assumed a special place in steamboat history. Not a little of this was planted and sustained by journalists of the time, for such stories made for lively commentary. In a world suddenly obsessed by speed, westerners were at the forefront of making, breaking and exploiting records. Their achievements, both heroic and tragic, have forever marked the era.

13

The Development of Steamboat Crews

Eastern and western steamboat crews were part and parcel of the massive industrial movement in the United States that took place in the 19th century. Harnessing steam provided the impetus for the development of factories that mass-produced everything from shoes to rifles. Cheap, unskilled labor departed farms for cities where working conditions were bad, and the standard of living was low. Children toiled side by side with adults from sunup to sundown. The concept of guilds applied only to craftsmen; unions for the great unwashed were unheard of when Robert Fulton maneuvered his awkward watercraft up the Hudson.

It was not surprising, therefore, that when the Fulton-Livingston monopoly was challenged and boats were constructed along the Ohio, the earliest men who swelled the ranks were the rough-and-tumble keelboatmen, who knew more than a thing or two about hard, backbreaking, low-paying jobs.

This early group divided into two distinct categories: officers and "hands." The captain, clerks, pilots, engineers and mate (first officer) belonged to the command structure. The captain, clerk and mate were most closely allied with the boat, and thus the interests of the ownership. Pilots and engineers tended to support management in matters of policy, but in other respects they acted as independent contractors. This unique position set them apart from the other officers and allowed them a freedom not found in the formers' ranks. Beneath them on larger vessels were the second and third mates, the second clerk, cub pilots and strikers. Then followed firemen, deckhands, rousters and cabin personnel.

Captains had a private stateroom, usually in the forward end of the texas. His habitual position was on the hurricane deck, and when giving orders, he spoke through a "speaking tube." Other officers had quarters similar to those used by cabin passengers, but theirs were less ostentatious.

By and large, those men used to commanding the flatboats and barges of the Mississippi gathered together what money they could and became captains or pilots; the latter group assumed the manual labor needed to load and stow cargo, chop wood and perform all the sundry menial tasks required aboard the boat.

The engineer held an officer's rank, but these individuals did not come from the rafting trade. Early engineers were often imported from the East Coast, but quickly this critical position was filled by local men.

From this point, subgroups developed: the firemen (stokers), the second and third mates, the second clerk (mud clerk), ship's carpenter, deck hands, wooders and those specifically hired to perform cabin duty: chambermaids, bartenders, servers, cooks and galley help.

Early boats of small tonnage—the transients and the packets—got by with a minimal crew: the captain, pilot, engineer and four or five hands. This number grew as the size of boats dramatically increased and more emphasis was placed on passenger comfort, but always with an eye toward economy: tramp owners operated close to the bone and wages constituted a large portion of their expenses. Job requirements were informal and many officers and hands, particularly the Upper Mississippi captain, performed multiple tasks, occasionally serving as pilot, clerk and always handling all matters pertaining to the passengers. It was not until the advent of "Floating Palaces" that ancillary positions were created. In 1852, the *Eclipse* had a 121-man crew, probably the highest number ever to work an antebellum steamboat. That number included 70 firemen and deckhands, 25 waiters and stewards, 5 cooks, 3 mates and 5 engineers. Wages for the entire crew, including officers, amounted to $4,605 per month.

In 1831, a typical steamboat of 235 tons employed a crew of 26. By 1843, the average western steamboat of 154 tons used a crew of 21. Boats working the Pittsburg to Cincinnati Trade in the 1850s carried as many as 50 crewmen. By 1851, the number of men employed on the Ohio—Mississippi Trades totaled 14,752, a very considerable number for the times.[1]

In the early development of steamboating, vessels did not navigate at night but pulled over to the riverbank, it being deemed too dangerous to proceed without adequate light. As more and more boats entered the trades, competition became more marked. By necessity, pilots learned the intricacies of nighttime travel and thereafter a boat operated 24 hours a day, putting in only when running downstream with a swift current, at periods of extreme low water, or when fog enshrouded the river. Travel through the hours of darkness required that a boat carry two pilots and two engineers who spelled one another at regular intervals. A boat might challenge icebergs, sawyers, hurricanes and earthquakes, but never fog. Unable to see the banks or the water ahead, not even the most skilled navigator dared risk his vessel (and incidentally, his life) on instinct, alone.

Specific job titles and duties were drawn from the maritime service, adapted as necessary to inland conditions. This evolution applied to the laws as well, and throughout the steamboat era, there was always the flavor of saltwater and sails thrown into the mix.

Nowhere was the commingling of steamboat and schooner more apparent—and more divergent—than the position of captain. If he did not own his boat, the captain was paid by the year, as opposed to other officers, especially the pilots, who were paid by the trip or the month. The largest salaries were offered by the "brag boats" that ran the more important trades, such as that between New Orleans and St. Louis.

Monthly Salaries (in Dollars)

Year	1837	1843	1849	1851	1855
Captain	150	75	50–120	~83	200
Clerk	50	50	40–100	75	150
Pilot	140	—	60–160	75	150
Engineer	125	40	50–100	75	150
Mate	—	30	30–60	75	150[2]

The Upper Mississippi trades generally paid better wages than those on the lower river trades. The monthly salary structure offered:

Year	1857
Captain	$300
Clerk	$150–200
Pilot	$500–550
Engineer	$150–200
Mate	$200[2]

Aside from the captain, most officers and crew found work only eight to nine months of the year. During the remaining months, which included the winter and periods of low water, all larger vessels and many smaller ones put up and did not operate. That meant the crews would have to live off their steamboat earnings or obtain employment ashore. For most, the winter months were the only times they had to spend with their families.

The "Old Man": The Upper Western Steamboat Captain

By and large, the Upper Mississippi, Missouri and northern tributary captain fit the bill so often described in novels, film and television series. These men tended to be jacks-of-all-trades, most familiar with the intricacies of navigation and fully capable of spelling the pilot, or even of standing as the second pilot aboard. The position of clerk was slower to evolve on the waterways above St. Louis, and thus the captain assumed the responsibility of hiring the principal pilot, engineers, firemen, mates, deckhands, roustabouts, the clerks (when needed) and other cabin employees.

These captains were well aware that their reputations might be the sole reason a shipper or passenger selected a boat, and they made it a point to present a respectable, civilized demeanor. This "gentlemanly deportment" was often remarked upon, and satisfied customers often published "cards," thanking that officer for a safe (and sometimes an exciting) trip.

Bragging about his ship did not hurt a captain's cause, either. A small article in the *Adams Sentinel,* August 10, 1835, gives a prime example:

> *Coming in strong.*— The Philadelphia Gazette gives the following as a specimen of the western superlative: A Kentucky steam boat captain dilating in a strain of exuberant commendation on the excellence of his craft, says: She trots off like a horse; all boiler; full pressure; hard work to hold her in at wharves and landings. I could run her up a cataract. She draws eight inches water-goes at three knots a minute, and jumps all the snags and sand banks.

The captain had to be a good judge of human nature, and beyond that, to him lay the responsibility of making a profit, either for the owner of the line service for which he worked or, as was often the case, for himself, if he owned all or a share of the boat.

The captain was required to see to the proper order of the boat — the crew as well as the passengers included. He handled disputes among his own men, firing those he deemed unfit for service and occasionally leaving them along the shore to fend for themselves if no town were close. Before the advent of labor unions, crewmen had little, if any, defense against charges, and were well aware their livelihood rested in the good opinion of the captain. In extreme cases when the crew were involved in a labor strike, it fell to the captain to negotiate their grievances. He set and paid wages, and withheld the minimal amount required by law to support the marine hospital.

Furthermore, the captain dealt with irate and drunken passengers, set and oversaw the rules for gambling and drinking aboard the boat, and stood as jury and judge for such crimes as stealing, picking pockets and cheating. When diseases such as cholera were discovered aboard, the captain made the determination of how to handle an often volatile situation and, in the case of deaths, saw to the disposal of bodies, either by taking them ashore for a quick burial or merely dumping them overboard. It was also his position to keep the books of such illnesses and deaths and report them at his next port of call. Not all captains were truthful in these matters, as deaths of any type reflected badly on the reputation of the boat and frightened away business. Without anyone to contradict his records, it was a simple matter to underreport the totals.

The Upper Western captain assumed almost total authority for the boat except for that accorded to the pilot. On boats without a clerk or those owned by the captain himself, he bargained for freight, raised and lowered fares as the situation demanded, granted free passage to visiting pilots (those not currently employed who desired to keep their river skills sharp by observing water conditions), purchased foodstuffs (and often set the menu), ordered and paid for fuel, and handled all ancillary charges, such as canal and wharf fees. In that light, he had to balance the ledgers with an eye toward ultimate profitability, for some trips during the season were highly lucrative, while others might actually be run at a loss.

A captain made the decision to leave port or to delay departure until a full cargo could be obtained; he also took responsibility for changing course, literally in midstream if he heard of better financial opportunities along the way. It was he who determined whether the challenge of a race be accepted and how it would be run; and with him rested the ultimate responsibility for any accidents which occurred as a result of such action, although in fact it was the engineers who suffered most from bad publicity over explosions.

The captain also decided when or if to retrieve a person who fell overboard. It was not uncommon to continue without bothering a rescue, on the supposition that the unfortunate was torn to shreds by the paddle-wheel and retrieving the body was a waste of time. On the other hand, letters and newspaper accounts appeared with some regularity describing the heroic efforts of a captain to save an unfortunate.

According to the rigid river standards of the day, however, every man aboard had his duties and his time on watch. When the captain was officially off duty, "the amenities and traditions of the river" dictated that he have no hand in the affairs of the boat, leaving them to the officer in charge, whether he be the mate or the chief clerk. The code of honor, as with the tradition at sea, also dictated that the captain be the last man to leave a sinking boat. As with other unwritten laws, these were seldom, if ever, broken. Breaking these rules could exact at terrible cost to the captain's reputation. There were few secrets among the river community and a man's honor — or lack thereof went with him to the grave.

There were many reports of captains performing heroic service at times of disaster. One of the most eloquent came from the Albany *Daily Advertiser* (reprinted in the Milwaukie *Sentinel,* February 18, 1840). The article is prefaced by a similar disaster with a far more tragic outcome:

> THE LEXINGTON: The burning of this steamboat is one of the most heart-rending disasters on record. Such an appalling loss of human life has seldom occurred. The event is imbittered [*sic*] by the reflection that a proper degree of presence of mind on the part of those having command of the boat might have saved most of the passengers. The small boats were launched while the steamboat was under full headway and immediately lost. They would have been sufficient to have carried off every person on board had they been skillfully managed.

The Albany *Daily Advertiser*, while discoursing on this subject, relates the following incident.

> Several years ago a disaster occurred on Lake Champlain similar in many respects to the burning of the Lexington. One stormy night as the steamboat Phoenix, with a full load of passengers and freight, was ploughing her way through the waters of Champlain, a fire broke out at midnight, and soon raged with fearful violence. The passengers roused by the alarm from their slumbers, and wailing to a terrible cause of pending destruction, rushed in crowds upon deck and attempted to seize the small boats. Here, however, they were met by the captain, who, having abandoned all hope of saving his boat, now thought only of saving his passengers, and stood by the gunwale of his vessel *with a pistol in each hand,* determined to prevent any person from jumping into the boats before they were properly lowered into the water, and prepared to receive their living freight. With the utmost coolness and presence of mind he supervised the necessary preparations and in a few minutes the boats were lowered away and the passengers received safely on board. They then shoved off and pulled through the darkness for the distant shore. As soon as this was reached, the boats returned to the steamboat and took off the crew, and as the captain supposed, *every living soul except himself.* But shortly after the boats had left the second time, he discovered under a settee the chambermaid of the Phoenix, who in her fright and confusion had lost all consciousness. Lashing her to the plank which he had prepared for his own escape, this gallant captain launched her toward the shore and was thus left alone with his vessel, now one burning pile. Having satisfied himself that no living thing remained on board his boat, and with the proud consciousness that he had saved every life entrusted to his care, he sprung from the burning wreck as it was about to sink beneath the waters and by means of a settee reached the shore in safety.

> The above is no exaggerated story. It is the simple narrative of one of the most heroic acts on record. We have only to add that the captain, who so faithfully and fearlessly discharged his duty on this trying occasion, is still in command of a noble boat on Lake Champlain, and is known to every traveler as Capt. Sherman, of the steamboat Burlington.

Reports of captains abandoning their stations were rare and often contradictory, as in the case of the steamboat *Anthony Wayne,* which sank on April 27, 1850. Prefacing an eyewitness account, the reporter of the Fort Wayne *Sentinel* (Indiana) of May 4, 1850, describes the *Wayne* as being a very old boat which had been repaired the past winter. She was owned by Mr. C. Howard of Detroit, valued at $15,000 and insured for $10,000. The writer goes on to note (italics added):

> *The Captain is much censored for going off with the life boat, capable of carrying 15 or 20 persons, taking only two others with him, and making no effort to take on board any more of the ill-fated passengers.*

An article reprinted from the *Mirror* in the same paper gives the following particulars:

> On Saturday night, April 27, the steamer Anthony Wayne arrived at Sandusky City from Toledo, and after taking on board twenty-five passengers, left at half past 10 P.M. for Buffalo. About half past one, while nearly opposite and six or seven miles distant from Vermillion, her boilers exploded. The explosion was most terrific. Her midships were literally blown into fragments—1 boiler thrown overboard and the other rolled upon the main deck. She filled and her hull went down in twenty minutes after the explosion in six fathoms of water. The upper decks separated from the hull when she went down, and continued afloat, remaining anchored to the hull by the tiller-ropes. The wreck was total, and the loss of life most appalling.
>
> The books and the trip sheet went down with the boat, and it is therefore impossible to tell with accuracy the number lost; but from the most careful estimation that can be made, we put the number of passengers on board at 63. Of this number, 25 are known to be saved, 11 wounded, 10 pretty definitely ascertained killed—leaving 17 (if on board) missing unaccounted for. The crew consisted of 30. Of these, 11 were killed, 4 wounded, and 15 saved uninjured. Making an aggregate of 93 persons on board, 40 of whom are saved unharmed, and 15 wounded—leaving 38 persons killed and missing.
>
> At the explosion, one of the small boats was blown overboard and lost. As soon as she was ascertained to be in a sinking condition, a life boat was launched from the upper deck and swamped. The large yawl was also launched at the same time, under the charge of the first mate, H.G. Edgcomb, who took into the boat with him 11 others of the passengers and crew. Others prepared rafts and boats and got on them. Capt. Gore jumped overboard and bailed out the life boat and after picking up the clerk, a deck hand and a fireman, and three passengers, and directing the first mate to steer for a schooner which had been seen up the lake, turned his course for Vermillion. Before leaving, the Captain and the Mate hailed those on the wreck, and learning that the hull had gone down, and that all remaining on the wreck were safe on the floating decks....
>
> The Wayne was an old boat, and the general impression is that she ought to have been pronounced unseaworthy long since. Her boilers, however, were new, having been made at the foundry of Wolcott & Savage, Detroit, but one year since. Her engineer, Jeremiah Elmore, had a wide reputation as one of the most competent and careful engineers on the Lakes, and has been known to quit boats because he pronounced them unsafe. It is said that he had been heard to complain on Saturday that the Wayne's engine was old. He was on duty at the time of the explosion and died at his post.

Along with Daniel Smith Harris, Joseph Throckmorton was one of the earliest and most notable of Upper Mississippi captains. Born June 16, 1800, in Monmouth County, New Jersey, he worked in the mercantile business in New York City. He moved to Pittsburg in 1828 and bought a portion of the steamboat *Red Rover*. Bringing her to St. Louis, he engaged in the upriver trade by June 1828.

Throckmorton worked the lead trade and also earned considerable monies delivering Indian annuities. In 1830, he formed the first cooperative of steamboat captains on the Upper Mississippi, working in conjunction with Captain S. Shallcross. Shallcross brought his boat, the *Chieftain,* loaded with freight and passengers as far as the Lower Rapids, where Throckmorton continued the journey. The following year, he formed a similar deal with Captain James May, of the *Enterprise*.

In 1833, Throckmorton wrote a letter to the St. Louis *Missouri Republican*, which sheds a fascinating look not only on a captain's life, but on the tricks of the trade perpetrated upon vessel owners.

> On loading a boat at a place of this kind [a wood stop], I discovered several black marks upon the deck, which, on examination, I found to be gunpowder, from a box which my men were about to store away as dry goods, which in part did contain dry goods, but in the middle concealed, was a considerable quantity of gunpowder, so carelessly placed, that it was strewn about the package. Now, I have only to request, that whenever any of my customers have powder to ship, that they will not conceal it, and thereby endanger our lives, but inform us of it. I am not a little surprised that so respectable a concern should attempt a thing of this kind, particularly as the freight would not have been more upon the article of powder than any other. I should suppose that shippers would have taken the *hint* after what has recently occurred on our western waters. At any rate, it is high time that *we* should. It is not my wish to complain, but it *is my wish* to run my boat with as much safety as possible; and I trust this gentle caution will be attended to.

The incident to which he referred was the explosion of the *Phoenix* in July 1832, which was steaming upriver from New Orleans. A fire broke out and the boat might have been saved but for a shipment of gunpowder, hidden in packages. The incendiary exploded, dooming the vessel.[4]

Gunpowder, whether smuggled aboard or properly consigned, proved a tremendous danger to steamboats. A year after the explosion of the *Phoenix*, the Hagerstown *Mail* (November 22, 1833), reported another disaster involving that explosive material:

> *Loss of the New Brunswick*—We learn by a gentleman who arrived here yesterday that this boat was burnt on her passage from New Orleans up to St. Louis, with a valuable cargo. She took fire among some chairs on deck, that were matted, and while the passengers were at dinner, which had burnt into the Ladies' Cabin, before it was discovered. The Boat was immediately run on shore, from the apprehension that she would blow up, the alarm of powder being on board, having been spread among the passengers; the passengers and crew scarcely got on shore, when she blew up, and is a total loss, vessel and cargo, with passengers' baggage. No lives lost.—*Poulson.*

Another explosion seven years later proved that no safe way had been determined to ship explosives. The following is from the Madison *Express,* March 7, 1840:

> The steamer *Belle,* which took fire eighty miles below St. Louis, was immediately run ashore, and though full of passengers, all escaped before her explosion, which was found to have been caused by *sixteen hundred kegs of powder on board.*

Throckmorton worked the "Grand Tour" trade and also continued working the Upper Mississippi routes. In 1842, he earned $10,000 on the *General Brooke* for five trips to the lead mines, and the following year, made fifteen trips, garnering the hefty sum of $25,000.

In 1848, Throckmorton became an agent for the Tennessee Insurance Company, in St. Louis, but then returned to his former occupation, captaining the *Genoa, Florence, Montana,* and *Columbia.* In 1870, he served as a United States engineer, working on improvements on the Upper Mississippi. He died in St. Louis in December 1872, having lost his considerable fortune along the way.[5]

The standard of rugged individualism among Upper Mississippi steamboat captains eventually became impossible to sustain, and most were compelled to join the powerful line services. Among the major employers were the expanding Keokuk Packet Company, the Minnesota Packet Company, the Northern Line, the White Collar Line and the Diamond Jo Line.

Where Anyone Might Set Himself Up As Captain: Skippers on the Lower Mississippi

As though there were an invisible dividing line drawn at St. Louis, the differences between Upper and Lower Mississippi commanders diverged at that point. While the lower portion of the Great River had its fair share of "hot" captains, vastly differing interpretations of the role became apparent. It is uncertain whether that stemmed from the fact that the trades between New Orleans and Ohio Valley cities developed earlier and there were far more boats on the water, or because a rapidly developing passenger trade occupied more of a captain's time and energy. But clearly, responsibilities divided between captain and pilot took on significant differences.

The greatest difference between the two sections involved river skills. While the Upper Mississippi captain had excellent, well-honed abilities and often assumed part-time piloting duties, those on the lower portions were primarily jacks-of-all-trades: they had a little knowledge of everything, but no specific skills in operation. At the dawn of steamboat navigation, captains came from the ranks of keelboaters and flatboatmen. As time went on, a man usually worked his way through the ranks, often starting as a deckhand, striker or steersman and graduating to mate before earning the right to command. The unique state of steamboating, however, wherein boats were relatively inexpensive, provided opportunities for a different type of captain: the man who had enough money to buy his own boat.

In an era when no master's license or certification was required, anyone buying a boat had the right to call himself "captain." Although those captains with solid experience often earned their reputations and thus their share of the freight and passenger trade, other considerations soon assumed great importance: speed and luxury accommodations. The man or men who bought a boat divided the best and highest-paid positions among themselves or their relatives, with no one the wiser as to their lack of qualifications. Without proper training, these captains ceded

Captain Samuel Dean, commander of the *Buckeye State*. The portrait was taken in Pittsburg in 1850.

A detailed illustration of the *Buckeye State* packet, which operated from 1850 to 1857. Wood stacked on the lower deck is visible, as are crew and possibly deck passengers at the bow.

responsibility to the pilots, so that an unwritten (but rigidly adhered to) code came into practice. The captain would take responsibility only at certain times, such as when the steamboat was pulling into the wharf or departing. At all other times during operation, the safe conduct became the exclusive province of the pilot.

Poor Judgment, and Judgment Rendered

The fact that so many individuals of varying skill and temperament set themselves up as captains invariably led to degrees of competence and judgment. While most western captains were well aware of the necessity of maintaining a solid reputation, some were not so conscientious. The *Belmont* (Wisconsin) *Gazette*, April 12, 1837, reported,

> We learn from a gentleman of this city who has just ascended the Illinois river, that a most melancholy occurrence took place, on Sunday the 18th, inst. about five miles from its mouth, where, through the obstinacy of the captains of two steamboats, one of their boats was sunk, the lives of all the deck Passengers, amounting to more than twenty, lost, and the freight and baggage entirely destroyed.
>
> The Captain of the Wisconsin, which was then ascending the river, had repeatedly stated that if he should meet the *Tiskilwa,* and her Captain would not give him a clear channel, he should run her down.—This, it seemed, provoked the Captain of the Tiskiwa, and he became obstinately determined not to turn out of his course. Both boats met about five o'clock in the morning, a time when all the passengers were in bed, and steered directly for each other till within a distance of only a few rods, when the captain of the Tiskilwa endeavored, but too late, to avoid the concussion; and turning a little out of direct course, thus gave a fair broadside to the ascending boat, which took her just behind her wheel, and she sunk in less than three minutes after she struck. The first notice of the extreme danger which the cabin passengers received, was the screams of those below, who were drowning; and without even time to put on their clothes, they merely escaped by jumping through the windows of the cabin, which, fortunately for them had been completely separated from the sinking boat by shock.
>
> The Captain of the Wisconsin is stated to have acted, even to the ladies, in a most brutal manner, having put them ashore *barefooted,* at more than a mile and a half from any habitation, and who [had] nothing but their night clothes on. Reports say that men were even worse treated, as he endeavored to prevent their getting on board the Wisconsin at all.
>
> A large sum of money belonging to one of the ladies named has been recovered subsequently, from the ladies' cabin; and one gentleman was fortunate enough to find his coat floating on the river; with his money, amounting about $4,000, in the pocket. *(Commercial Advertiser)*

There is no record of what happened to the unnamed captain of the *Wisconsin,* but a report in the *Adams Sentinel,* February 5, 1838, cites a case in which the captain's culpability was decided without much ado:

> Cincinnati, Jan. 23: The Grand Jury for the Court of Common Pleas for Hamilton County, at their sitting last week, found a bill against the Captain and Engineer of the Steamboat *Home,* for manslaughter, caused by the collapsing of the flue of the boat, a short time since, while at the city wharf.
>
> The trial occupied the time of the Court all day yesterday, and was concluded in the evening. After the examination of a large number of witnesses, and a charge from the President Judge in favor of the accused, the subject was committed to the jury, who returned a verdict of *not guilty,* without leaving the box. *(Whig)*

Evolving Traditions

Tradition at sea dictated that a captain's word was law: the master made all decisions concerning the ship and they were followed without question. Anything less was considered mutiny. As maritime law developed, it legalized ancient custom and courts upheld the captain's ultimate authority aboard his vessel unless an inability to command was proven beyond a shadow of doubt. This concept did not translate perfectly or easily to masters of the inland waterways. As more and more "owner-captains" entered the river trades, many men did not wish such unlimited power. Complicating matters, the evolving law was always vague on such authority. Because rivers crossed multiple state lines, legislators often sought to enact rules and regulations singular to their territory. This caused many legal questions, most of which the courts were hesitant to address. That left it up to common practice, and eventually unwritten standards of conduct developed, created by the rivermen themselves.

As it evolved, the captain had no authority (or only quasi-legal authority) to order the pilot to do his bidding. He could not, for instance, order the pilot to navigate through conditions the pilot felt were too dangerous; nor could he countermand an order the pilot gave to the engineer. In most cases, if a captain were to attempt to advise the pilot on the boat's operation, his action would be considered highly improper, perhaps even inciting a mutiny. Captains who involved themselves with navigation had difficulty recruiting hands and earned bad reputations along the river.

As recorded by George Byron Merrick in his description of the later steamboat era:

> Only in cases where such interference was necessary for the safety of the boat was [his intervention] deemed permissible; and a captain who so far forgot himself as to interfere, lost caste among all classes of rivermen, high and low.[6]

Once the boat was in the water, the captain was expected to devote his energies to seeing to the well-being of the passengers, and in that capacity he often had his hands full. The captain took his regular tour of duty on deck, and in case of emergency, was expected to direct the crew in getting passengers to safety. In his absence, the second mate usually stood in his stead. On larger boats, the captain customarily intervened in the handling of the steamboat only to the extent of directing policy. His primary duties involved securing cargo, maintaining good relations with merchants and keeping an eye on the profit and loss ledgers, for he was likely either a co-owner or worked directly for the ownership, with his primary responsibility being their welfare.

Lacking federal guidelines and the questions arising out of which, if any, states held sway over passengers and crew committing crimes on open waterways, it fell on the captain's shoulders to enforce rules and conduct for those aboard his vessel. Popular right (if not legal authority) granted him the power to judge those accused of crimes, but in this duty, unlike a sea captain, he openly sought the assistance of passengers for final judgment and determination.

By 1817, rules were posted on the cabin wall of the second *New Orleans*, stating:

> At noon, every day, three persons to be chosen by a majority of the passengers shall form a court to determine on all penalties incurred and the amount [of fines] collected shall be expended in wine for the whole company after dinner. For every transgression against good order and cleanliness, not already specified, such fine shall be imposed as the court in their discretion shall think fit.[7]

Courts were generally held on the cabin deck. Systems varied from boat to boat, and in instances where a crime was violent or especially heinous, the commanding officer could take it upon himself to inflict punishment. However, if the passengers disagreed with the sentence, they proved — again by common consent — the ultimate deciders.[8]

Steamboat "Justice"

An extreme case of steamboat passengers defying the captain and taking the law into their own hands comes from the *Adams Sentinel*, May 21, 1838, which reprinted an article from the *American Sentinel*:

The St. Louis Republican of the 30th ult. gives the particulars of the drowning of a free negro, a cook on board the steamboat Pawnee, on her passage up from New Orleans to that city under the sentence of that most unmerciful of judges, Judge Lynch. The circumstances are substantially these: — On Friday night previous, about ten o'clock, a deaf and dumb German girl was found in the store room with the cook. The door was locked, and at first he denied that she was there. The girl's father came, he unlocked the door, and the girl was found secreted in the room behind a barrel. He was accused of having used violence to the girl, but how she came there did not very clearly appear. The Captain was not informed of this during the night. — The next morning some four or five of the deck passengers spoke to the Captain about it; this was about breakfast time. — He heard their statements and informed them the negro should be safely kept until they reached St. Louis, when the matters should be examined, and if guilty, he should be punished by the law. Here the matter seemed to end: the Captain after breakfast returned on deck, passed the cook's room, and returned up to his own room; immediately after he had left the deck, a number of deck passengers rushed upon the negro, bound his arms behind his back and carried him forward to the bow of the boat. A voice cried out, "throw him overboard" and was responded to from every quarter of the deck — and in an instant he was plunged into the river. The Captain hearing the noise, rushed out in time to see the negro float by. The scene of trying and throwing him overboard scarcely occupied ten minutes, and was so precipitate that the officers were unable to save him. Several of those engaged were identified and some of them arrested and lodged in jail in St. Louis, where, we trust, they will be treated as murderers should be.

"The grandest position of all": Steamboat Pilot

The great American writer Mark Twain wrote those words in 1883, and he ought to have known, for he apprenticed as a cub pilot in the years before the Civil War, and earned his license as one of the Mississippi River's elite.

He further stated, with obvious nostalgia for a bygone life, that everyone had a master but the pilot:

> The captain could stand upon the hurricane deck, in the pomp of a very brief authority, and give him five or six orders while the vessel backed into the stream, and then that skipper's reign was over. The moment that the boat was underway in the river, she was under the sole and unquestioned control of the pilot. His movements were entirely free; he consulted no one, he received commands from nobody, he promptly resented even the merest suggestion. Indeed, the law of the United States forbade him to listen to commands or suggestions, rightly considering that the pilot necessarily knew better how to handle the boat than anybody could tell him. So here was the novelty of a king without a keeper, an absolute monarch who was absolute in sober truth and not by a fiction of words.

Twain went on to describe the life he once led:

> In those old days, to load a steamboat at St. Louis, take her to New Orleans and back, and discharge cargo, consumed about twenty-five days, on an average. Seven or eight of these days the boat spent at the wharves of St. Louis and New Orleans, and every soul was hard at work, except the two pilots; they did nothing but play gentleman up town, and receive the same wages for it as if they had been on duty. The moment the boat touched the wharf at either city, they went ashore; and they were not likely to be seen again till the last bell was ringing and everything in readiness for another voyage.[9]

Captains, clerks and occasionally the mate, who worked the entire season on one boat, were often paid by the year; engineers were paid by the month. Pilots, however, were hired and paid by the trip. The reason for this was that if the captain/owner decided to change trades, he would need a pilot familiar with that stretch of water. As the era progressed, pilots became specialized and therefore tended to stay in one particular area, whereas officers went with the boat wherever she steamed. However, on larger or more stable steamboats, pilots were attached to the vessel, even during the winter months or periods of low water when the boat was in dock. A pilot's good reputation attracted money, and money meant success. The results were mutual, for established navigators earned a considerable fortune for their labors. Mark Twain observed that the salary was $400 a month, an almost inconceivable splendor.

> Few men on shore got such pay as that, and when they did, they were mightily looked up to... Lying in port under wages was a thing which many pilots greatly enjoyed and appreciated.

And as for a pilot's ultimate authority aboard a boat, Twain has the last word.

> One day on board the "Aleck Scott," my chief [pilot], Mr. Bixby, was crawling carefully through a close place at Cat Island, both leads going, and everyone holding his breath. The captain, a nervous, apprehensive man, kept still as long as he could, but finally broke down and shouted from the hurricane deck,—
>
> "For gracious' sake, give her steam, Mr. Bixby! give her steam! She'll never raise the reef on this headway!"
>
> For all the effort that was produced upon Mr. Bixby, one would have supposed that no remark had been made. But five minutes later, when the danger was past and the leads lead in, he burst instantly into a consuming fury, and gave the captain the most admirable cursing I ever listened to. No bloodshed ensued; but that was because the captain's case was weak; for ordinarily he was not a man to take correction quietly.[10]

On occasion, however, pilots took their exalted status too far. The *Adams Sentinel,* November 9, 1835, reports the following:

> *Disgraceful Conduct.*—The Rodney (Miss.) Telegraph states, that the pilot of the steamboat Mazeppa, on her way from Louisville to New Orleans, was so enraged at the owner's taking in tow a flat boat, that he purposely ran the boat against a snag, in broad daylight, some miles below New Madrid, by which 3 negroes, (a woman and 2 children) and 23 horses on the flat boat were killed.

The Pilot: King of Rivermen

On steamboats, the pilot and engineer possessed the greatest amount of skill and expertise. The pilot was a regular officer, but confined himself to the pilothouse, where he reigned as sole master. Unlike the position of helmsman at sea, the steamboat pilot laid in the course and directed passage of the boat. The position required absolute attention to detail and a prodigious memory, for he steered during the hours of darkness. Working with limited visibility required the pilot to remember shapes of riverbanks, possess the ability to differentiate shadow from substance and wind shoals from actual shallows, set the speed in unfavorable conditions, and maneuver over or around obstructions.

Vigilance was the critical factor in determining the safety of the vessel, and if the steersman became diverted for whatever reason, the boat could easily fall victim to harm. At no time would he ever leave the pilothouse until relieved, and the only other personnel allowed to handle the great wheel was a cub apprentice.

The steamboat *Hope*, Vicksburg, Mississippi, 1860. The elevated pilot-house reveals glass walls, which were a development to give the pilot a 360-degree view of the river.

Pilothouses ranged from the grim to the grandiose. On a tramp vessel, it might be cheap and dingy, a battered, cramped, rat-trap affair. On the other end of the spectrum, pilothouses of Floating Palaces were more representative of the navigator's exalted status. Mark Twain described the *New Orleans* facilities as being a sumptuous glass temple; room enough to have a dance in, with showy red and gold window curtains. The furniture comprised an imposing sofa with leather cushions and a back to the high bench where visiting pilots sat and spun yarns. He adds there were

> bright, fanciful "cuspidors" instead of a broad wooden box filled with sawdust; nice new oil-cloth on the floor; a hospitable big stove for winter; a wheel as high as my head, costly with inlaid work; a wire tiller-rope; bright brass knobs for the bells; and a tidy, white-aproned, black "texas-tender" [waiter] to bring up tarts and ices and coffee during mid-watch, day and night.[11]

If a pilot's skill and value do not come as a surprise to the television aficionado who grew up watching such series as Darren McGavin's *Riverboat* or Jack Kelly and James Garner's *Maverick*, perhaps Twain's description of a mid–19th-century navigator will:

> Two or three of them wore polished silk hats, elaborate shirt-fronts, diamond breast-pins, kid gloves, and patent-leather boots. They were choice in their English, and bore themselves with a dignity proper to men of solid means and prodigious reputations as pilots.[12]

A pilot's skill came from practice, as opposed to the technical prowess of an engineer, and his specialty lay in handling the boat and reading the river. Prior to the spread of the telegraph (before the 1850s), pilots gathered socially to discuss navigation and, in that way, passed along river conditions to those making similar trips. There were also published guides, called "navigators," which listed the more common obstructions and techniques of navigation on the Ohio and Lower Mississippi. The most famous was Zadok Cramers', *The Navigator*, published in Pittsburg. It went through twelve editions in the period 1801–24. Others included J. C. Gilleland's *The Ohio and Mississippi Pilot* (Pittsburg, 1820), Samuel Cumings,' *Western Navigator* (Philadelphia, 1822), reissued over a period of 30 years under the title *The Western Pilot*) and George Concilin's *New River Guide* (Cincinnati, 1843).[13]

Outside the Pilot House: Taking Care of Business

The pilot's official duties were confined to the pilothouse, but occasionally, as demonstrated in this article from the *Sandusky Clarion*, May 21, 1849, he felt obliged to take a hand in other duties:

A Reckless Scoundrel

Yesterday morning about half past 4 o'clock, on the arrival at the wharf of the steamer Isaac Newton, a man named Martin Brennan went on board the boat and commenced, in a most reckless manner, knocking down the passengers and crew. Ald. Harcourt who is freight agent to the People's line, went up to him and requested him to desist or he would, as a magistrate, commit him. Brennan seized the alderman by the neck and arm, and being a very powerful man, forced him to the gang-plank, and then violently threw him into the river, crying that "he would kill him and take his life." Ald. Harcourt sank twice and would inevitably have been drowned had not Mr. J. Newton and several others thrown him ropes, with which and by the aid of one of the hands in a small boat, he was taken out of the water much exhausted. He came on deck wet and dripping, and again arrested Brennan, the ruffian once more attacking him, when George Lester, the pilot of the boat, came to his assistance, and knocked the big burly fellow as "flat as a flounder." He struggled violently, but a Hyer-like blow on the cheek settled him, and Mr. Lester procured some rope-yarn and despite his struggles, coolly bound him and put him into a cab to be taken to a jail. Justice Parsons examined him and made out a full commitment on the charge of assault and battery with intent to kill. There are other charges against Brennan, and it is the intent of Justice P. to bring him to full and speedy justice. (*Alb. Knick. 9th inst.*)

The "Cubbie Factor"

Pilots were not born but made. To become a pilot required between two and three years of apprenticeship. A youth wishing to become a pilot paid a master $500 to $1000, which was either paid upfront or taken out of the first wages he received after graduation. The "cub" was taught not only the intricacies of the river, but also the important customs associated with the exalted position to which he aspired.

"Turning out" was the expression used when a pilot reported for duty; the expression "off-watch" referred to the pilot handing over the wheel to his partner. Custom, borrowed from maritime practice dictated that when one pilot relieved another, the latter would put on his gloves and light his cigar, while the former briefed him on river conditions. Arriving late was the one breach of etiquette pilots never tolerated in one another, and if such were the case, the pilot going off-watch refused to speak to the other. Each pilot ran the same part of the river going downstream as he had going up the river. However, when conditions were unfavorable, both stayed and worked together.

The cub, or "steersman," followed the master pilot to whom he was articled from boat to boat. When the teacher thought the cub needed additional experience, he shipped him out to a pilot working a larger or smaller boat, but maintained ultimate control of the cub's education. After acquiring all the knowledge deemed necessary, and proving that to the satisfaction of any two licensed pilots, the cub filled out an application directed to the United States Inspector General. Before the advent of regulations, no questions were put to the applicant, and no further proof was needed, it being felt that the testimony of two competent pilots proved sufficient. Not all pilots were amenable to trainees, however, and often refused to "learn" a cub on the grounds that too many pilots ruined the remuneration for the whole.

Prior to the Steamboat Act of 1852, there were several attempts to organize pilots. The Louisiana Pilots Benevolent Association was established in 1841, with an aim of raising standards, but was equally "a combination or monopoly got up for the purpose of extorting from steamboats an unfair price" for labor.[14] Several years later, the Little Rock Pilots Association forbade pilots from attempting to set and demand higher wages. None of these groups proved particularly effective.

In 1853, the first year the regulations dictated by the act of 1852 went into effect, strict licensing laws made it much more difficult to obtain a pilot's license. More pilot associations were created, and these, according to Mark Twain, were highly effective. He describes an almost cult-like environment wherein member pilots refused to have anything to do with nonmembers, and utilized specialized locks to prevent outsiders from reading navigational reports. The Pilots Benevolent Association declared a moratorium on creating cub pilots for five years believing a surplus of pilots was driving down wages. Standards were set whereby an apprentice had to be at least eighteen years of age, of respectable family and of good character. He had to pay $1,000 in advance for his training, and after completing the course, he required more than half the membership's signature on his application for a license.

"Well, blow me down!": The Steamboat Engineer

After the captain and the pilot, the steamboat engineer was the next officer in line. Although critical to the safe operation of the vessel, the engineer received lower wages for his work than the more heralded pilot and captain and received little popular acclaim, while absorbing much, if not all, the blame for an explosion. This was often unfair, as the engineer was merely an employee, hired by the captain and responsible to him. If the commander desired more speed than was prudent, or if the pilot required additional power to get the vessel over a bar, it was the engineer's duty to comply. Mistakes in judgment of those over him were seldom regarded as mitigating factors.

The engine room was a cramped, noisy, constantly moving and fiercely hot environment in which to labor. Grease coated everything; sparks flew from open furnaces; steam escaped leaky valves. The engineer's orders often came fast and furious, and were often contradictory.

As steamboats first worked the western waterways, individuals with specialized mechanical skills were sought from the East Coast, but soon men with an eye toward adventure—and danger—joined the ranks from machine shops and smithies. They learned the position on the job, and early explosions were rightly attributed to their lack of skill and, occasionally, lapses in oversight. With experience came technical expertise and a good engineer was worth his weight in gold. As with cub pilots, those desiring to learn the trade apprenticed aboard a boat, starting out as a striker. A striker learned the craft of engineering from the bottom up, first by cleaning and manning the boilers. Those who proved competent were taught the intricacies of running and repairing the machinery.

Repairs constituted an important part of an engineer's duties. While major problems with the machinery required expensive layovers at a boatyard, many smaller problems were handled on the fly. The American Fur Company first introduced spare parts and blacksmithing tools as standard equipment after it began running steamboats to its trading posts on the Upper Mississippi. The benefits of this soon became apparent, and the idea was copied along all the inland waterways. Engineers thus honed their mechanical skills and became adept at jury-rigging nearly anything needed aboard.

Breakage and the wearing of necessary parts was the result of constant use under harsh, 24-hour a day conditions. Minor leaks from valves, joints and pipes were either ignored or patched; more serious problems, such as blown-out cylinder heads, broken shafts, damaged wheel arms and connecting rods were more difficult to handle. More significantly, the constant shifting of water in the boilers due to uneven weight distribution or erratic river levels often exposed red-hot plates to air. Immediate correction did not always alleviate the danger, and explosions were the result.

Additionally, sudden orders to stop or start the engines, reverse the wheels, or change speed hundreds of times a day increased the wear on the machinery as well as the tension on the crew. Communications from the pilothouse by a series of bells (the "jingler") or the speaking tube required the engineer to adjust the rate of firing to meet the demand for power, always with an eye toward maintaining adequate steam for use in emergencies.

Although the salary was more than a mechanic could earn on shore, it was small compared to the dangers, for engineers' names were often prominent on casualty lists. That did not seem to keep applicants away, however, and many thrived under the less-than-ideal conditions.

The most hot-blooded among this group were called, appropriately enough, "hot engineers," and they were the most sought after by river captains. A hot engineer was a man with a taste for speed, one who did not care how hard he drove the engines, as long as they provided the steam needed when the orders came. And they often came when the boat was involved in a race. The engineer was just as caught up in the rivalry and competition as the officers and passengers. The fastest boat earned the most lucrative contracts and a marketable name for the crew, whose wages could increase accordingly. Not only that, but an engineer from a brag boat was important within the circles he frequented, and appreciated the high esteem in which he was held.

In the unfortunate circumstance of a boiler explosion, popular opinion held the engineer accountable, as typified by an article run in the Hagerstown *Mail,* May 7, 1830. It goes into great detail about the sinking of the *Chief Justice Marshall,* under Captain Ford, which took place while the boat was just getting underway. (See a separate account of the steamboat *Chief Justice Marshall* striking a rock and sinking in chapter 11. It is unclear whether this is the same boat, as names were frequently used many times, often by the same owners. Both accounts detail accidents of boats working the same trade. The article "presumes" and "conjectures" the cause; italics added.)

> Of the cause of the disaster, we have no certain information. *It is presumed to have arisen from too great an exhaustion of water in the boilers — owing, it is conjectured, to the incompetence of the engineer.* The old engineer, who has had charge of the engines for several years, was dismissed a few days since, and a new one engaged, who offered his services at a reduced salary.

In truth, there were many mitigating factors, other than "incompetence." The article goes on to add (not, however, as mitigating evidence), that

> It is proper to mention also that in order to accelerate the speed of the boat, a new boiler had lately been put in below, and it was that which had burst.

An article in the *Republican Compiler,* November 29, 1836, left no doubt who was to blame:

> Distressing Steamboat Accident. We learn from the Cincinnati Whig of the 18th inst. that a most distressing accident occurred on board of the steamboat Flora, Capt. R.D. Chapman, on the 17th while on her way from Louisville to Cincinnati. — The boat had approached to within thirty miles of the latter place, when the pipes which connect the two boilers together, commonly called the "connecting pipes," suddenly broke, or separated, causing the death of one man, and the scalding and mutilation of thirteen others, all cabin passengers but one.
> As soon as the noise, which the accident occasioned, was heard by the persons in the cabin, most of them unfortunately ran to the door, on which being opened, enabled the scalding steam to rush in and perform its work of destruction....
> The boat was towed to Cincinnati the day after the accident, by another steamboat, & several of the wounded persons taken to the Hospital.
> The Cincinnati Gazette states that "the accident is imputed to the gross negligence of the engineer." *Balt. Amer.*

Factors which also came into play when explaining boiler explosions should have, but often did not, exonerate the engineer:

(1) Engineers did not have the authority to assert their opinions over a captain's desire for speed
(2) Defects in manufacturing, worn equipment and depreciation were not their fault
(3) The lack of adequate or tested safety valves and steam gauges made their job more difficult than any other aboard a steamboat
(4) The boat's owner did not care to spend money repairing equipment
(5) No one fully understood the mechanisms of what caused boiler explosions
(6) Inadequate training and the need for engineers placed men in a position they were unable to handle[15]

The derisive term for an inadequately trained engineer was a "three-month engineer," and owners were often criticized in the press for their greed and disregard for passenger safety by hiring such individuals. This attitude was augmented by advertisements in the paper, declaring that "any man after two months' experience" was capable of handling the machinery (1812); another, in 1813, declared a learning period of two weeks was sufficient. Clearly, they were incorrect, but it brought into the fold many to whom great responsibility fell, with little actual background.

Like the pilots, engineers formed associations to improve their occupation. In 1842, unions were formed in Cincinnati and St. Louis. The *Tioga Eagle* (Wellsboro, Pennsylvania), reported on November 16, 1842:

> WESTERN ENGINEERS.— The steamboat engineers of Cincinnati have formed an association, the subject of which is declared to be "to regulate and promote the interest, character and respectability of that class of men known as 'steamboat engineers.'" They hope by means of their association, to remedy the evils that arise from defective boilers and ignorant and careless engineers.

Engineers' associations undertook investigations of steamboat accidents and, like the pilots' associations, lobbied Congress for the licensing of qualified men. These groups were often looked at with skepticism by captains, who viewed them only as being a means whereby a member could command a higher salary. In any case, they were largely ineffective, and during the 1840s, when a glut of engineers arose, a great war to undercut one another for the limited positions available caused turmoil in the ranks.

This was brought under control by the Steamboat Act of 1852, which mandated strict regulations and licensing procedures for engineers. The year after, a new association of engineers was created at Pittsburg, Wheeling, Cincinnati and Louisville. This was more successful; the number of members rose substantially from previous years, and they adopted wage schedules and uniform practices for the profession.[16]

The Cussing Officer: The Mate

The mate was the lowest-ranking officer aboard a steamboat, and also the lowest paid. He worked directly under the captain in the general oversight of the boat and, like the captain, typically possessed the least specialized skills. The steamboat mate's role differed dramatically from mates who served at sea. Those officers were required to have a thorough working knowledge of navigation and all aspects of ship operation, whereas the steamboat mate's duties were limited to handling the crew and performing common labor. He saw that the freight and fuel were stowed, ensuring proper weight distribution throughout the deck, hold and guards.

In cases of emergency, the mate might assist in sparring or warping operations when getting the boat over a bar, or in performing repair work following an accident. He also spelled the captain on deck, at those times assuming the authority in routine matters. As an officer, he would be expected to help during a crisis, assisting passengers to abandon the vessel in case of fire or sinking.

His greatest ability came in the use of colorful language, which he frequently employed at the top of his voice. To augment verbal commands, the mate often urged the deck hands to greater effort by dominating them with brute force and intimidation, usually with a piece of wood or a capstan bar in his hands.

The Mate in Command

The disaster of the steamboat *Pulaski* was one of the worst on American waters. A description of the tragedy from the Bangor (Maine) *Daily Whig*, of June 27, 1838, gives a fascinating insight into the tragedy and the role played by the mate.

> *Loss of steam packet Pulaski, with a crew of thirty seven, and one hundred or one hundred and sixty passengers.* On Thursday, the 14th instant, the Steamboat Pulaski, Captain

Dubois, left Charleston for Baltimore, with about 150 passengers, of whom about 50 were ladies.

About 11 o'clock in the night, while off North Carolina coast, say 30 miles from land, weather moderate and night dark, the starboard boiler exploded, and the vessel was lost with all the passengers and crew except those whose names are enumerated among those saved in the list to be found below.

We have gathered the following fact from the first mate, Mr. Hibberd, who had charge of the boat at the time. Mr. Hibberd states that about 10 o'clock at night he was called to the command of the boat, and that he was pacing the promenade deck in front of the steerage-house; that he found him — If, shortly after, upon the main deck, lying between the mast and side of the boat; that upon return to consciousness, he had a confused idea of having heard an explosion, something like that of gunpowder immediately before he discovered himself in his then situation. He was induced, therefore, to raise and walk aft, where he discovered that the boat midships was blown entirely to pieces; that the head of the starboard boiler was blown out, and the top torn open; that the timbers and plank on the starboard were forced asunder, and that the boat took in water whenever she rolled in that direction.

He became immediately aware of the horrors of the situation, and the danger of letting the passengers know that the boat was sinking, before lowering the small boats. He proceeded, therefore, to do that. Upon dropping the boat, he was asked his object and he replied that it was to pass round the steamer to ascertain her condition. Before doing this, however, he took in a couple of men. He ordered the other boats to be lowered, and two were shortly put into the water, but they leaked so much in consequence of their long exposure to the sun, that one of them sunk, after a fruitless attempt to bail her. He had in the interim taken several from the water until the number of ten. In the other boat afloat there were eleven. While they were making a fruitless attempt to bail the small boat, the Pulaski went down with a crash, in about 45 minutes after the explosion.

Both boats now insisted upon Mr. Hibbard's directing the courses to the shore, but he resisted their remonstrances, replying that he would not abandon the spot till day light. At about 3 o'clock in the morning they started amidst of the wailings of the hopeless beings who were floating around in every direction, upon pieces of the wreck, to seek land, which was about 30 miles distant. After pulling about thirteen hours, the persons in both boats became tired, and insisted that Mr. Hibberd should land. This he opposed, thinking it safest to proceed along the coast, and to enter some of its numerous inlets; but he was at length forced to yield to the general desire, and to attempt a landing upon the beach a little east of Stump Inlet.

He advised Mr. Cooper, of Ga. who had command of the other boat, and a couple of ladies with two children under his charge, to wait until his boat had first landed, as he apprehended much danger in the attempt, and should they succeed, they might assist him and the ladies and children. There were eleven persons in the mate's boat (having taken two black women from Mr. Cooper's). Of these, two passengers, one of the crew, and the two negro women were drowned, and six gained the shore. After waiting for a signal, which he received from the mate, Mr. Cooper and his companions landed in about three hours after the first boat, in safety. They then proceeded a short distance across Stump Sound [?], to Mr. Redd's of Onslow county where they remained from Friday evening until Sunday morning, and then started for Wilmington. The mate [?] and two passengers reached here this morning (10th June) about 9 o'clock.

Passengers saved in the two yawls: Mrs. P.M. Nightingale, servant and child, Car.—Island; Mrs. W. Frewer and child, St. Sindons, Geo.; J.H. Cooper, Glynn, Georgia; Capt Pooler-son, Wm. Robertson, Savannah, Georgia; Elias L. Barney, N.C.; [name illegible]; S. Hibberd, 1st mate Pulaski, W.C.N. Swift, New Bedford; [name illegible], Munich; Charles B. Tappen, New York; Gideon West, New Bedford, boatswain; B. [name illegible], Norfolk, steward.

Persons drowned in landing: Mr. Fird, of Bryan of Georgia; an old gentleman from Buffalo, New York, and recently from Pensacola; a young man, name unknown; Jenny, a colored woman; Priscilla, a colored woman, stewardess.

In this case, the list of those who survived was far shorter than the list of those who perished.

One subject seldom discussed in the matter of steamboat disasters is the aftereffects of those who perished. While it would seem no more was involved than identifying the body or place of interment, the *Adams Sentinel,* January 27, 1840 (nearly a year and a half later), reveals another aspect of death, the sometimes complicated setting of estates.

The Charleston Courier states that the Court of Inquiry in that city has been engaged in the trial of a case of deep interest, arising out of a suit instituted between the representatives of the late H.S. BALL and Lady, of that city, two of the unfortunate victims in the explosion and wreck of the steam packet *Pulaski,* in order to settle the question of survivorship between the husband and wife, who shared an ocean-grave on that appalling occasion. All the harrowing incidents of that dreadful calamity were disclosed by the evidence, so far as they could be collected from the surviving witnesses and sufferers; and a beautiful model of the boat was placed in evidence before the Chancellor, to illustrate the evidence & the argument.

The Man with the Reputation: Steamboat Clerk

Perhaps no one but the captain had more interaction with the public than the steamboat clerk, and his reputation for providing assistance and accommodations reflected favorably (or less so, depending on the individual) on the image of the vessel. On less reputable boats, the clerk, along with the captain, was held accountable for whatever deceptions, misstatements or fraud were practiced on passengers.

The clerk served as the business manager, keeping ledgers, maintaining accounts with merchants, checking for spurious currency and serving as the agent who actually sold the passage tickets and assigned rooms. He solicited cargo and made certain it got aboard, collected payment for such, checked manifests when unloading, prepared the waybills, paid for the wood and fixed rates and fares. On some boats, he was even responsible for the hiring and firing of the non-skilled crew. His working area was located forward on the boiler deck and adjoined the main saloon opposite the bar. Known simply as "the office," it was where routine business matters were handled.

The head clerk held an officer's title and typically ranked next to the captain and pilot in salary and prestige. On boats of smaller tonnage the captain often acted as his own clerk, but on larger vessels, the clerk handled most financial transactions, including paying the crew. He was responsible directly to the owners, and received little or no interference from the captain. The position was similar to a ship's purser, but the steamboat clerk held more responsibility.

The head clerk was assisted by a "mud clerk," so called because he had the duty of receiving and delivering freight on the open wharves in all weather. The mud clerk stood alternate watches to his superior and performed whatever passenger service might be required. His salary was half that paid the head clerk.

Besides the clerk, second engineers and second mates were also considered officers, although of the lowest grade.

An Officer of the First Class: The Steward

Although commonly believed to be an officer of the second class, the steamboat steward actually belonged to the more exalted class, and although he stood no regular watch as did the other officers, his duties required him to be available at any and all times. In salary, he ranked between captain and pilot, and a man with a considerable reputation might earn pay equal to the chief officer. The steward earned his pay by providing three meals a day at the stated hour, offering passengers a wide variety of entrees, side dishes and desserts. The work was arduous and stressful.

On "crack boats" with large crew lists, the steward had two assistants, with meat, vegetable and pastry cooks, bread makers, waiters and pantrymen under his command. He was responsible for seeing to it that meals were set on time, the menu varied and the cabins set to rights. It was also his responsibility to wake passengers sleeping on the cabin floor — those who paid full fare on overbooked voyages, as well as the cabin crew and servants.

14

Deckhands: A Distinctive Class of Casual Workers

Deckhands, or non-skilled laborers, comprised more than half the crew of a steamboat. These men were hired by the one-way trip, and were engaged or dismissed at will, with many going from boat to boat over the course of a season without ever knowing where their next job would be. The work was physically demanding and the position little regarded. Once the boat was loaded, however, they had few set tasks and this group did not stand regular watches as did the officers.

As the age of steamboating began, early crews came from the rivermen who had worked the keel and flatboats. Most were older men with a reputation for brutality and lawlessness. They required a heavy hand to oversee their labor, and fights, often bloody and fierce, were not uncommon. The salary was minimal and most were compelled to work the four or five winter months ashore in order to make ends meet. As a transient class, they had little opportunity for a stable life, and those with families were away far more than they were home.

Until the 1840s, most deckhands were American born. After immigration became more of a factor (increased by the Irish potato famine of the late 1840s), the Irish and Germans took over a considerable portion of the jobs. In ten years, the Irish population working the steamboat trades increased dramatically to 40–50 percent. In the upper regions of the country, immigrants reached St. Louis by 1820. The line of penetration reached Keokuk by 1830, and ten years later, immigrants appeared as far as the lead-mining regions of Galena and Dubuque. By the 1850s, the new labor class extended to the boundary of Iowa and to the frontier just beyond St. Paul.

Many of these travelers came across the Cumberland Road that reached as far as Columbus, Ohio, by 1833. By the mid–1840s it had extended into Illinois. A smaller number arrived via the Ohio River to the Mississippi and went up to St. Louis. From that city, most migrated west or north.

The population of foreign-born crews increased from 63 percent in 1850 to 67 percent in 1860. By this time, the profile of the typical deckhand had dramatically altered. Long gone were the older, tougher keelboatmen, replaced by very young immigrants who were desperate to land any paying job. In 1850, 60 percent of unskilled steamboat labor

A post-bellum photograph from 1867 of the *New Era*. Freight lines the shore waiting to be loaded. During the Steamboat Era, manual labor was required to load and unload cargo; no mechanical devices were used in the process until after the Civil War.

were under 30 years of age, and in some trades, the number rose as high as 68 percent. Of that number, a full 25 percent were under twenty-five years old.[1]

Aboard a steamboat, the deck hand performed all the hard labor, was called to wood when necessary, and also manned the capstan and pumps. In antebellum times, no mechanical devices were used to bring cargo aboard. In the late 1840s, small hoisting engines (called "mickeys," or "niggers") were used to move heavy freight from the hold, but not to shift cargo from deck to shore, or vice versa. That work was all performed on the backs of the rousters.

All Fired Up: The Stoker

The position of the fireman, or stoker aboard a steamboat was hardly an enviable one. Those hired for this somewhat specialized position were paid slightly more than a deck hand, but that hardly compensated for the constant exposure to red-hot boilers and the extremes of temperature blowing in from open furnace doors.

The firemen's labor was divided into three shifts, with a man working four hours twice in twenty-four hours: once in the day and once at night. This was all anyone could stand, for the act of constantly stoking the fires was backbreaking. Even the sturdiest broke down, despite the common practice of plying them with brandy. Customarily, a bucket of alcohol was delivered every four hours and each man filled his cup. During races, especially, when prodigious amounts of steam were required, they were plied with spirits to numb their brains and strengthen their limbs. During the famous race between the *Die Vernon* and the *West Newton* in 1852 (see chapter 12), Captain Ford supplied his Negro firemen with whiskey toddies to stimulate a maximum effort.

Firemen were also the primary wooders on a steamboat. They were called out at any hour of the day or night to go ashore and lug back heavy logs, often in adverse weather conditions. Not only was this heavy-duty labor, it was also dangerous, for the narrow, unsteady ramps from boat to shore soon became slick with mud or river water. Falling into deep water could often be a death sentence.

The most dangerous aspect of a fireman's life, however, was the omnipresent threat of explosion. Those working closest to the boilers were either killed outright or badly scalded in case of explosion, and the anonymous designation "fireman" was often seen on the scrolls of the dead after such disasters.

The *Republican Compiler,* on May 14, 1830, reprinted a typically tragic note from the Baltimore *American*:

Explosion of the Steamboat Huntress

The Cincinnati Gazette states, that a Steamboat accident occurred on the 4th ultimo, at a place on the Ohio river about 14 miles above Smithland.—The Steamboat Huntress had put to shore to leave a passenger, and care was not taken to let a sufficient quantity of steam escape to secure the safety of the engine; and as the boat put off from the shore the explosion took place. Three persons were killed—one engineer, one fireman and the Cook; two other hands on the boat jumped over board, though very badly scalded; no other serious injury sustained.

Another such brief notice came from the Hagerstown *Mail* of November 27, 1840:

Steamboat Accident.—The New Orleans Bee says the steamboat Persian exploded her boilers on the night of the 7th instant, at about three miles from Napoleon. The first engineer, second mate, two firemen and seven deck passengers were killed; twenty-four deck passengers were badly scalded, and four missing.

Transferred Hotel Staff: The Cabin Crew

"Floating Palaces" were equated to the finest hotels, and the level of service passengers received on steamboats was similar. Staff included cooks, stewards, waiters, cabin boys and chambermaids, as well as barbers and bartenders. Personal attendants were also available for a price.

Although not officers, first cooks and first stewards received a salary similar to second engineers and second mates. Wages paid to the rest of the cabin crew were less than what a deckhand received, although the cabin crew had better working conditions. Rousters slept on the main deck alongside the deck passengers, using boxes or crates for beds, whereas cabin boys and waiters generally slept in the saloon on carpeted floors, along with servants brought aboard by passengers. Meals were provided, but the cabin crew usually had the benefit of eating leftover fare from breakfast, supper and dinner served in the main cabin.

Additionally, the cabin crew had a somewhat more stable working arrangement than the deckhands, being hired by the season rather than the trip. Well into the middle decades of steamboating, cabin staff were native-born white Americans, with a diverse age and gender assortment of older men, boys and some women.

The Slave Question

Not routinely chronicled in steamboat lore, slaves were an integral part of a boat's crew. Owners early on discovered that they could loan out a slave for the season and earn a fee for services rendered. This became an important source of income, and with the ready acceptance of Negroes working on boats plying the southern trades, their numbers swelled. In fact, many captains preferred blacks for what they saw as their "docile" temperament over the "quick-tempered Irish" or the "slow Germans," and found them less expensive than a free emigrant. By the turn of the century, slaves were a principal source of labor on boats operating the Lower Mississippi and the New Orleans-to-Memphis trades, as far north as Louisville or St. Louis.[2]

Captains taking their boat above those cities often found it advisable to leave their slaves at Louisville or Covington to be held until their return (*Beverly v. Captain and Owner*, Louisiana Annual Report), In 1850, there were no records of free Negroes working the boats, but on 56 boats, a total of 221 slaves were listed, about one-eighth of the total crew.[3]

Slaves were seldom used on Upper Mississippi boats, although occasionally a freed black was found employed as a fireman or, more commonly, as part of the cabin personnel. Passengers from northern states were far less tolerant of seeing men in bondage, although George Merrick points out that in his experience west of the Mississippi, from Iowa to the Gulf, and even as far east as Illinois, there was no question on slavery. It was "an 'institution,' as much to be observed and venerated as any institution in the country."[4]

Following that perspective, a slave was worth $800–$15,000 and considered too valuable to be used on boats running the Upper Mississippi, where cold weather, strenuous labor or maltreatment could result in serious injury. The Canadian border was also beguilingly close — too close. If a slave escaped, the owner's loss would be complete.

The disadvantages of slave labor, which apparently did not deter their use in some fashion or another, came from the fact that when times were slow and the regular crew discharged while the boat put up, the captain had no such option with his hired Negroes. They were contracted through the season, and their owners had to be paid, regardless of work performed — or in this case, not performed. The boat owner also bore the brunt of reimbursing the owner if a slave were injured or killed. If a white crewmember were lost, it was of no financial concern to the captain, and on older boats especially, where the risk of explosion was greater, the Irish were the preferred firemen.[15]

African Americans held in bondage were also transported by steamboat, occasionally with tragic consequences as two articles relate. The first, from the *Adams Sentinel* of May 13, 1833, reports:

> *The Reaper*—The Louisville Herald of 17th April says: We learn from the Cincinnati Gazette, that the Steam Boat Reaper was sunk in seventy feet of water; that Mr. J.H. Wood, one of the owners, the engineer, pilot, mate and *eight slaves chained together,* lost their lives.

The Hagerstown *Mail*, November 22, 1833 (from a story in the New Orleans *Bee*), gives an account of the catastrophe of the steamboat *St. Martin*, which was destroyed on May 4 after fire spread to some bales of cotton near the furnace:

> Mr. J.F. Miller, who himself escaped most miraculously, offered two thousand dollars to anyone who would save a young slave whom he had with him; but seeing the impossibility of any such an attempt, the flames having progressed to an alarming extent, no one could be found to accept the offer. The slave perished in the river, having thrown himself over board, after extending his hands toward heaven as if in supplication.

Slavery in America was always a troubling problem to those unaccustomed to the "peculiar institution." Frances Trollope wrote in her *Domestic Manners of the Americans:*

> The effect produced upon English people by the sight of slavery in every direction, is very new, and not very agreeable, and it is not the less painfully felt from hearing upon every breeze the mocking words, "All men are born equal." The condition of domestic slaves, however, does not generally appear to be bad; but the ugly feature is, that should it be so, they have no power to change it. I have seen much kind attention bestowed upon the health of slaves; but it is on these occasions impossible to forget, that did this attention fail, a valuable piece of property would be endangered.[6]

She adds:

> There is something in the system of breeding and rearing negroes in the Northern States, for the express purpose of sending them to be sold in the South, that strikes painfully against every feeling of justice, mercy, or common humanity.... The same man who beards his wealthier and more educated neighbour with the bullying boast, "I'm as good as you," turns to his slave, and knocks him down, if the furrow he has ploughed, or the log he has felled, please not this stickler for equality. There is a glaring falsehood on the very surface of such a man's principles that is revolting.[7]

Captain Basil Hall, in his *Travels in North America*, complained bitterly of conditions he found throughout the country. Many Americans took exception to his views and wrote long rebuttals, including this extraordinary paragraph defending the continued existence of slavery.

> There is no topic, as is well known, which has furnished so many sarcasms against the United States, as the existence of a practice so utterly at war with that universal freedom, which their popular institutions are supposed to guarantee. Under the pressure of these reproaches Americans have taken the trouble to trace with great care the history of the rise and progress of this evil, and have established, by the clearest evidence, that it was placed there against the earnest remonstrances of the colonists — that it was fixed on us at a period when we formed a component part of the British Empire, and that the earliest efforts of the States, so soon as they became independent, was directed to mitigate, and in some of them actually to extirpate it.[8]

Charles Dickens put the matter more honestly and succinctly:

> Slavery is not a whit more endurable because some hearts are to be found which can partially resist its hardening influence.[9]

The Great Southern Fear: Servile Insurrection

Utilizing slave labor aboard steamboats presented one major obstacle: the fear that slaves might escape. In such a case, the boat's owner would be liable for the loss to the slave's owner. In 1841, the Ohio Supreme Court declared that a slave was free from the moment he touched the soil of that state with the consent of the master: "If a slave, hired on board a steam-boat should escape, the master or owner of such a steam boat may be made responsible for his value."[10]

Questions arose as how best to prevent trouble, and various actions were taken. In 1857, a case came before the law courts in Kentucky concerning the risk of losing slaves in the ports of free states. At issue was the question of whether an employer had the right to iron (chain) or otherwise confine slaves during that period the vessel was docked. The jurists decided, "If that should become common, the practice of hiring slaves on steamboats would be at an end" (*Meekin v. Thomas*).

It does not appear, however, that escaping slaves proved a pressing issue, and when some owners adopted the policy of permitting slaves to keep their Sunday wages, work on a steamboat appealed more to them than plantation labor.

John A. Murel and His Mystic Clan

Whether "docile," or living in a relatively "favorable" domestic condition, whites lived in constant fear of slave uprisings. Perhaps one singular individual best describes conditions of the times. John A. Murel (also spelled "Murrell" and "Murrel," or by the corruption, "Merel"), known to contemporaries as the "Great Land Pirate," was an outlaw of considerable repute.

Born in Middle Tennessee, and described as a "d — d likely fellow, tall and well proportioned," he dressed "in the Methodist order," perhaps because his father was a Methodist preacher, but more likely because he was often said to preach "d — d fine sermons" while his cohorts stole his parishioner's horses.

Much was written of Murrell's exploits, primarily in the yellow-jacket genre (popular press) such as an 1835 pamphlet, the title page of which proclaimed: "A History of the Detection, Conviction, Life and Designs of John A. Murel, Together With His System of Villany, and Plan of Exciting a Negro Rebellion. Also a Catalogue of the Names of Four Hundred and fifty five of His Mystic Clan Fellows and Followers, and a Statement of Their efforts for the destruction of Virgil A. Stewart, the young man who detected him, to which is added a biographical Sketch of V.A. Stewart," by Augustus Q. Walton, Esq.

Believed to be in some, or all parts, fictitious, the assertions nevertheless captivated an eager audience and made infamous both the villain and his alleged deeds. Subsequently, the *National Police Gazette* ran a multipart article entitled, "The Life and Adventures of John A. Murrell, the Great Western Land Pirate." The series began on September 12, 1846, and ran every week until April 24, 1847. A book by the same editors was published in 1847.[11]

It was, however, Mark Twain, writing in *Life on the Mississippi* (1883) who lent credence to the legend:

> There is a tradition that Island 37 was one of the principal abiding places of that once celebrated "Murel's Gang." This was a colossal combination of robbers, horse-thieves, Negro stealers, and counterfeiters, engaged in business along the river some fifty or sixty years ago.

Noting that cheap histories of Murel were available for sale by train boys at the time (from which he took his own account; note the spelling of the name), Twain dismisses any comparison between the land pirate and Jesse James, noting:

Murel was his equal in boldness, in pluck; in rapacity; in cruelty, brutality, heartlessness, treachery, and in general and comprehensive vileness and shamelessness.

The Walton account makes fascinating reading nearly 200 years after the fact. The author described in minute detail, down to quoted conversations, the bandit's wicked acts. In an early exploit, Murel booked passage on a steamboat for Natchez with a stolen slave. The captain, when told by an informer that Murel had contraband, confiscated the property in hopes of a reward. Murel, prepared for such a contingency, produced false papers claiming ownership. To exact revenge, the informer "found his way to the bottom of the Mississippi, and his guts made into fish bait."

The same fate awaited the hapless slave. After being sold and escaping several times (a well-tried scheme to gain money at the expense of the black man, who was promised freedom for his part), the fellow was "taken out of the reach of all pursuers" by being drowned in the river, "where his carcass has fed many a tortoise and catfish."

Developing a devoted following, called the "Mystic Conspiracy," Murel's band was divided into two groups: the grand council, who designed and perfected plans, and the second, the "strikers," who knew nothing more than what immediately concerned them, and carried out the villainy. Murel boasted 400 in the grand council and nearly 650 strikers.

Murel's grand scheme, as Stewart explained, was to "excite a rebellion among the Negroes throughout the slave holding states." Setting December 25, 1835, as the date, he planned on having an army of slaves stand up and rebel against their masters. While they "commenced their carnage and slaughter," Murel's men would rob banks. The principal target, although by no means his only one, was New Orleans.

Fortunately for the country, young master Stewart turned Murel in to the authorities. He was tried and convicted of the lesser crime of horse stealing, for which he received a ten-year sentence.

Stewart's tale, along with the rest of the pamphlet, caused an uproar, not so much over the heinous crimes of horse stealing, counterfeiting and murder, but over a much graver issue to southern men: servile insurrection. Terrified by the threat they perceived in *The Life and Adventures of John A. Murel,* numerous lynch parties formed throughout the South. In the summer of 1835, "vigilance committees were organized where the Negro population was most numerous, and where, of consequence, the slaveholding class was more sensitive to the cries of alarm ... the male population patrolled the surrounding country. The committees of safety sat daily, and some persons suspected of abetting the alleged insurrection were brought before it, while others, 'whose guilt seemed to be fully established, were hung without ceremony along the roadsides or in front of their own dwellings.'"

A contemporary account states:

> There is no doubt that the State at this time was overrun with highway robbers, Negro-stealers, and Black legs, of which organization Murel was a member; that some abnormal people were impelled to foment insurrection by the doctrines of abolition, as others have from time to time been impelled to assassination by political and religious doctrines.

Speaking before the Young Men's Lyceum of Springfield, Illinois, on January 27, 1838, Abraham Lincoln detailed what had happened in Mississippi:

> ... this process of hanging [went on] from gamblers to Negroes, from Negroes to white citizens, and from these to strangers, till dead men were seen literally dangling from the boughs of trees upon every roadside; and in numbers almost sufficient, to rival the native Spanish moss of the country, as a drapery of the forest.[12]

Thus, the Great Land Pirate caused the havoc he anticipated, when he predicted:

> I will have the pleasure and honor of seeing and knowing that my management has gutted the earth with more human gore, and destroyed more property than any other robber who has ever lived in America, or the known world.

Only not quite in the way he imagined.

Living and Working Conditions of the Deck Crew

During the development of the steamboat structure, much care and consideration was given to handling and storage of freight, passenger accommodations and proper quarters for officers. Lost in these concerns was any sort of living space for the deckhands. It thus evolved that the crew were to be housed in whatever area was not taken up by deck passengers in the cargo area. On some boats, tiers of bunks were constructed, but in general practice, the hands ate, lounged and slept wherever they found room on the main deck.

No utensils were furnished, because "in all cases," the men stole them when leaving the boat, and some were "so filthy in their habits, that to furnish them a bed was out of the question."[13] Meals were served in tin or iron pots, and the deckhands used a piece of hardtack or a shingle as a makeshift spoon.

In his book *Old Times*, George Merrick describes the scene of deckhands eating from the leftovers of the well-fed passengers:

> Each steamer carried forty or more hands or "rousters." For them, the broken meat was piled into pans, all sorts in each pan, the broken bread and cake into other pans, and jellies and custards into still others — just three assortments, and this, with plenty of boiled potatoes, constituted the fare of the crew below decks. One minute after the cry of "Grub-pile!" one might witness the spectacle of forty men sitting on the bare deck, clawing into the various pans to get hold of the fragments of meat or cake which each man's taste particularly fancied.[14]

Another author added that German deckhands would take bad food with some complaint but little protest; the Negroes were compelled to suffer in silence; and the Irish would speak out vigorously, on occasion leaving the boat in protest.

A Heavy Hand: Discipline Aboard Steamboats

Officers were accorded respect and well-paying jobs, but the deckhand received neither. Nor was his life aboard made easier by working conditions, particularly when discipline was involved. Although commonly asserted that the German roustabout could take severe punishment and the Negro much more, and the Irishman could not stand a blow from the mate, treatment was meted out with a heavy hand to all.

While it might have been argued that early, rowdy deck crews — formerly rowdy keelboatmen — were used to rough treatment and required severe guidance, as the era progressed and more and more American immigrants, farmer boys and foreigners entered the service, they represented a different class of worker. The methods of discipline did not seem to lessen, however, and beatings, clubbings and even knife wounds were inflicted by the mate, often to maintain or assert his own authority.

Of all steamboat officers, the mate had the most violent reputation. When unknown bodies were found face down in the river ("floaters"), and death came from violence, they were "supposed to be deck hands."[15] This did not mean the mate's control over life and death was absolute, however, for steamboats made many stops along the way and redress to local officers might be made. Victory was far from certain, however, for in a rare instance where a deckhand pressed charges for injury, the fellow-servant doctrine was used by the plaintiff. A federal court judge ruled that "steamboats are not liable for injuries received by a workman, though acting under the legal instruction of a superior officer."[16]

Considering the brutal treatment of deck hands, George Merrick wrote:

> It may be and was asked by Eastern people, unused to river life, "Why do men submit to such treatment? Why do they not throw the mate into the river?" The answer is caste. They were used to being driven, and expected nothing else, and nothing better, and they would not work under any other form of authority. As I stated at the beginning, they were of the very lowest class. No self-respecting man would ship as a deck hand under the then existing conditions.[17]

A hard life, indeed, which garnered little sympathy. And likely asked for none.

Shore Leave

Due to the strenuous labor associated with a deckhand's work aboard a steamboat, two trips "on end" was as much as the regular man could tolerate, and he was then required to take time off to recover.

There were many entertainments available for a rouster with money in his pocket. The saloons served whiskey and mixed drinks, while red light districts offered what was slyly called "horizontal refreshment." Gambling and cockfights drew crowds, and theaters in river towns were supported chiefly by "gentlemen of the navigation." Cheap boardinghouses lined the wharves, and if a man did not care to spend the money, he often slept outdoors, in or around the wharves.

Mark Twain described a "fashion-freak" sport in New Orleans, so called because all the people associated with mule racing were "people of fashion." Amid cheering crowds, thirteen mules set off as one after several false starts. The animals would then sprint away, each mule and each rider having conflicting opinions on how the race should be won. The mile heat was run in 2:22, with eight of the thirteen mules ending up somewhere other than the finish line.[18]

For those less disposed to pleasures of the flesh, missionaries and religious groups proliferated around the waterfronts. They distributed pamphlets such as "The Boatswain's Mate," "The Drunkard's Looking Glass," and "Contemplations on Eternity." Many of these could not have come as a surprise to rivermen, for it was a common practice among steamboat clerks to carry a large assortment of religious tracts to distribute along the way. The crews of "small-fry rascals" (skiffs and rafts) would call out to the boat crew, "Gimmee a pa-a-per!" and the clerk would toss over a file of New Orleans journals. Next went over neat bundles of religious tracts, tied to shingles.

> The amount of hard swearing which twelve packages of religious literature will command when impartially divided up among twelve rivermen's crews, who have pulled a heavy skiff two miles on a hot day to get to them, is simply incredible.[19]

The Sabbatarian Movement urged captains not to put out on Sundays or load cargo on the Sabbath, and often preachers could be heard delivering open-air sermons to any willing to listen.

A prime example of the religious fervor exhibited by some comes from the *Adams Sentinel,* November 28, 1836:

> Five Thousand Sabbath Breakers.— On Sabbath, the 23rd of October, nine steamboats left Buffalo for the west, carrying about four thousand passengers.— The owners of the boats, the hands that navigate them, and the various attendants at the different ports, making a total of 5000 Sabbath breakers. As much as we rejoice at the increase of business and prosperity in our city, we confess we have some fearful forebodings when we see the law of God thus disregarded and the institutions of religion trampled on.— "Righteousness exalteth a nation, but sin is a reproach to any people." *Buff. Spec.*

Orren Smith was one of those captains who believed in keeping the Sabbath, and he never ran a minute after midnight Saturday. He would tie up his boat and rest until 12:00 the following night, when he would resume his route. Smith often invited traveling ministers aboard to read the service; if none were available, he would read it himself. "But the captain's reverence and caution did not save his boat, and she sank below La Crosse in the autumn of 1854."[20]

Bethel societies constructed special chapels especially for those working the river, but these were sparsely attended, as the boatmen showed reluctance to "exchange the songs of obscenity for the songs of grace."[21]

Little or no complaint was raised about improving temporal or religious life aboard a boat, however (other than from physicians and newspaper editors), and perhaps it was just as well. George Merrick noted:

> The mention of St. Louis and St. Paul lent the only devotional tinge to steamboat conversation in the fifties. Without this, there would have been nothing religious about that eight hundred miles of Western water.[22]

The same was likely true for the 1,000 miles below those upper river ports.

It was not until the Act of 1871 that the lives and working conditions of steamboat crews were addressed.

15

Steamboat Gothic: The Florid and the Ornate

The American steamboat began life as a "floating teakettle," and no one had anything positive to say about the *Clermont*'s lines, her grace or her beauty. She was designed as a functional vessel propelled by steam power, and although she had cabins for passengers who braved the new mode of public transportation, it is doubtful anyone marveled at her external appearance or noted in their journals about the luxurious accommodations.

When Fulton and Livingston brought their model west and built the new vessels for their line running from New Orleans to Natchez, the pair had their eye on passenger comfort, but as the service developed into the dual freight-passenger style, more thought was given to the stowing of cargo than people. Aesthetics were left aside, as designers concentrated on placing the redesigned engines and creating more room for cargo. It took twenty years for a standard configuration to develop.

The second of Fulton-Livingston's boats was the *Vesuvius* (1814). The ladies' cabin was placed in the hull with the engines and the cargo; the gentleman's quarters was built on deck and "elegantly fitted up as a cabin." Two years later, Henry Shreve's steamboat *Washington* (1816) had the entire passenger cabin built within the hull, taking up nearly all of one deck. The women were placed in the stern. The captain's cabin and the bar were forward; between bow and stern was the saloon, with berths lining either side for the men.

The men's area was separated from the main saloon by curtains or room dividers. The bunks that lined the walls in double tiers were little more than removable shelving, taken down in the morning to enlarge the main cabin. If the bunks were permanent, the curtains, often of a fine, attractive material, served to screen them from view, and provided some privacy while dressing. This section was uncarpeted and women were allowed there only after it had been reconverted into the dining area.

As passenger traffic became more intense, the steamboat design expanded and a second level (upper boiler deck) was added exclusively for their use. The women were placed toward the rear, with the forward section given to deck passengers. Men's sleeping quarters continued aft on the main deck. Eventually, the men's cabin was moved to the boiler deck, forward. The setup, which became the normal arrangement, thus solidified: gentleman's staterooms forward by the bar; an open saloon used for dining at mealtimes and games, lounging and entertainment during the rest of the day and evening; women's cabins at the rear. When private staterooms became more common, curtains and room dividers were no longer necessary.

It was felt that a stateroom for women at the back offered a safer position, farthest from the engines and boilers. However, the placement subjected them to the most intense vibration from the paddlewheels and also constant noise. Married men traveling with their wives stayed in the women's section in private staterooms; no unattached men were ever allowed there. If one thought to visit a lady and was caught, he was unceremoniously escorted out and often tried by the boat's jury as a morals offender. Punishment usually involved a rapid departure off the boat.

The design of the second story, and the subsequent creation of the gallery, or veranda, that ran nearly the complete way around the cabin (except where it met the wheelhouse), matched the guard on the deck below, and served as a promenade. Deck passengers were relegated to the lower or main deck, and thus kept separated from the quality.

Better Than a Five-Star Hotel

Frances Trollope described her accommodations aboard an early steamboat:

> We embarked on board the Belvidere, a large and handsome boat; though not the largest or handsomest of the many which displayed themselves along the wharves.... We found the room destined for use of the ladies dismal enough, as its only windows were below the stern gallery; but both this and the gentleman's cabin were handsomely fitted up, and the former well carpeted.[1]

Charles Dickens, upon first entering his cabin aboard the steam-packet *Britannia* on January 3, 1842, was so shocked at the cramped, "not as advertised" cabin, he wrote of the hope that

> this room of state, in short, could be anything but a pleasant fiction and cheerful jest of the captain's, invented and put in practice for the better relish and enjoyment of the real state-room presently to be disclosed.

It was not, however, and he determined that

> deducting the two berths, one above the other, than which nothing smaller for sleeping in was ever made except coffins, it was no bigger than one of those hackney cabriolets which have the door behind, and shoot their fares out, like sacks of coal, upon the pavement.

Alas for great expectations. But it was no worse than the saloon:

> Before descending into the bowels of the ship, we had passed from the deck into a long narrow apartment, not unlike a gigantic hearse with windows in the sides; having at the upper end a melancholy stove ... while on either side, extending down its whole dreary length, was a long, long table, over each of which a rack, fixed to the low roof, and stuck full of drinking-glasses and cruet-stands, hinted dismally at rolling seas and heavy weather.[2]

Upper Mississippi steamboats lagged somewhat behind their gaudier cousins to the south, and it was not until the 1840s that speed and beauty became as important as capacity. Adaptations were slow to catch on, however, for they were not as necessary in the local trades as for those running the more lucrative Ohio River and the New Orleans routes

In 1834, a passenger noted that of the western puffing high-pressure steamboat could be heard from miles ahead.

> The cabin was plainly but substantially furnished, and kept very clean. There were no state rooms; but two tiers of bunks, containing the beds, ran along the side of the boat and were separated at night from the saloon by curtains. The fare was substantial, plentiful, and good, and the officers were pleasant and gentlemanly.[3]

Four years later, conditions remained basically the same. Rooms were spacious, and included a washstand and other necessities of the toilet. In 1839, Joseph Throckmorton, captain of the *Malta* (a boat valued at $18,000) added spring mattresses for the comfort of his passengers, but provided no heat, which presumably kept many in their bunks.

The Malta

Built: 1839 in Pittsburg
Length: 140 feet
Tonnage: 114
Beam: 22 feet
Hold: 5 feet
Equipment: double engine with a fire pump and hose

In 1845, Robert A. Reilly, captain of the *Wiota*, relocated the stairs running each side of the boat upward from the boiler deck to the front. This made ease of travel on the vessel more practical, and was soon adopted by others.

The Wiota

Built: 1845 in Elizabethtown, Pennsylvania
Owners: Robert A. Reilly, Henry Corwith, William Hempstead of Galena
Tonnage: 219
Length: 170 feet
Beam: 25 feet, 6 inches
Hold: 5 feet, 3 inches
Cylinders: 18 inches in diameter
Stroke: 7 feet
Wheels: 22 feet in diameter with 10-foot buckets

In 1847, a traveler wrote of the *Red Wing*:

> The walls, the ceiling, the beds, are all uniformly painted white. Even in the ladies' salon there are none of the chandeliers, the lamp globes, the gilded scrolls and arabesques, the pianos, the sofas, and the couches which made the Lakes steamships so pleasant. A few red tables and yellow chairs, that is the total — except that in the ladies' salon there are some of those rocking-chairs which seem to be a sort of *sine qua non* of feminine existence everywhere in America.[4]

The Red Wing

Built: 1846 in Cincinnati
Tonnage: 142 tons
Length: 147 feet
Beam: 24 feet
Hold: 4¼ feet high

The *War Eagle* (the Second) boasted that

> the cabins are furnished, with just enough of the gilt work to give them a cheerful appearance. All the modern steamboat improvements have been attached, and the barber shops, wash room, &c. are on a liberal scale. She will draw but 24 inches of water, light [unloaded], and only five feet when loaded to the guards. The carpets, of the finest velvet, are from Shilito & Co.'s; the furniture from S.J. Johns; mirrors from Wiswell's; and the machinery by David Griffey.[5]

The War Eagle *(the Second)*

Built: 1854 in Cincinnati
Cost: $33,000

Passenger cabins continued to improve, with the introduction of featherbeds, wardrobes and the ever-present American rocking chair. By the 1850s, a typical stateroom was usually about six feet square, a luxury far surpassing anything encountered while traveling in a cramped stagecoach or sleeping in a wayside hotel.

The *Messenger,* a packet running the Pittsburg-to-Cincinnati line, carrying forty cabin and numerous deck passengers, was typical of its day. A married man and his wife shared a tiny stateroom in the ladies' cabin, which contained two berths. It opened out onto the main saloon. A second door, this one made of glass, provided access onto the veranda.

As was typical of steamboats, the *Messenger* was advertised to start every day for a fortnight, and did not leave until the captain had a profitable cargo.

Charles Dickens describes the boat, which he had previously noted lacked any mast, cordage, tackle or rigging, commonly found on ships.

> There is no visible deck, even; nothing but a long, black, ugly roof covered with burnt-out feathery sparks; above which tower two iron chimneys, and a hoarse escape valve, and a glass steerage-house. Then, in order as the eye descends towards the water, are the sides, and doors, and windows of the state-rooms, jumbled as oddly together as though they formed a small street, built by the varying tastes of a dozen men; the whole is supported on beams and pillars resting on a dirty barge, but a few inches above the water's edge; and in the narrow space between this upper structure and this barge's deck, are the furnace fires and machinery, open at the sides to every wind that blows, and every storm of rain it drives along its path.[6]

English travelers to the United States were taken with the majesty of the steamboat in varying degrees. Frances Trollope considered them greatly superior to those she had seen in Europe, but added, "The fabrics which I think they most resemble in appearance, are the floating baths (*les bains Vigier*) at Paris." Dickens (well familiar with Trollope's

15. Steamboat Gothic   181

The side-wheeler *Alton* in a post-bellum image. Note the decoration between the two chimneys, which was typical of the later era.

writing) concurred, likening them to huge floating baths. He adds that the primary difference is that the American boats have so much out of water, with decks filled with cargo.

Americans, for their part, could hardly believe that cabin passengers could be transported for so long a journey, provided with a stateroom and fed so well at a price less than one would expect to pay at a luxury hotel. In fact, that was a primary selling point of steamboat travel: not only were passengers treated to all the accommodations and foodstuffs normally the exclusive realm of the well-to-do, but they were also treated to the extraordinary sightseeing opportunities lacking in a stay at a five-star hotel.

The "Salon" or "Saloon": Steamboat Accommodations

While staterooms and amenities continued to improve and be expanded upon, the primary source of wonder came from the main cabin, or saloon, running the entire length of the boat. It was in this chamber that steamboat owners expended the most money and consequently derived the most pride.

Saloons gradually expanded from a forty-two-foot length (1816). By 1852, a length of 300-feet became the average. The overhead height of the cabin increased from six to between ten or fourteen feet in late antebellum boats. This forced designers to raise the ceiling beyond the hurricane deck. Glass windows and a profusion of skylights and stained glass admitted dazzling shafts of light. The overall effect, while stunning, was mitigated somewhat by the narrow width (typically no wider than twenty feet), occasioned by the construction of the sleeping chambers to either side, the need to maintain a narrow hull for speed and to keep the weight of the upper structure within the perimeter of the hull for stability.[7]

Within this enclosed world, passengers spent most of their time, and it is the gilded saloon that truly gave western steamboats their reputation. Nothing was too ostentatious, nothing too overdone. Amid a background of white paneling rose mighty columns adorned with scrollwork. They reached to the ceiling where huge arches crisscrossed one another. Chairs of red plush lined the walls, gold was painted on wainscoting, and heavy curtains hung beside mighty tapestries.

After meals when the long table was cleared, music was often performed. Early steamboats utilized the talents of crewmembers playing fiddles. As music proved a great inducement to passengers, larger boats employed small orchestras for dancing and listening pleasure.

American Gothic, Gingerbread and Gilt

When the structure of steamboats began to emerge from the simple "open engine on a keelboat" style, more attention was paid to appearance, as owners discovered that looks played a huge role in prestige, whether in gaining the most favorable freight or in luring the better-paying travelers.

The exterior of the steamboat throughout the era was painted white with few deviations. Towering overhead were the coal black chimneys and the black painted ironwork. More adventurous owners had the trim painted blue, but the majority of color came from the gilt of the finials (ornamental projections or ends), and the decorations of the paddle-wheel housings. These areas gave rise to artistic displays of bold and fancy lettering and scenes depicting anything from a representation of the boat's name to pastoral settings, patriotic battle scenes and biblical figures. By the 1850s, spreader bars stretching across the chimneys were formed into elaborately petaled flowers, and later, always with an eye toward making a dollar, advertisements were displayed in the most visible locations.

Native tastes ran toward the extreme, and the preponderance of gingerbread (described as tawdry, gaudy or superfluous ornament), jigsaw and pseudo–Gothic (pointed arches, ribbed vaults, piers and buttresses) quickly dominated the exterior and interior of the steamboat. United States travelers loved their uniquely American vessels and as the era developed, owners reached for more and more lofty heights of decoration. The result was a profuse combination of styles, techniques and tastes, with an accent on bright light, stained glass windows and red plush.

Mark Twain, cub pilot, viewed his first big New Orleans luxury steamboat with wonder:

> She was as clean and as dainty as a drawing room, when I looked down her long, gilded saloon, it was like giving through a splendid tunnel; she had an oil-painting, by some gifted sign-painter, on every state-room door; she glittered with no end of prism-fringed chandeliers; the clerk's office was elegant, the bar was marvelous; and the barkeeper had been barbered and upholstered at incredible cost. The boiler deck (i.e., the second story of the boat, so to speak), was as spacious as a church, it seemed to me; so with the forecastle; and there was no pitiful handful of deck-hands, firemen, and roustabouts down there, but a whole battalion of men. The fires were fiercely glaring from a long row of furnaces, and over them were eight huge boilers! This was unutterable pomp.[8]

The western steamboat became the symbol of everything that was beautiful and elegant, and the standard by which hotels, spas and even mansions were judged. Not infrequently, someone was likely to observe, "Why, this is almost as pretty as a steamboat!" and the phrase gained widespread acceptance both east and west. Beauty came at a price. The Madison *Express,* September 19, 1840, notes:

> The boats on the Lake are both commodious, rapid and beautiful, $4000 being paid sometimes for painting and ornamenting a single steam packet.

Planned Obsolescence

One of the most fascinating and least-discussed aspects of the steamboat era was the ready acknowledgment by owners and builders that the great floating palaces were constructed for the moment rather than the long haul. This fell into two categories: the hull and the interior decoration.

The hull was subjected to incredible wear and tear from the constant striking of submerged logs, running into sandbars or merely worn down from the silt-laden current. Constant upkeep, including expensive repairs, was required to maintain the integrity of the wood. When the outside was penetrated by snags or other obstructions, repairs were made on the river, but often survived no longer than passage to the next boatyard. A rent that proved too costly to fix meant scrapping the boat. In these instances, age was

An early sketch of a "floating palace" from 1850. It is interesting because of the early date but is not typical of steamboats. It is unclear from the illustration where the paddle-wheels are placed. The illustration is credited to "Spalding & Rogers." As fancy as brag boats were, they were one of the original American examples of "planned obsolescence."

no factor: even boats on their maiden voyage were subjected to the dangers of the river and many instances are chronicled describing such tragedies.

Worse was the wear and tear on steamboat engines and boilers, which operated twenty-four hours a day for the length of the trip. Intense heat and stress quickly eroded working parts. Valves stuck, steam pipes burst, boiler heads became dangerously worn and mud clogged everything. Even frequent cleaning and maintenance did not significantly prolong the usefulness of the machinery.

These factors took a toll on the boat, making its life expectancy fewer than five years. After that time it became dangerous to operate; many were sold as scrap and the rest passed from owner to owner, working their way down the line and thereby lessening their usefulness and reputation. The same held true for the working parts. If a boat were sunk, there was little point salvaging the engines or boilers because exposure to water quickly corroded the metal. This is not to say engines and, to a lesser degree, boilers, were not hauled up and reinstalled in other boats, but this was a dangerous proposition. Many such recycled components caused explosions, destroying not only the vessel but passengers' lives, as well.

The principal object of planned obsolescence, however, was the superstructure: the upper decks. The decks, their columns, railings and guards were built of light, unsubstantial wood. One primary reason for this was weight. In order to maintain balance and allow the boat not only to float but to be capable of speed, a builder would take any opportunity he could to save on weight. While the outside of a boat appeared sturdy and substantial, that was more a trick of the eye than reality.

An article from the Milwaukee *Sentinel,* July 23, 1839, remarks on the flimsy construction when reporting on a tragic accident:

> *Steamboat Collision — Loss of Life* — Mississippi boats are frail things. The steamer Danube ran into the Macferland on the night of the 16th inst near Helena and sank the latter in five minutes. So slightly are the upper works of the river boats constructed that the concussion broke loose the cabin of the Macferland, which floated off! There was 110 passengers on board, six of whom were drowned. The loss were deck passengers except one, the cabin boy. The boat was from New Orleans freighted for Cincinnati.

The gaudy, much heralded decoration of a boat was equally subject to the stresses of time. Exterior paint peeled and faded — even numerous reapplications failed to restore the glitter of a new boat. Inside, the expensive Oriental and Brussels carpets were quickly worn by innumerable feet passing over them, to say nothing of the great American sport of spitting, a complaint of numerous foreign visitors. Frances Trollope perhaps best described their feelings:

> But oh! that carpet! I will not, I may not describe its condition; indeed it requires the pen of a Swift to do it justice. Let no one who wishes to receive agreeable impressions of American manners, commence their travels in a Mississippi steamboat; for myself, it is with all sincerity I declare, that I would infinitely prefer sharing the apartment of a party of well conditioned pigs to the being confined to its cabin. I hardly know any annoyance so deeply repugnant to English feelings, as the incessant, remorseless spitting of Americans.[9]

There were other causes of wear and tear, including the national pastime of whittling. Since steamboat passengers reflected a remarkable conglomeration of people, from princes to sod farmers — anyone, in fact, who had the price of a first-class fare — individuals from all walks of life congregated together in the main cabin. Among them were frontiersmen, whose habit of plying the knife blade to anything that fell within reach resulted in the mutilation of chair handles, sofa legs, wooden lamp bases and tables. Two or three years of rough treatment rendered them unattractive at best and valueless at worst.

Gail Hamilton writes in *Wool-Gathering* (Boston, 1867) that steamboats had:

> an indefinable sham splendor all around, half disgusting and wholly comical. The paint and gilding, the velvet and Brussels, the plate and the attendants show bravely by lamplight, but the honest indignant sun puts all the dirty magnificence to shame.

Pieces of matching china were broken, crystal chipped, silverware bent, and gilt, the most luxurious and prestigious decoration, worn away. Constant use flattened the cushions of sofas, original and reproduction artwork suffered from exposure to the elements and the haze of cigar smoke, floorboards warped, walls of thin plaster buckled, and linens became stained. In the sleeping quarters, mattresses sagged, pillows absorbed noxious odors, and footboards were marked by gentlemen's shoes and boots, despite admonitions that such be taken off before lying down.

It did not take many months before a floating palace became a haggard scow. Passengers seeking the ultimate in travel quickly learned which boats to shun, and once-famous vessels were soon relegated to the auction block, or dismantled and sold piecemeal.

All this was recognized by owners and accepted as a hard fact of the business. While furniture could be replaced and new dinnerware purchased, there was little point. Not only would that require a tremendous expenditure of cash, such outward improvements did nothing for the structure of the vessel itself. It was therefore deemed more prudent to abandon the boat and purchase another, setting in motion the continual reemergence of newer and sleeker crafts.

16

"My satisfaction was complete": First-Class Passengers

And when I found that the regiment of natty servants respectfully "sir'd" me, my satisfaction was complete.[1]

Mark Twain's delight at being "sir'd" was not the exclusive realm of cub pilots, but actually represented the overall attitude of the crew toward the first-class passengers. Paying cabin fare on a steamboat meant moving into the upper echelons of society (however temporarily). Captains tipped their hats, rousters bowed, servants scrambled for bags, chambermaids made the beds and cooks prepared meals with ingredients most westerners probably had never tasted.

A trip on a steamboat, regardless of the destination, was a holiday in itself, and in fact, with fares seasonally lowered by competition, tourists had a chance to see the country at a price cheaper than staying at a first-class hotel. Americans loved to travel, and almost as soon as steamboats set out upon the inland waterways, passengers eagerly booked sightseeing tours.

The "Fashionable Tour"

Like Robert Fulton, George Catlin was born in Pennsylvania. And like the great inventor, Catlin studied art as a boy. In 1823, at the age of twenty-seven, he moved to Philadelphia, Pennsylvania, to practice law. Later, Catlin found great success in portrait painting and miniatures, eventually producing likenesses of such influential people as De Witt Clinton and Dolly Madison. It was the inspiration of a visiting delegation of Native Americans, however, which changed the course of his life.

Leaving eastern civilization behind, Catlin went west to St. Louis, determined to "use my art and so much of the labors of my future life as might be required in rescuing from oblivion the looks and customs of the vanishing races of native man in America.[2]

Catlin met William Clark, a native Virginian who had already gained a respect for the Indians during his military service in Ohio and Indiana. Clark had furthered his knowledge and reputation among the Native Americans by traveling through the Rocky Mountains with Meriwether Lewis as a co-leader of the government-sponsored Corps of Discovery Expedition in 1803. They returned to St. Louis on September 23, 1806, with massive amounts of zoological and botanical specimens, as well as having accumulated numerous stories, facts and legends of the Native Americans.

President Thomas Jefferson appointed Clark a brigadier general of military affairs and superintendent of Indian Affairs for the Louisiana Territory. He fought against abolishing the Factory System. In 1813, he received the appointment as governor of the newly created Missouri Territory and maintained that position until 1821, when the area was granted statehood. Clark ran for governor of Missouri and was defeated; a year later, in 1822, he lost another important fight when the Factory System was abolished.

Clark continued as superintendent of Indian Affairs and brokered many successful treaties, exerting his influence to prevent the Sioux and other tribes from joining the British in the War of 1812. Clark, known as the "Red-haired Chief," was an important and respected man among the Indians, and his meeting with George Catlin helped lead to the artist's acceptance as the pair traveled up the Mississippi on a diplomatic mission.

Using St. Louis as his base, Catlin took five trips upriver between 1830 and 1836, visiting over fifty tribes. In 1838, he made a voyage up the Missouri to Fort Union, where he sketched and painted eighteen tribes, including the Pawnee, Omaha, Cheyenne, Crow and Blackfoot. His collection eventually reached 600 works of art, capturing the likenesses of prominent individuals, as well as depicting village life with scenes of games, everyday life and religious ceremonies.

George Catlin was not only taken with the Native Americans, but he developed a life-long appreciation of the rugged country in which they lived. Describing his passage by steamboat, he observed:

> The scenes that are passed between Prairie du Chien and St. Peters, including Lake Pepin, between whose magnificently turreted shores one passes for twenty-two miles, will amply reward the tourist for the time and expense of a visit to them.

Already having earned a reputation from his western artwork, Catlin extolled the wonders of America's new lands and encouraged visitors to come and spend their holidays on the frontier. So successful was he in promoting the trip, that tourists flocked to the Upper Mississippi, taking part in what became known as the "Fashionable Tour." Catlin described his suggested itinerary:

> A trip to St. Louis; thence by steamer to Rock Island, Galena, Dubuque, Prairie du Chien, Lake Pepin, St. Peters, Falls of St. Anthony, back to Prairie du Chien, from thence to Fort Winnebago, Green Bay, Mackinaw, Sault de St. Mary, Detroit, Buffalo, Niagara, and home.[3]

In 1838, George Catlin returned east and toured with an exhibition of his paintings, hung steamboat salon-style, side by side and one over another. He received great acclaim for his work and eventually took the exhibition to London, Brussels and Paris. His talks on the great American wilderness inspired many Europeans, and they journeyed west to travel on steamboats through the scenic countryside.

Steamboat operators were quick to realize the benefits of tourism, promoted as early as 1826 by Captain D. F. Reeder. Reeder took his 120-ton boat, the *Lawrence*, from Fort Snelling (known as Fort St. Anthony until 1824) on a pleasure trip to the Falls of St. Anthony, loaded with local civilians. With a military band playing dance music, the group had a festive time amid the Indians, who came out to view their frolics. In 1823, the Italian explorer Giacomo Constantine Beltrami also extolled the beauty of the scenery, raising excitement for such a trip.

Once the Blackhawk War eliminated (or nearly so) the threat of Indian attack, the Summer Jaunt became a popular destination for locals, easterners and tourists from the Continent. To accommodate them, "excursion" steamboats took on a new look, and from this point, Upper Mississippi boats began to catch up with the frills and luxuries associated with Lower Mississippi boats.

Towns along the route vied for tourist money and newspaper writers often promoted certain boats in which they had an interest. Adverts for trips were published adjacent to articles flattering to captains or boats, boasting of their records, gentlemanly attributes and the splendors of the accommodations.

With the Fashionable Tour in full swing by 1835, steamboats that had previously depended on freight spent considerable time and money upgrading. Private staterooms were added and a new emphasis was placed on speed and novelty. One such boat plying the tourist trade was the *Smelter,* owned by Captain Daniel Smith Harris and his brother, Robert Scribe Harris. Built in 1837, in Cincinnati, the *Smelter* was a side-wheeler and one of the first boats on the Upper Mississippi to be built with a cabin or "boiler deck."

Renowned as one of the fastest and most luxurious of the Upper River steamboats, the *Smelter's* owners decorated the boat with festive evergreens and, when "rounding to at landings, or meeting with other boats, fired a cannon from her prow to announce her imperial presence."[4]

Other "tourist boats" included:

The Pizarro

Built: 1838 in Cincinnati
Tonnage: 107 tons (144 tons burden)
Cost: $16,000
Length: 133 feet
Beam: 20 feet
Owners: Captains D. S. Harris and R. S. Harris
Style: Commodious, with staterooms
Equipment: Fire engine and hose

The Brazil

In Service: 1838–1841
Owner: Captain Orrin Smith
Style: Two berths in each stateroom
History: Served as an expedition boat until 1841; sunk in the upper rapids at Rock Island in 1841; a total loss*

The Brazil *(aka the New Brazil; or Brazil, the Second)*

Built: 1838 in Cincinnati
Tonnage: 166 tons

Captain Robert Scribe Harris, brother of Captain Daniel Smith Harris. Together, they owned some of the fastest steamboats on the Upper Mississippi.

---

*The Cincinnati Star, *April 23, 1840,* notes "that a rumor was rife in that city that the steamboat* Brazil *has blown up somewhere on the lower part of the Ohio."*

Length: 144 feet
Beam: 22 feet
Hold: 5 feet
Style: 30 staterooms
Equipment: 22-foot paddle-wheel, double engine, 3 boilers
Owner: Captain Orrin Smith

The *Adams Sentinel,* July 8, 1839, gives a detailed account of one trip, extolling the enjoyment of the Fashionable Tour:

> *A Delightful Trip.*—The Pittsburgher says:—The steamboat Pennsylvania, under the command of Captain Stephen Stone, arrived at our port a few days since, from a trip to the Falls of St. Anthony. The party who went on board of her, amounted to about 60, and were composed of ladies and gentlemen, a large portion of whom were from this city. They had every thing that could contribute to their enjoyment—good music, good company, plenty of dancing, and fine spirits. On their outward voyage they left St. Louis on the 27th of May and arrived at the Falls on the 2nd June. The Falls are non-accessible by steamboat navigation; and in order to reach them, the company had quite a novel and romantic excursion of seven miles over a beautiful prairie, in wagons supplied by the politeness and hospitality of the commandant at Fort Snelling.
>
> We understand there is a perpendicular fall in the water of about eighteen feet. But there is a smooth rapid for about two miles above the falls, making the whole descent about sixty feet.
>
> The party visited the Indians in the neighborhood of the Falls. They saw them in their wigwams and villages.—The Indians, to the number of several hundred, entertained the company with a dance called the Buffalo dance, in which we imagine there was a good deal more life and animation displayed than in one of our cotillions. When the company was about to take their departure, the Indians, in full costume, commenced the war dance!

Summer Jaunts

Holidays were always a special event aboard steamboats, especially Independence Day. In 1835, the *Warrior* steamed into Fort Snelling, firing the customary twenty-one-gun salute to the waiting passengers, including George Catlin and bands of Sioux and Chippewa. So pleased were the alighting Easterners to see peaceful Native Americans, they eagerly paid the Indians a small stipend to perform dances and play games.

The first known Fourth of July celebration aboard an Upper Mississippi steamboat above Alton was in 1828, when the *Indiana* went down the Fever River. American flags flew everywhere, including from the villages of the Fox Indians. Adding excitement to the celebration of the Nation's Birthday, the *Pizarro* took passengers from Galena to Cassville at a cost of five dollars. The advertisement ran:

> There will be a splendid band of music on board the boat, and arrangements will be made for dancing;—four sets of cotillions can be accommodated. A superb dinner will be served up, with the best lemonade and liquors. Every attention will be paid, to make the day one of comfort—and to celebrate the glorious day in a style worthy of the Far West.[5]

Two leading citizens of Alton, Illinois, offered excitement of a different sort in 1841. The pair chartered the *Eagle,* having her "repaired and fitted up for the occasion." For the price of $1.50 each, passengers were to be brought to Duncan's Island, just below St. Louis, where four Negroes were to be hanged. The boat promised to "drop along side, SO THAT ALL CAN SEE WITHOUT DIFFICULTY," and to "reach home the same evening."[6]

June 5, 1854, was the date selected for a grand outing to celebrate the completion of the Chicago and Rock Island Railroad. As this was the first railroad to reach the Mississippi, it was felt appropriate that dignitaries and passengers travel from Chicago to the river by rail and then take a pleasure cruise upstream to St. Paul. Five steamboats were ultimately selected to transport the celebrants; their captains little realized that by the end of the decade they would have cause to rue the Iron Horse's appearance.

The *Golden Era, G.W. Spar-Hawk, Lady Franklin, Galena* and the *War Eagle* set out with two additional craft: the *Jenny Lind* and the *Black Hawk*. Passengers imbibed freely, and two or more boats were occasionally lashed together, enabling pleasure-seekers from one to pass over to another. Dancing to the music of two full bands was the order of the evening, and when the revelers were not otherwise engaged, eating was the prime attraction.

Breakfast was served at 7:00 A.M., after which the tables were cleared and preparations were undertaken for the main meal of the day. Fresh stores were bought along the way, including game, eggs, vegetables and speckled trout, while lambs, pigs, hens, turkeys and ducks were taken aboard and slaughtered as needed. Cows on the lower deck provided fresh milk. Delicacies included canned oysters and lobsters from the East, and desserts were nearly unlimited, including puddings, pies, ice cream, custards, jellies, fruits, nuts, cakes and ices. Tea, coffee and river water were served as beverages to go along with cold and warm breads.

Good Food, Bad Manners

Aside from the fascination of breathtaking scenery, eating was a prime way to pass the time. Breakfast was usually served at 7:00 A.M., "dinner" (lunch) at 12:30 P.M., and supper at 6:00 P.M.

A typical menu served to passengers traveling from New Orleans in the early 1830s was described as

> *Breakfast*, beefsteak, fowls, pigeon or chicken fricassee, ragout, plates of cold steaks, baked pork or turkey, small platters of ducks, chickens or other fowl, plates of cold sliced meat, potatoes, rice, corn, etc., with meal or rice pudding, tarts and rum and water; *supper*, very much like breakfast.[7]

Travelers were advised to bring their passage tickets to the table and prominently display them before eating, as proof they had the right to partake of the repast. They were, not surprisingly, requested to turn the tickets in at the completion of the voyage.

Menus continued to increase in variety. In 1849, "one soup, five kinds of fish, *six* kinds of boiled meats (several with sauces), eleven entrees consisting of meats and baked dishes including such delicacies as fricasseed kidneys and spiced pig's head, nine roasts, five kinds of game (including venison, wild turkey, and squirrel), fifteen kinds of pastry and dessert, and some fruits and nuts" were provided.[8] In addition, fruit, nuts, raisins, bread and cheese were often set out on sideboards for in-between meal refreshment for those not overstuffed during regular table sittings.

Charles Dickens described the fare served aboard a canal boat taking him to Pittsburg.

At 8:00 A.M., after the berths had been taken down and the table set, passengers were offered tea, coffee, bread, butter, salmon, shad, liver, steak, potatoes, pickles, ham, chops, black-pudding and sausages. (He notes that after the table was cleared, waiters then served as barbers.) Dinner, Dickens added, was breakfast again, except without the tea and coffee, and supper and breakfast were identical. On his way to Cincinnati, Dickens was served a mighty spread: slices of beet-root, shreds of dried beef, yellow pickles, maize (Indian corn), apple-sauce and pumpkin to go along with roast pig and preserves.[9]

Finding and preparing food that appealed both to the palate and the sense of novelty was a constant challenge. On January 1, 1851, the Cincinnati *Gazette* described a feast fit for a (French) king. The repast consisted of sixteen entrees, of which eleven were:

> Calves Head a la Tortul
> Giblets a la Glassey
> Boneta la Paysanne
> Veal a la Chasseur
> Bone Ducks a la Chamford
> Pula a la Anglais
> Chickens a la Diable
> Macaroni Parisienne
> Galatin a la Saute
> Venison a la Crapsden
> Quails a la Du Price

It was the habit of pastry chefs to prepare a special treat for passengers once a trip. On one trip, diners were served thirteen different desserts, presented in slender glass goblets the size of vases. Inside were custards, jellies and flavored creams, offered side-by-side with pies, puddings and ice creams. An extra set of glass and china was carried aboard especially for this once-a-trip banquet. It is no wonder that George Merrick's comment on the meals served aboard the Minnesota Packet Company applied to all luxury boats:

> The fact was, that most of the passengers so served had never in all their lives lived so well as they did on [the best steamboats!][10]

That said, foreign visitors often — and loudly — criticized American eating habits. Dickens observed that diners helped themselves by sticking knives and forks in their food, cramming the utensils down their throats and diving back for more.[204] Perhaps Frances Trollope's most famous observation came in her description of a steamboat dining table.

> The total want of all the usual courtesies of the table, the voracious rapidity with which the viands were seized and devoured, the strange uncouth phrases and pronunciation; the loathsome spitting, from the contamination of which it was absolutely impossible to protect our dresses; the frightful manner of feeding with their knives, till the whole blade seemed to enter into the mouth; and the still more frightful manner of cleaning the teeth afterwards with a pocket knife, soon forced us to feel that we were not surrounded by generals, colonels, and majors of the old world*; and that the dinner hour was to be any thing rather than an hour of enjoyment.[12]

Americans were also accused of partaking of their meals in silence. Eating was occasioned by no conversation, laughter or cheerfulness. If anyone spoke, the topic centered around politics, accompanied by a generalized cursing and swearing. Making the meal

---

* A sarcastic reference to the southern habit of all gentlemen being addressed by military title.

less appetizing were the smells of gin, whisky, brandy and rum from the bar, and the scent of stale tobacco.

Similar to the gingerbread and jigsaw of exterior decoration, steamboat meals were rather more numerous than properly prepared. George Merrick offered the sanguine saying that on the river if the boat owner wished to save on food expenses, he first took the passengers on a tour through the kitchen, the inference being that if the passenger saw firsthand how the meals were prepared, he would give them a wide berth. The main (starboard) galley was usually a hive of activity and of doubtful, if acceptable, cleanliness, for live poultry, lambs and hogs were slaughtered there. The pastry kitchen, however, was considered the showplace of food preparation, and occasionally ladies were taken there to observe the baking.[13]

Along the same vein, if a captain wished to save money on food, he made a point of serving the most elaborate meal on the first night aboard. This was a time most passengers suffered from seasickness and were unable to partake of the feast. The captain completed his obligation of offering one particularly bountiful meal, although it hardly enhanced the reputation of his boat.

The vast meals served aboard provided travelers with more than they could possibly eat, but like nearly everything else on the river, there was a way to mitigate financial losses. After the first-class passengers had finished, the leftover food was given to the cabin staff and the deck hands. While entrees, desserts and breads were mixed together, the scraps served as a tremendous break from their promised fare of boiled potatoes, crackers and dried meat.

Money Well Spent

Cabin fares depended on the season, the destination, and most importantly, the number of boats available. Steamboat skippers were not averse to charging huge prices when they thought they could get it, and were equally willing to sink to cut-throat fees in the face of competition.

In 1845, fares from Galena were posted as follows:

Destination	*Miles*	*Cabin Fare*
Dubuque	27	$1.00
Potosi	42	1.50
Cassville	57	2.00
Prairie du Chien	90	3.00
Fort Winnebago	220	5.00

In 1852, competition was so drastic, that steamboat ledgers took a severe beating. Three years later, profits soared, with the Minnesota Packet Company declaring a dividend of $100,000 for the season. Single boats cleared as much as $40,000.[14]

Passengers traveling in lean years took full advantage of lower prices, and in flush times they paid the penalty of higher fares. For the average person, this could mean an increase or decrease of one to two dollars or more per trip, which represented a considerable change for the 19th-century American.

On the Lower Mississippi, as steamboat efficiency improved and travel times significantly decreased, fares followed.

New Orleans to Louisville (1,400 miles)

Prior to 1818	$100–125
1825	$50
1828	$35
1830s	$25–30

New Orleans to St. Louis, Louisville and Cincinnati

1850s	$12–15 on an average boat
	$25 on a "crack" boat

Pittsburg to Cincinnati (500 miles)

1825	$12
1840s	$5

Louisville to St. Louis (580 miles)

1835	$12
1847	$5 (high-water season)
	$20 (low-water season)

Cincinnati to Louisville

1819	$8 downstream
	$12 upstream
1825	$4 downstream
	$6 upstream
Early 1850s	$1.50–2.00 either way[15]

Just as with freight, passenger way fares were higher than through fares.

In the 1840s and 1850s, the average price for first-class passenger travel averaged once cent per mile on a typical steamboat. Prices for luxury boats were always higher, although the money collected from ticket sales usually did not cover the cost of their extravagant meals.

Steamboat passage was significantly less expensive than overland stagecoach travel. In his *Travels in North America* (1827–28), Basil Hall gives the statistics for an overland route from Alabama to Georgia for three adults and one child as costing ninety-one cents per mile, while the same group was able to take passage aboard a steamboat from New Orleans to Louisville for only eight cents per mile. Comparatively, in 1859, it cost $80 to cross the Atlantic from New York to England or France on the North Star Line and $100–120 on the Vanderbilt Line.[16]

An advertisement run in the Alton (Illinois) *Telegraph*, May 31, 1837, gives an idea of early steamboat travel. Interestingly, it states "meals extra," showing the state of Upper Mississippi river transportation at that stage:

St. Louis and Alton Daily Packet. The superior fast sailing steam boat ALPHA, Harris *master,* will perform daily trips between St. Louis and Alton, leaving St. Louis at 9:00 a.m. and leaving Alton at 3 o'clock p.m. Merchants, farmers, produce dealers and others in the interior can be accommodated into and from St. Louis by the above conveyance at less expense than by land.

N.B. Passengers visiting St. Louis by the *Alpha* can be accommodated with Board.

RATES

Passage from St. Louis to Alton	$2.00
" " Alton to St. Louis	$1.50

MEALS EXTRA

Charge for dinner	50 cents
" breakfast and tea	50 cents
" lodging on board	25 cents

Gillespie & Co., Agents at St. Louis
Townsend & Co.   "   Alton
May 31, 1837

The "Necessaries" of Life

If the three principal requirements for life may be classified as food, shelter and clothing, a fourth need is not far behind: a place to eliminate bodily waste. This, of course, leads to the problem of disposal. In the history of human advancement, how to deal with such matters was spotty and characterized by long gaps in improvements. Moreover, the entire process was highly dependent on social mores.

As early as 3000 B.C.E., the Mesopotamia palace of Sargon I, king of Sumeria, had six primitive toilets installed. The Minoan civilization (2000 B.C.E.) actually had a system of flush toilets that worked on a sophisticated rainwater pipe system under the force of gravity. The Romans constructed the largest sewer system in the world and not only utilized public bathrooms, but also afforded wealthier citizens private chambers hooked up to the sewer system. Seats were cut from marble with no dividers for privacy; water flowing underneath carried away the waste. After the fall of the Roman Empire, it was nearly 1,000 years before the concept was revitalized.[17]

In 1596, an Englishman by the name of Sir John D. Harrington "invented" the water closet, by merely making improvements on ancient designs. He placed the waste apparatus in a small room within the home, but the concept did not catch on. The first patent for a flush toilet was given to Alexander Cummings in 1775 but, as with Harrington's work, the idea did not gain popular acceptance. People continued to utilize the outdoor privy, which typically comprised a narrow shed inside which an elevated seat had been placed over a hole. When the pit became full, it was covered up, the shed moved and a new hole dug. In later years, lime was often used to eat away the waste, but as lime cost money, most families found it easier just to start anew.

For city dwellers, the chamber pot was the primary means used for elimination. These were emptied by tossing the contents out a window, leading to a great stench and a proliferation of flies and other insects. In the 1700s, those away from home used the pit toilet, dug at the outskirts of town. These were open to anyone, without distinction of gender or modesty. If a person desired more immediate remedy, for a small fee street vendors would provide a cloak and a bucket. Alleys, side streets and stairwells were also used by the less discriminating, adding to the overall smell of urban cities.

There were no public sewer systems, so letting pigs run loose to eat the waste was the usual remedy. Since the awareness of disease-breeding organisms was unknown, entire cities were often contaminated by the infiltration of germs into the soil. This in turn contaminated the public water systems, leading to outbreaks of water-borne diseases that could annihilate hundreds of people within days or weeks.

In 1827, conditions in the thriving port town of Cincinnati were "devoid of nearly all the accommodation that Europeans conceive necessary to decency and comfort." Francis Trollope wrote:

> No pump, no cistern, no drain of any kind, no dustman's cart, or any other visible means of getting rid of the rubbish, which vanishes with such celerity in London, that one has no time to think of its existence

She was advised to have her servant dump the "refuse" in the middle of the street (laws forbade throwing it to the sides), and the pigs would soon clean it off.[18]

The idea of privacy while completing one's business was a relatively new development, arising from the Victorian era. Prior to that time, defecation was often performed in public by kings, princes and generals with an invited audience; those so honored to view the sight were accorded a great favor.

In 1739, a Parisian restaurant provided separate toilet facilities for those attending a dance; in 1824, the first public toilet facilities appeared in the capital. Twenty years later, the French also reintroduced men's public urinals, but again, most were open, without an eye toward privacy. Perhaps the greatest contribution to the acceptance of public facilities, however, was the Crystal Palace Exhibition in England in 1851. Due to vast numbers attending the event, there was, by necessity, a dire need to provide a means for convenient relief. George Jennings installed the toilets and it was reported that nearly 800,000 people paid a small fee to use these restrooms. The success of the endeavor prompted a new awareness in convenience and privacy.[19]

The Steamboat Crapper

The Steamboat Era dawned at the beginning of public consciousness concerning privacy in the matter of elimination. Whether referred to as the water closet, the necessary, the head, the loo, the outhouse, the bathroom, the privy, the "jake," or the "crapper," early accommodations were primitive. Most men simply urinated over the side of the boat. Water closets were built into the wheelhouse or over the stern, and consisted of a seat under which a hole allowed waste to fall into the river.

Thomas Crapper (born September 1836, in Yorkshire, England; died January 27, 1910, in London), is often credited with designing and patenting the first modern flush toilet. His father, Charles, interestingly enough, was a steamboat pilot who apprenticed his fourteen-year-old son to a master plumber in Chelsea. After serving his apprenticeship, Thomas served three years as a journeyman plumber before setting up his own shop.[20]

Based on the work of Crapper and others, the flush toilet became significantly more important. Always with an eye toward innovation, the wealthier and more luxurious steamboat lines introduced such apparatus on their vessels, providing the first exposure most Americans had to the concept. The first commercially produced toilet paper was manufactured in 1857 by a New Yorker named Joseph C. Gayetty. The "Therapeutic Paper" contained an abundance of aloe and was sold in packs of 500 sheets for the price of fifty cents. (Before this time, "bum fodder" was anything from washable cloth to newspaper — or not used at all.) There was no difficulty identifying the designer of the Therapeutic Paper, for Gayetty had his name printed on every sheet.[21]

It has been suggested that the slang word "crapper," denoting a bathroom, was derived from Thomas Crapper's surname. Authorities differ on this, some suggesting the word was in popular, if not polite, use before the plumber and his flush toilet became synonymous. It is possible the term is a corruption of the Dutch *Krappe*, the Low German *Krape* (meaning a vile and inedible fish), or the Middle English "crappy," which meant chaff, or waste material.[22] If true, it is certainly an odd coincidence that the common slang just happened to coincide so neatly with the inventor's last name.

Washrooms aboard a steamboat were hardly more advanced than the toilet facilities. Early spaces provided for the purpose of cleaning the body were crude affairs, constructed beside the ladies' and gentlemen's cabins. Most held no more than several tin basins placed on a bench and filled with river water. In lieu of hand towels, common "jack" or roller towels were provided. In addition, some washrooms offered a comb, a brush and a commonly-shared toothbrush.

Charles Dickens remarks that a looking glass for shaving, nailed to the wall, was provided on the boat he took from England in 1842. Traveling from Washington to Richmond, he observed:

> The washing and dressing apparatus for the passengers generally, consists of two jack-towels, three small wooden basins, a keg of water and a ladle to serve it out with, six square inches of looking-glass, two ditto of yellow soap, a comb and brush for the head, and nothing for the teeth.

When he opted to use his own comb and brush, the gentlemen in the fore–cabin were annoyed about his "prejudices."[23]

The time a passenger spent in the washroom was limited, as those waiting to get into the cramped chamber grew impatient. There is little doubt those inside tarried long, however, as the enclosed spaces were heavily infested with vermin and insects. Men either went unshaved (few men in the mid–1800s shaved more than twice a week), or visited the boat's barber.

17

Steamboat Diversions

The excitement of a steamboat trip began at the wharf. In large cities such as New Orleans, where steamboating dominated the waterfront, excitement was always in the air. Bands often played music for the entertainment of those watching the travelers depart, jugglers tossed balls, magicians performed tricks, and hucksters of all ages hawked fried breads, fruits and nuts. Boys sold newspapers and paperback "penny dreadfuls," slaves in the southern markets carried carpetbags and heavy boxes, merchants hurried last-minute shipments aboard, and everyone made a valiant effort to avoid tripping over legions of feral cats and dogs, or stepping in the massive accumulation of trash and feces clogging the passageways.

Across a two- or three-mile waterfront, cotton bales were piled three high, and barrels, hogsheads, sacks and crates blocked the thoroughfares, creating numerous narrow, twisted walkways, nearly impossible to navigate. Mates aboard departing boats swore, Negro deckhands sang while they worked, whistles blew, smokestacks erupted with black clouds, the waters churned and everyone tried to shout louder than the next in a vain attempt to be heard by the person standing next to him.

Crowds gathered around the steamboat chalkboards, discussing posted cargo rates set only until a neighboring clerk reduced his prices. Passengers scurried up ramps and made their way to the clerk's office to buy their tickets; well-wishers waved hats and hands. When the last warning bell sounded, those departing did so in a rush, pushing and shoving their way down the plank, just as late arrivals elbowed and swore their way up and past the visitors.

In New Orleans, departure time was set between four and five o'clock. Numerous vessels left simultaneously, and up and down the line, boilers were fired, made hotter still by the addition of pitch pine. Flags blew in the breeze as the great boats backed up, straightened and moved into the river. With all hands on deck, paddlewheels turning and the singing of the crew, it was a glorious sight, repeated time and time again in cities dotting the Mississippi.

Once passengers got settled and unpacked with the assistance of chambermaids, they left their staterooms by doors leading out to the promenades and gathered to watch the passing scenery. For those on their first trips, there was much to see: curving shorelines covered with low-hanging trees, colorful shrubs and native flowers, and always the odors of combusting wood and coal mixing with the fresh smell of paint, tar and oil. It was always a thrill to observe the white-churned water, the passing current against the bow and the flocks of birds driven skyward by the moving monsters.

Gambling in the United States

Once the novelty (and the seasickness) wore off, gentlemen quickly retired to the saloon for a game of cards, or to watch others stake their money. The acceptance of gambling aboard steamboats reflected the temper of the times and the wild frontier in which it gained acceptance. Gambling had always been a part of the American culture, but its popularity waxed and waned according to place, religious influences and a sense of fairness. In the early Colonies, lotteries were an accepted means of raising money. The Virginia Company of London, the financier of Jamestown, was authorized by the Crown to raise money by this means to support the venture. These proved so popular that they were eventually banned because it was feared too much money was leaving England.

This did not stop the Colonies, and all thirteen used this means for raising revenue, to the point where it was considered a civic duty to buy a ticket. George Washington, Benjamin Franklin, John Hancock and other notable Americans sponsored the idea of lotteries to support public institutions, and their success helped establish Harvard, Yale and Princeton universities. They did not always pay off, however, as the 1823 lottery for the beautification of Washington proved. This private lottery, passed by Congress, might have done great things for the city, but unfortunately the organizers absconded with the funds.[1]

Lotteries continued in use on both a public and private nature, such as the "scheme" to raise money for the Union Canal in 1828, as advertised in the Adams *Sentinel* on February 20:

Only 14,000 Tickets
Union Canal Lottery

The Scheme Contains

1 prize of	$10,000	Total $10,000
1	4,000	4,000
1	2,500	2,500
1	1,880	1,880
2	1,000	2,000
2	600	1,200
4	400	1,600
8	200	1,600
39	50	1,950
39	40	1,560
39	30	1,170
78	15	1,170
390	10	3,900
4446	5	22,230

5,015 prizes amounting to $56,760[2]

In 1835, D. Carver's "ever Lucky Office" offered for sale tickets on the state lotteries of Delaware and Maryland for Christmas presents. Top prizes for those schemes were $8,000 and $20,000, respectively.[3]

Gambling also took the form of horse racing, with the first track constructed on

Long Island in 1665. Card and dice games led to casino houses, and this idea quickly spread west, where such cities as New Orleans adopted the style.

Poker had been established in New Orleans by the French, who played a game called Poque in the early 1800s. It was similar to draw poker, and quickly found favor with the influx of immigrants to the new American state. The first gaming casino in New Orleans was established in 1822 by John Davis. Open around the clock, it featured gourmet food, liquor, roulette wheels, card games and "painted ladies." The main area of gambling became known as "the Swamp." Situated near the waterfront, it quickly acquired a nefarious reputation. Those who survived their initiation here proved adept at fleecing steamboat passengers, where the accommodations were far better and the rewards greater.[4]

During the 1830s, gambling fell into disrepute. Professional card sharps were blamed for stunting economic growth, interfering with business, endangering the streets, committing numerous crimes and debasing morality. In 1835, a vigilante mob hanged five gamblers in Mississippi. Finding no safe haven on land, the blacklegs took to steamboats, finding there a more tolerant attitude. The period between 1840 and 1860 represented the heyday of steamboat gambling, and the advent of its long and glorious history.[5]

"Games for money strictly forbidden"

Many boats carried the warnings "Games for money strictly forbidden," or stated that men playing cards did so at their own risk. While it was popular along the waterfront for a captain to profess his distaste for cards and insist that no card playing was allowed on his boat, few, if any, followed that pronouncement with any practical action. As long as stakes were kept within reason and the players caused no trouble, they saw no reason to prohibit the western passion. The explanation was simple.

> The reason boats went one eye blind on this business was because the professional gamblers were quite a percentage of the passengers, very liberal in paying for the best rooms, tipping the cabin boys and good liberal customers at the bar; they, in connection with the non–professional, in some cases very rich planters ... were the majority of the passengers whose will made law.[6]

One of the best descriptions of the steamboat gambler came from George Merrick, who saw enough of them in his tenure as cub engineer and pilot to have formed a thorough idea of the breed.

> The professionals who traveled the river for the purpose of "skinning suckers" were usually the "gentlemen" who displayed the greatest concern in regard to the meaning of this caution [play at your own risk], and who frequently expressed themselves in the hearing of all, still less for money; but if they did feel inclined to have a little social game it was not the business of the boat to question their right to do so; and if they lost their money, they certainly would not call on the boat to restore it.[7]

Most accounts of steamboat gambling refer to small pots, and Merrick declares that on the Upper Mississippi, the usual "take" of a professional gambler was three hundred dollars a week. That, however, was not an inconsequential sum during the mid–1800s, and certainly elevated a man to the ranks of the "elite." This was a euphemistic term denoting the status conferred by money. Many travelers who objected to games of "chance," or who objected to losing, thought otherwise. Mark Twain fell into this category, observing the profession while traveling to California by sea:

There were three professional gamblers on board — rough, repulsive fellows. I never had any talk with them, yet I could not help seeing them with some frequency; for they gambled in an upper-deck state-room every day and night, and in my promenades I often had glimpses of them through their door, which stood a little ajar to let out the surplus tobacco smoke and profanity. They were an evil and hateful presence, but I had to put up with it, of course.[8]

Gamblers typically traveled in pairs, avoiding others of the profession so as not to intrude on their territories. They were consummate actors, pretending not to know one another until introduced by an innocent third party. They frequently assumed disguises and played roles to make themselves appear innocuous to the uninitiated. Depending on the part assumed, they dressed accordingly and could speak the language of the job they professed to represent.

Similar to the dime novel accounts of the time, blacklegs presented a gentlemanly deportment, tipping their hats to the ladies and using refined language in mixed company. One of the slickest tricks of the trade was to give the impression of being drunk. This led their opponents to believe they had an "edge," the expectation that an inebriated gambler would bet more recklessly.[9] In fact, many sharpers kept a private bottle at the bar, usually nothing more than river water tinted with burned peaches to give an authentic look. Bartenders and waiters were aware of the deception and would take their cut at night's end.

Cheating has always been synonymous with gamblers, perhaps primarily so with those working the steamboat trades. According to George Merrick, they rightfully deserved the reputation. He offers a fascinating account of the fine art of "stripping," or shaving the deck. The high cards — king, queen, jack and ten-spot — were placed between two thin sheets of metal, the sides of which were very slightly concave. Then the edges of the cards were sliced away with a razor. The result was undetectable to the eye, but readily apparent to an experienced hand.[10] When dealing, the confidence man could dexterously pass the pasteboards of highest value to himself or his partner, and in that way, win the heaviest pots. They did not practice this art on every hand, for that would be a giveaway to even the greenest novice. Typically, they lost a number of hands at the outset to demonstrate their fallibility, then won the later games as Lady Luck appeared to shower her good fortune on them.

New packs were called for whenever "luck" changed, or to convince the players or onlookers that the cards were legitimate. The gamblers made a display of breaking the seal, thus proving no tampering had been performed. Losers were not always good sports, however, and occasionally a fight would break out. A gambler carrying a "lady's gun," or derringer, might carry the day, but he did not always escape justice. If the captain were called to settle a claim of cheating, he had the option of putting the gambler ashore at any place along the river or convening a court. Whether or not there was sufficient proof, juries were seldom lenient, and it usually went badly for the gambler. Not only was he forced to give back his winnings, but he might be whipped for his crimes. The ultimate punishment was to toss him overboard, where he was sucked under the water and ripped to shreds by the huge paddlewheels.

Passengers were not the only ones given to a taste for cards. A report in the Wisconsin *Democrat,* June 23, 1838, relays what may be a unique, but ultimately fascinating look at a slice of steamboat gambling:

EXTRAORDINARY CIRCUMSTANCE OF GAMBLING. It is well known upon the western waters, that the firemen and other hands employed upon the boats spend much of their idle time in playing cards. Of the passion for gaming, thus excited, an instance has been narrated to us, upon most credible authority, which surpasses the highest wrought fictions of the gambler's fate. A colored fireman, on a steamboat running between this city and New Orleans, had lost all his money at *poker* with his companions. He then staked his clothing, and being still unfortunate, pledged his own freedom for a small amount. Losing this, the bets were doubled, and he finally, at one desperate hazard ventured his full value as a slave, and laid down his free papers to represent the stake. He lost, suffered his certificates to be destroyed, and was actually sold by the winner to a slave dealer, who hesitated not to take him at a small discount upon his assessed value. When last heard of by one who knows him, and informed us of the fact, he was still paying in servitude the penalty of his criminal folly.

According to Hoyle

While poker is typically the game portrayed in the fictionalized accounts of steamboat gambling, the conventional European games of whist, brag and euchre were often played alongside western favorites such as Three Card Monte, Faro and Brag. The actual game of poker possibly originated from a 16th-century Persian card game called "As Nas," played with a 25-card deck of five suits, similar to five-card stud.

Although it may have been traditionally associated with white men, gambling also attracted Hispanics, blacks and Chinese, particularly in San Francisco, where the Gold Rush provided much hard money for betting. Women, too, were more than able to hold their own. Among the most famous were Calamity Jane, Poker Alice and Madame Mustache.[11]

Hardly any film set on the Mississippi or an episode of *Maverick* went by without one of the gamblers being questioned about a certain rule of the game. The reply always started out, "According to Hoyle..." and included some obscure justification that invariably resulted in winning the hand. The implication was that, according to "Hoyle's rules," the established bible of poker, Bret and Bart were correct.

The expression "According to Hoyle," was, indeed, a common one throughout the mid–1700s and beyond, but it bore little relevance to the subject at hand, for Edmund Hoyle (1672–August 29, 1769) never wrote a word about poker. Hoyle's fame actually originated with the game of whist, for which he was a tutor to wealthy Londoners. He wrote a short booklet on the game and it became so popular that unauthorized copies quickly circulated. To fully establish himself as the author (and collect his royalties), Hoyle published *A Short Treatise on the Game of Whist* in 1742. This proved so successful that he followed it with several other books, on backgammon, chess, quadrille, piquet and brag. In 1750, a compendium of these smaller works were released as *Mr. Hoyle's Games Complete*, and it soon became the standard used on the Continent and the United States.[12]

"Fast and elegant steamer"

This phrase was a common one, frequently appearing in advertisements for Upper Mississippi boats. While speed could be proven, beauty was left to the eye of the beholder.

Mark Twain sarcastically noted that oils produced by "some gifted sign-painter" adorned the cabin doors of the most elegant steamboats.[13] George Merrick seconded the

A very handsome engraving of the *Mississippi*. This is a famous image often used on paper money and steamboat documents. "Crack" boats such as this provided luxury accommodations for first class travelers. Inside the saloon, gentlemen would amuse themselves by drinking and playing cards.

motion by stating that those artists who could not find a more profitable venue for their art gladly sold their pictures to steamboat operators. Furthermore, these artists were happy to ply their talents by decorating the thirty-foot paddle-boxes owners were so eager to have resplendent in paint.[14]

From the brushes of the less fortunate portrait and scenery artists came "figures of historic size," ranging from beautiful women (modestly and becomingly clothed), with or without bundles of wheat and reaping hooks. Boats with "Eagle" in the name all bore strikingly similar avians; sunbursts were always popular; and occasionally, the figure for whom the boat was named found his or her likeness spread over the wheelhouses.

The more luxurious the boat, the better the artwork. The *Northern Light*, belonging to the Minnesota Packet Company, had oil paintings of the Falls of St. Anthony, Dayton Bluffs and Maiden Rock in panels in the cabin, and the aurora borealis on her paddle-box. The artist was said to have received $1,000 for his work. Most of the commissioned works, at least on less pretentious vessels, were done by lesser persons.[15]

The Northern Light

Built: 1856 in Cincinnati
Style: Side-wheeler
Length: 240 feet
Beam: 40 feet
Hold: 5 feet
Cylinders: 22 inches
Stroke: 7 feet
Boilers: 8; 46 inches diameter, 17 feet long
Wheels: 31 feet in diameter
Buckets: 9 feet; 30 inches dip
History: Sunk in 1862 by striking shore ice that tore out the stern

Music and dancing were other modes of enjoyment offered the first-class passenger. These amusements came into their own with the advent of the Fashionable Tour, and reached their peak in the 1840s and 50s, when luxury travel was at its height. Smaller boats got double their money's worth by having cabin hands display their talents, but on larger crafts, captains hired professional musicians. These ran the gamut from pianists to fiddlers to small orchestras. Brass bands occasionally made an appearance, but were expensive and fell out of favor with boat owners.

The wonder boat of her day was the *Eclipse,* renowned throughout the Lower Mississippi as being in a class by herself.

The Eclipse

Built: 1852
Cost: $120,000
Tons: 1,117 tons
Length: 350 feet
Hull: 36 feet (overall, 76 feet)
Depth: 9 feet
Cylinders: 36 inches, 11-foot stroke
Paddle-wheels: 41 feet × 15 feet

The *Eclipse* boasted a 300-foot-long cabin decorated in the Gothic and Norman styles. Serving the Louisville trade, she had a 122-person crew to work the huge vessel and attend the passengers. Nothing was too good for those traveling on her, and in antebellum days, few, if any boats were her equal.

On the other end of the scale, at least in terms of entertainment, was the *Excelsior.* Captain Ward was the first to introduce the "steam piano" (called, amusingly, by George Merrick and his compatriots, a "calliope"). Music was produced by a "barbaric collection of steam whistles" and played at all hours of the day and night. Several others operators copied this style, but as Merrick points out, no passenger ever shipped out twice on a vessel with such a noisemaker, so it soon fell out of fashion.[16]

The cabin orchestra, comprised of between six and eight Negro cabinmen, was the most enduring draw and the most popular. Instruments included the violin, banjo and guitar, and were accompanied by the melodic voices of the musicians. The cabinmen were occasionally permitted to pass the hat after a performance and allowed to keep the extra they made. Boats with the best orchestras quickly earned a good reputation, and passengers patronized them more than any others.

Augmenting word-of-mouth, Negro bands often played as the boat pulled into the wharf, drawing crowds with their spirituals (plantation melodies), occasionally with lyrics amended to reference peculiar traits of the crew. Anything which set a boat apart was viewed in a positive light, and captains were eager to pay these men extra for services rendered.

When and if the thrill of gambling and dancing wore off, passengers conversed among themselves. Every four years, the subject turned to presidential politics and the upcoming election. An article from the *Adams Sentinel,* April 2, 1838, describes how one group found diversion.

A Vote on the Western Waters.—A slip from the office of the Louisville Journal contains the following:

The steamboat Gen. Gaines arrived here yesterday from New Orleans. On her passage up, a vote was taken on board of her upon the question of the next Presidency. The results stood: for Clay, 37, for Van Buren 3, for Harrison 4, for Webster 1. Of the 37 votes for Clay, 6 were from Virginia, 18 from Kentucky, 6 from Mississippi, 1 from New Hampshire, 4 from Indiana, 1 from Arkansas, and 1 from New York. Of the 4 for Harrison, all were from Ohio. The 1 vote for Webster was from Virginia.

Interestingly, the correspondent did not give a breakdown of the three votes for Van Buren.

The Fleet Ages

Not all hardships were reserved for deck passengers and not all steamboats of the later years were in the same class as the *Eclipse*. Boats that survived river obstructions, the devastation of fire, boiler explosions, or merely outlived their usefulness as pleasure craft were sold from owner to owner. Fresh coats of paint covered a multitude of sins, and passengers, either unaware of a vessel's history or hoping to save a few dollars in fare, took their chances. While it would be unfair to say all regretted their lack of foresight or thriftiness, many came to rue their decision.

Accommodations lacked the clean linen of "A" boats, and mattresses were as likely to be filled with corn husks as feathers or spring mattresses. Sheets were soiled and ripped; the rooms smelled of previous occupants and whatever decorations were left had a coating of grease and grime. Chair cushions sagged, curtains were limp, and floors, scored by countless boots, were warped from long exposure to humid weather.

Conditions in the salon were no better. The gilt had faded, the paint peeled. Layers of tobacco juice and spilled drinks, added to accumulations of dust and debris, presented a gloomy picture. Dining tables rocked on uneven legs, chair joints creaked and plates were cracked and chipped. Patterns, once so carefully set out, were mismatched; tarnished silver trays displayed dents; pitchers had broken handles, knives were dulled and spoons bent out of shape.

Likewise, the food served on these relics was far below the standards promised in printed advertisements. An owner might promise "luxury suites" and "gourmet fare," but such was not the case. Instead of canned lobster and fresh chicken, poorly prepared salt beef, cornbread and stale pastry were provided, and only in limited quantities.

Adding to the discomfort, the carpets were threadbare and often soggy from leaking roofs, mold grew along the sideboards and an all-prevailing dampness permeated the boat. The walls of the vessel, once firmly anchored, buckled; stairwells swayed and the fancy promenades were reduced to cluttered, dirty walkways, hardly conducive to pleasant conversation, while noisy engines prohibited long, contemplative naps.

As a boat aged, the insulation which masked some, but not all of the vibrations, disintegrated, making the entire vessel shiver from the effort of straining machinery. Tremendous heat from the main deck made summer temperatures unbearable, and the lack of heating on early and late season trips forced passengers out of their cabins. They huddled around the small stoves by the bar, those too close nearly scalded and those further back almost frozen. In fairness, seasonal extremes of cold were a common complaint on most boats, especially those working the northern trades, but the older or more dilapidated the boat, the worse conditions were.

Nor did paying a premium fee guarantee that the steamboat would arrive on time. Captains desperate for cargo often made side trips or delayed departure from small towns, hoping to pick up a commission before setting out. In such cases there was nothing a traveler could do but wait it out, or depart, forfeiting the rest of his passage money in the process.

Older boats also tended to break down more frequently, necessitating stopovers along the riverbank for repairs. If a pilot were unlucky enough to get the vessel stranded on a bar, it was unlikely that the captain would pay the fee to have it towed. That meant waiting for the crew to eventually work it off, or for the river to rise. A delay of several days or even weeks was not unheard of. If this happened, rations fell to dangerous levels and passengers were compelled to go hungry or make their way to shore and hunt for their dinner.

Among the crew, there was little regard and less sympathy for the passengers' plight. Men working these "scows" were used to hardship and paid a minimum wage. It was a common joke that before a passenger paid his fare he was sent to the safest part of the boat, and that afterward, he was on his own. It may have been amusing to rivermen, but hardly so to those bearing the brunt of the joke.

Even on well-maintained steamboats, odors from livestock and crowded belowdecks quarters added to the general discomfort. Straw used beneath horses, cattle and swine was occasionally swept into the river and replaced, but no toilet facilities beyond pails were provided for the hands or the deck passengers, and those were emptied only if a traveler took it upon himself. Without washing facilities, body odor quickly became intolerable and clothes were seldom washed.

Insects, too, were a severe problem, particularly on boats navigating the southern waters. Mosquitoes, gnats and flies swarmed everywhere and bedbugs bit with a ferocious appetite. Boats rarely provided netting around the bunks, and even those that did were unable to prevent their passengers from being bitten when outside that scant protection.

Motion sickness was also a frequent problem for which there was little effective remedy. Unless a doctor happened to be traveling aboard, passengers or crew in need of medical attention were either treated by the captain or hurried ashore at the earliest opportunity. The most effective cure known to 19th century sufferers was to chew the root of the ginger plant.

Obnoxious traveling companions, drunks and bullies also made life miserable for the steamboat passenger. The handling of these undesirables was left to the captain or the clerk, but unless violence was involved, most were ignored, if not tolerated, by the unfortunates aboard.

18

"Just the bar' necessities, ma'am"

The scene: Dodge City, Kansas
The year: Anytime from 1867 to 1876
Interior: The Long Branch Saloon
Wide Angle: MARSHAL DILLON walks in through the batwing doors. MISS KITTY meets him; they exchange pleasantries before sitting at a private table where DOC and FESTUS interrupt their argument to stand, Festus pulling out the chair for her. SAM, the bartender, brings them a tray with four beer mugs; he distributes the drinks.
Angle at bar: Two COWBOYS enter, saunter to the bar. One orders a bottle of Red Eye; pays $2.00 in coin. Sam slides the bottle and two glasses across the counter. Cowboys drink.

A fabricated, but all-too-typical scene from *Gunsmoke,* played out countless times with little variation over a twenty-year span. With a different cast of characters, the same scene might have been taken from any number of films, or television series, including *Riverboat*, in which Darren McGavin played Captain Grey Holden of the *Enterprise*. Anyone who grew up watching such fare would be justified in believing that was the way all 19th-century folks took their alcoholic refreshment: either in a glass with foam spilling over the sides or poured out of a bottle containing rotgut whisky.

They would be mistaken.

In the 1800s, Americans were more likely to imbibe sweetened mixed drinks or take a glass of fine Madeira wine than gulp warm carbonated beverages or shove moonshine down their gullets. Alcohol was an accepted part of life, typically served at breakfast (where it was called an "elevator"), dinner and supper, with all members of the family taking part. In fact, alcohol was considered liquid food, inasmuch as rye and corn whisky came from grain, wine from fruit and beer from barley.

In 1810, Louisville shipped 250,000 gallons of whisky up the Ohio River. By 1822, that output had grown to 2,250,000 gallons, a considerable boon to steamboat operators and crews, for they were often plied with hard liquor as an incentive to work harder. In the 1830s, Americans consumed more than five gallons of liquor per capita annually, the highest rate in the history of the country. Beer was always available and locally brewed, but did not achieve the widespread popularity it later enjoyed until there was a great influx of German émigrés, around 1850.[1]

Alcohol was enjoyed with any number of additives, most notably sugar, and treasured for its medicinal benefit in cordials and tonics. It was used as a painkiller, to soothe jangled nerves, bring down fevers, warm cold bodies and even as a treatment for dyspepsia (indigestion), from which so many suffered. The *aqua vitale*, or "water of life," was, ironically enough, probably better to drink than well water, which was often contaminated (a cause of cholera), or milk, which unscrupulous distributors polluted with such noxious materials as chalk, flour, stale egg whites, magnesia, molasses and burnt sugar.[2]

In 1829, intoxicated celebrants enjoying Andrew Jackson's inaugural ball nearly destroyed the Executive Mansion. This did not seem to dampen the spirits of their fellow Americans, however, for alcoholic consumption reached an all-time high the following year.

Bending the Elbow

Americans have long had an obsession with "bending the elbow," and in 1780, a "Moral and Physical Thermometer" was produced, giving the merits of drink in order of effect, from least to most powerful:

Water
Milk and water; vinegar and water; molasses and water
Small beer
Cider
Wine
Porter
Strong beer
Punch (weak)
Punch (strong)
Toddy
Grog
Flip
Slings
Bitters infused with spirits
Morning drams
Pepper in rum[3]

A well-stocked bar, found in any number of saloons and drinking parlors, as well as the more glamorous steamboats, would include all or many of the following:

Syrups

Ginger
Gum (arabic) (white gum arabic from the acacia tree, sugar and water)
Lemon
Orgeant (bitter almond extract)
Pineapple

Raspberry
Simple syrup (2½ pounds sugar to one pint water, boiled)
Strawberry

Sugars

Black strap molasses
Brown sugar
Capillaire (orange-flavored)
Fine sugar
Granulated sugar
Honey
Loaf sugar (sugar formed into bricks, crushed or powdered for use)
Powdered sugar
Rock candy

Essences and Tinctures

Aromatic tincture (ginger, cinnamon and orange peel)
Essence of cinnamon
Essence of cloves
Essence of ginger
Essence of peppermint
Oil of cinnamon
Pineapple oil
Tincture of capsicum (hot pepper)[4]

Travel narratives of the time are filled with descriptions of strange alcoholic beverages. There were "cobblers," a typically American mixture of what essentially was a mint julep without the mint, using wine instead of higher-proof brandy or gin. "Crustas" may have arisen in the French Quarter of New Orleans, developed by Santina, the owner of a popular Spanish saloon. Sugar was encrusted around the rim of a red-wine glass, then a lemon was pared to achieve one continuous strip of peel. It was placed in the glass whole, then liquor, bitters and juice were added. "Skins" were drinks made with boiling water and high-proof alcohol. The origin of the name was obvious: drink it too fast, and the beverage took the skin off your throat.

Victorians always appreciated their gin and bitters, and often used an ingredient called tansy. Also known as cow bitters or button bitters, it came from a weedy, fragrant plant native to England and cultivated in the Northern Hemisphere. Today, it is considered slightly poisonous.

"Fixes" were one of the most popular iced drinks, apparently a creation of the 19th century. Fixes were made with applejack, gin, brandy or rum, mixed with fruit syrups, lemon juice and sugar and poured over ice. It is said "flips" were the favored drink among seamen, presumably when they were ashore. A flip was a hot beverage made with ale, wine or liquor, mixed with eggs, sugar, and spices. British bartenders used a "flip dog" to prepare the drink, which was basically a red-hot iron poker plunged into the glass to heat the mixture.

An extremely rare photograph of the interior of a steamboat, taken in 1860. It reveals the Ladies' Cabin. The *Planter* was built at Wheeling, Virginia, and measured 343 tons. She was captured in 1863, and used by the Union Army. In 1875, she became part of the U.S. Navy.

In the middle of the 1800s, the word "lemonade" as often referred to a drink containing alcohol as it did to the lemon-and-sugar drink. Raspberry syrup was frequently used, along with freshly squeezed lemons, as were orange slices. Other recipes were even more removed from the modern conception.[5]

Charles Dickens mentions "Turtle, and cold Punch, and Hock, Champagne, and Claret" in his book, *American Notes*. "Hock" referred to Hockheimer wine, a sweet white wine produced from Riesling grapes in the Rhiengau area of Germany. Champagne probably arrived in the west in the 1850s. Mumms and Grande Marque (De Launay & Co.) were among the first and most favored. Claret was the English term for French Bordeaux wine, and recipes for "cold punch" were innumerable.

A Punch for the Road

The only description Dickens gives of his social punch was "cold." Just as today, punches were often used as celebratory drinks (in this case, the occasion of his departure for America), but creating the hot or cold beverage was as complicated and individualistic as each celebration.

"Punch" appears to have been mixed by many cultures, over many centuries. The origin of the word may have come from the Hindu *paunch,* the Persian *punj,* or the Sanskrit *pancha*. All mean "five," from the original ingredients used: lime, sugar, spice, water and fermented palm juice.[6]

As the drink was popularized in Europe and North America, lemon became the essential ingredient after alcohol, and bartending guides go to great lengths to explain how to obtain the necessary "ambrosial essence" of the fruit. Another tip was that when making hot punch, "You must put in the spirits before the water; in cold punch, grog, &c., the other way."

PUNCH A LA ROMAINE
(for a Party of 15)

1 Bottle Wine (Red or White)
1 Bottle Rum
Juice of 10 lemons
Juice of 2 Sweet Oranges
2 Pounds Powdered Sugar
1 Orange Rind
10 Egg Whites

Take the juice of 10 lemons and 2 sweet oranges, dissolve it in two pounds of powdered sugar, and add the thin rind of an orange. Run this through a sieve, and stir in by degrees the whites of 10 eggs, beaten into a froth. Put the bowl with the mixture into an ice pail, let it freeze a little, then stir briskly into it a bottle of wine and a bottle of rum.[7]

Punch bowl drinks were the most popular class of beverages in the mid–1800s, and variations went by many names:

Mull: a punch made from boiled wine, sugar, spices &c.
Nectar: any sweet drink, usually served cold
Negus: made from hot water, port or sherry, spices and sugar. Named after Colonel Frances Negus (d. 1732) who is said to have been the inventor
Bowls or cups: Used alternately to signify punch
Shrub: from the Arabic shrub, meaning drink
Nogg or Nog: from the English "noggin," a large pot in which the egg-laden beverage was prepared.[8]

Farther along in his travels, Dickens describes a saloon in Boston:

The bar is a long room with a stone floor, and there people stand and smoke, and lounge about, all the evening: dropping in and out as the humor strikes them.* There too

---

Contrary to American bars, an English bar "was neither a room for customers to gather in, nor a counter to drink at, the notion of stand-up drinking being an innovation borrowed from the counter takeout or stand-up-(cont.)

the stranger is initiated into the mysteries of Gin-sling, Cock-tail, Sangaree, Mint Julep, Sherry-cobbler, Timber-Doodle, and other rare drinks.[9]

The names of the drinks have local associations, as it was a common practice for bartenders to create new combinations with exaggerated names for the entertainment of the patrons, as well as to establish their own reputations. Gin slings (also known as "whisky slings") were wildly popular throughout the country; "sangaree" is an English corruption of the Spanish word *sangria;* and "the origin of Timber-Doodle" is lost to posterity.

GIN SLING

¼ Pint Gin (or whisky)
Thin peel of an Orange
Juice of 2 Oranges and 1 Lemon
Sugar

Soak the thin peel of an orange or lemon in one-quarter pint of gin or whisky. Add the juice of two oranges and one lemon; sugar to taste; add one pint of pounded lake ice; serve with straws.[10]

"Slings" came to America by 1807, and were traditionally made of gin, bitters, lemon juice, ice and sugar. Later, other types of liquor, fruit and spices were added. Hot slings came into use in the 1860s, and were very similar to hot toddies.

The "julep" was not an American concoction, but traced its origin back to the Persian *gul-ab* or Arabic *julab,* indicating rosewater or a cooling drink containing opium. Reaching England by 1400, the drink was alternately called julap, juloup, jewlip and juleb. It was brought to the Colonies in the 17th-century. The basic ingredients of the mint julep were fixed between 1760 and 1770. Captain Frederick Marryat, who hated the Mississippi River with such passion, clearly had kinder words for American mixed drinks:

> The mint julep [is], with the thermometer at 100 degrees, one of the most delightful and insinuating potations that was ever invented....

The famous southern drink was not made with bourbon until after the Civil War. The finest Cognac was the liquor of choice, or wines such as Madeira or claret were used. Juleps were imbibed through a straw or a piece of hollow pasta.

MINT JULEP

1–1½ Wine Glasses Cognac Brandy
1 Dash Jamaica Rum
2–2½ Tablespoonful Water
1 Tablespoonful White Sugar
3 or 4 Sprigs Fresh Mint

Dissolve one tablespoon of white pulverized sugar into two and one half tablespoons of water. Take two sprigs of fresh mint and press them well in the sugar and water, until the flavor of the mint is extracted; add one wine glass of Cognac brandy, and fill the glass with fine shaved ice, then draw out the sprigs of mint and insert them in the ice with the stems downward, so that the leaves will be above, in the shape of a bouquet; arrange

---

and-drink 'dram shops.' Rather, the bar was generally a small room near the entrance to the pub ... where the landlord could greet customers."[11]

berries, and small pieces of sliced orange on top in a tasty manner, dash with Jamaica rum, and sprinkle with white sugar on top. Place a straw [horizontally across the rim of the glass] and you will have a julep that is fit for an emperor.[12]

Other recipes call for the juice of one orange and one tablespoon crushed rock candy, or one half gill of maraschino.[13]

Sangarees date back to 1736, and were reputed to be the most popular drink in the Spanish Caribbean. The varied recipes were brought to America by seafarers. An early description included: Madeira wine, lime juice, sugar and nutmeg, served at room temperature. In the United States, it originally was a punch-like beverage serving a group, but later was made in individual portions.

BRANDY SANGAREE
1 Wine Glass Brandy
½ Wine Glass Water
1 Teaspoon Port Wine
1 Teaspoon Sugar

Dissolve the sugar in a little water. Fill the glass ⅔ full of shaved ice, shake up well, strain into a small glass and dash a little port wine so that it will float on top.[14]

Other recipes call for the substitution of gin or sherry as the main liquor, with a little nutmeg, and slices of pineapple.

"Sours" appeared on the scene around 1850, calling for brandy or gin with the juice of ¼ teaspoon lemon and sugar, served over ice and decorated with orange, pineapple and berries. "Toddies" were adopted by British soldiers, who drank a palm liquor called *tari tadi* in India. As there were few palm trees in England, the fermented palm juice was replaced by rum, whisky, applejack, gin and brandy, sweetened with sugar and spiced with nutmeg. The word *tadi* was corrupted to "toddy," and the drink appeared in the Americas by 1790. Apple toddy became very popular, and many saloon hot water urns were fitted with apple roasters to supply the baked apple needed for the recipe.

"Another round of beers, Sam"

The terms "beer" and "ale" were used interchangeably through the 1860s. Along with porter and stout, they were served at room temperature, or heated and used for mixed drinks. If Miss Kitty actually served "Cold Beer on Draught," she either used a mechanical ammonia and ether ice-making machine invented before the Civil War, (but rarely used) or she kept an icehouse filled with that precious commodity.

Ale, porter and stout were imported into the United States from the British Isles in an unrefrigerated, unpasteurized state, and were available in bottles or kegs. Bitter ale had bitter hops, and "old ale" was aged for smoothness and character.

American beers were lagered (cellared) beers, clear, light and effervescent. They were brewed from malt, hops and water, called "wort." Sometimes, corn, grits and cracked rice were added for flavoring.

Carbonation was added after lagering by the injection of carbolic acid gas, made from a system whereby marble chips were exposed to a stream of sulfuric acid. The resulting gas was piped into the beer. Less expensive beers used carbonate of soda and cream of tartar to create gas.

Porter was a heavy, bittersweet, dark beer made from charred malt; stout is porter with a higher alcoholic content. Ginger beer was a sparkling drink or mixer made in Britain and Ireland of fermented ginger, sugar, yeast, cream of tartar and water. Ginger beer was available in both alcoholic and nonalcoholic versions.

The Steamboat Bar Room

Alcoholic beverages were not served in conjunction with steamboat meals. Gentlemen wishing to drink did so at the bar, and paid for their drinks out of their own pocket. The bar was located at the top (bow) end of the saloon, set somewhat apart from the open area used for card playing, conversation and, when cleared, for dancing.

There was no shortage of customers at the bar, for whisky was classified as a necessity of western life. The sale of liquor garnered the captain a modest take, depending on the number of passengers aboard; those partaking included men, women, children and Indians, "presupposing that a certain percentage of the passengers' money would find its way into [the bar-keep's] tills, regardless of age, sex, or color."[15]

Merrick states that there was a saying on the Upper Mississippi, that if a man owned a bar on a popular boat, it was better than possessing a gold mine. "The income was ample and certain, and the risk and labor slight." When the captain or owner did not want to be involved in the business, he sold leases on the bar to an independent concessionaire for an annual fee of $1,500. Occasionally, financiers would buy the leases on a number of boats, or on entire lines. "Billy" Henderson of St. Louis went from owning one bar on the *Excelsior* to buying all the bars on the entire Northern Line.[16] In the decades before the Civil War, the operator cleared $400–$600 per month on the sale of spirituous beverages. Just as in barrooms across the East, favorite drinks were brandy smashes, gin slings, whisky cocktails (favored by easterners) and mint juleps (favored by southerners).

Brandy Smash

1 Wine Glass Brandy, Gin, Rum or Whisky
1 Tablespoon Water
1 Tablespoon white sugar
Fresh mint leaves

Take two sprigs of fresh mint and press them well in the sugar and water, until the flavor of the mint is extracted; add one wine glass of Gin, and fill the glass two-thirds full of shaved ice, then draw out the sprigs of mint and insert them in the ice with the stems downward, so that the leaves will be above, in the shape of a bouquet. Lay two small pieces of orange on top, and ornament with berries in season.[17]

Mark Twain had a much more generous remembrance from his days on the river. He wrote in *Life on the Mississippi*, "In the old times, the barkeep owned the bar himself and was gay and smarty and talky and all jeweled up and was the toniest aristocrat on the boat; used to make $2,000 a trip."[18] Presumably, a very long trip. Or a very thirsty one. His guess is likely exaggerated.

Bartenders tended to be well dressed and well mannered, with a gift for gab and an agreeable personality. Aside from knowing the recipes of mixed drinks, they usually possessed the talent of "manufacturing" liquor for those who patronized the smaller boats.

The "judicious admixture of burnt peach stones, nitric acid, and cod-liver oil, superimposed upon a foundation of Kentucky whiskey three weeks from the still" passed for a choice brand of French brandy "on less discerning palates."[19]

The precise origin of the word "cocktail" is unclear, but it is speculated it is a corruption of the French *coquetier* (a type of egg cup, sometimes used as a drinking glass) or *coquetel*, a wine cup used in Bordeaux.[20] It has been used to describe a category of mixed drinks since at least 1806.

In the mid–19th century, cocktails were basically tavern-bottled mixtures of wine, bitters and fruit juice, taken on outdoor excursions. Because many recipes contained the extracts of roots, bark and herbs, they were held in esteem as having medicinal properties, and were prescribed as early morning tonics "to fortify the inner man." (Of note, the dry martini did not come into use until the 1890s.)

BOTTLED BRANDY COCKTAIL

⅔ Quart Brandy
1 Pony Glass Bitters
⅓ Quart Water
½ Pony Curacao
1 Wine Glass Gum Syrup

To make a splendid bottle of brandy cocktail, use the above ingredients. The author has always used this recipe in compounding the above beverage for connoisseurs. Whiskey and gin cocktails, in bottles, may be made by using the above recipe, and substituting those liquors instead of brandy.[21]

Around the Next Bend

Steamboat passengers and crews consumed a great deal of alcohol, but the boats were also responsible for delivering such refreshment to all compass points directly adjoining a river or tributary. Due to the varied necessities for mixing, serving, sweetening and flavoring hard spirits, shopkeepers, tavern owners and saloon proprietors needed extensive supplies. Pioneers, miners, fur trappers and soldiers all wanted their libations, and cargo lists throughout the decades are filled with such orders, making the alcohol trade small but lucrative.

One typical tavern owner's order contained the following items for the needy westerner:

2 barrels peaches	3 barrels whisky
8 sacks dry apples	1 barrel brandy
1 barrel crackers	1 barrel old rye whisky
2 boxes tobacco	1 keg (10 gallons) gin
1 box smoking tobacco	1 keg port wine
1 box oysters	1 keg dark brandy
1 keg St. Croix rum	
1 box bar tumblers	1 keg peach brandy
1 box decanters	1 keg (5 gallons) Holland gin
1 box pint flasks	1 keg lemon syrup
1 box currants	1 jug peppermint
1 jug bitters	

The American Temperance Society to a "T"

The year 1826 saw the birth of the American Temperance Society, although there had been stirrings since at least 1773, when John Wesley, a Methodist minister, advocated the idea of making the distillation of whisky illegal. More significantly to the mainstream medical movement that would later join forces with the religious elements, Benjamin Rush (physician to the great Washington and friend of Jefferson and Franklin), published a pamphlet entitled "An Inquiry into the Effects of Ardent Spirits," in 1784. "Ardent spirits" in this case referred to alcohol distilled from fermented grain, as opposed to wine and beer. These drinks he considered healthful if consumed in moderation. In that, he had the support of the Pilgrims. During their arduous voyage across the Atlantic on the *Mayflower,* they drank a dark, rich porter and, once settled in the new land, made beer, hard cider and berry wine.

Rush was the first American to describe systematically habitual and heavy drinking as a disease, which he termed "a disease of the will" that led to physical addiction and destroyed the body. He concluded that this disease led to dropsy, diabetes, epilepsy and palsy, to say nothing of dyspepsia. His "Moral and Physical Thermometer," referenced earlier in the chapter, advocated water as the path to "health and wealth."[22]

Philadelphia was the center of the temperance movement, which quickly gained members from across the spectrum of those advocating moral and social change. Among the many notables was Reverend Sylvester Graham, father of the Graham cracker and advocate against the evils of masturbation. Armed with the belief that sin was a matter of choice, "Christ's Soldiers" spread the word far and wide, even to the distant reaches of the Mississippi, where captains were implored to prohibit drinking aboard their boats among crew and passengers alike. A note in the *Daily Sentinel and Gazette* (Milwaukee) on November 12, 1846, could easily have been used as a persuasive warning:

> *The Cause of the Great Britain's Disaster* — The following article, in which the late disaster of the steamer Great Britain is, upon good authority apparently, attributed to the use of intoxicating drinks by the officers of the steamer.... [One of the passengers aboard states that] if the officers of the steamer had been temperance men, the accident would never have occurred. We have seen and conversed with a clergyman who has resided within ten miles of the spot where the vessel went ashore, and from the published statement of the nature of the disaster, he gives it as his opinion that nothing but carelessness unknown in navigation could have allowed its occurrence. He is perfectly familiar with the locality, and thinks the officers must have been under some unexplained hallucination.

As with any populist movement of substantial success, politics reared its ugly head as members of the Know-Nothing Party joined the ranks. Blatantly racist and staunchly against immigration, believing the Irish and Germans were little more than drunkards, they argued that American voters were being bribed with alcohol, based on the fact many polling places were local bars and taverns.

The early feminist movement was also allied with the temperance movement, for it was no great leap for women of the day to associate the consumption of alcohol with beatings, loss of family income and abandonment. This sentiment was aptly described by Frances Trollope, who noted of a family in Washington, "The gentleman, indeed, was himself one of the numerous tribe of regular whiskey drinkers, and was rarely capable of any work; but he had a family of twelve children, who, with their skeleton mother" worked their fingers to the bone.[23]

In just a few short years, the idea of taking alcohol in moderation was altered to the promotion of total abstinence, and the abandonment of "the traffic," or sale of alcohol. On membership roles, a "T" stood for a pledge of total abstinence, and from this developed the new word "teetotaler." In Boston, the American Temperance Society (ATS) grew into a powerful force, and purists were dubbed "Ultras."

Not to be outdone, the temperance movement also had its drinks.

Jersey Cocktail

2 Dashes Bitters
Apple Cider
1 Teaspoonful Powdered Sugar or Rock Candy
Lemon Peel

Dissolve the sugar with the bitters in a tumbler, fill one-third full of shaved ice, and the balance with cider. Shake well, and serve with lemon peel on the top.[24]

Drink for the Dog Days

A bottle of soda water poured into a large goblet, in which a lemon ice has been placed, forms a deliciously cool and refreshing drink. But it should be taken with some care, and positively avoided while you are very hot.[25]

While visiting Cincinnati in 1842, Charles Dickens witnessed a temperance society meeting, which began with a procession of "several thousand men; the members of various 'Washington Auxiliary Temperance Societies,' marshaled by men on horseback." Irishmen formed their own distinct group with green scarves, carrying the national flag, as well as a portrait of Father Matthew. "There was the smiting of the rock, and the gushing forth of the waters," and a man with a hatchet symbolically smashing the head of a serpent about to spring from a barrel of spirits.

Significantly, there was "a huge allegorical device, borne among the ship-carpenters, on one side whereof the steamboat Alcohol was represented bursting her boiler and exploding with a great crash, while upon the other, the good ship Temperance sailed away with a fair wind, to the heart's content of the captain, crew and passengers."[26]

Whether fortunately or unfortunately, the temperance movement made little headway on steamboats along the western waters, and alcoholic spirits continued to be served as a matter of course.

19

The Nameless Masses: Deck Passengers

Possibly the most telling of all comments directed toward those who paid the cheapest fares aboard steamboats came from a casualty report after a tragic accident.

None but deck passengers were thought to be lost.*

No names were given (deckers, as they were called, signed no register) and few tears were shed for the luckless men, women and children who perished in the calamity. On the upper decks, passengers from all walks of life ate and conversed together, and — for a time at least — put aside social prejudices. Paying first-class fare elevated the western frontiersman to the level of the eastern banker; an army lieutenant's wife might sit beside the wife of an English nobleman; children from Boston romped with those from Atlanta. For a day or a week or a fortnight, they all shared the novelty of the river, the fare of kings and perhaps the excitement of a steamboat race. Such neighborly good feelings, however, stopped at the stairs leading down from the boiler to the main deck. There, an invisible line kept the "quality" away from the "quantity."

Deck passengers, who often outnumbered cabin passengers by a 4:1 ratio, were excluded from the good times and the camaraderie of their fellow man. Prohibited, and at times restricted by ropes, they huddled below between boxes, sacks, barrels and cotton bales, sleeping on any open space not already assigned to horses, cattle and rousters.

Cabin to Deck Passenger Ratios

Decade	*Route/Trade*	*Ratio Cabin/Deck*
1821–26	Louisville	491/2,506 (cumulative)
1824–29	Cincinnati	270/1,402 (cumulative)
1840–50s	Cincinnati	96/150 (average)
1850s	Pittsburg	100/125 (average)[1]

Although the humble, cumulative price they paid for transport amounted to greater profit for the boat's owner than all the more exalted cabin fees, these emigrants, laborers, farmers, clerks and their families were considered as mere chattel, or "freight with legs."

---

*Adams Sentinel, *August 21, 1840.*

More a necessary evil than a recognized stratum of humanity, the difference between those paying $25 a head and those paying $3.50 represented a divide greater than education, breeding and manners could account for. Into this low-class group also belonged free blacks, for Negroes were not permitted to buy first-class passage even if they could afford to do so.

The concept of deckers originated very early along the river. Before steamboats, those unable to pay full fare often worked their way downriver on keelboats, being treated as, or working with, the crew. Keelboatmen themselves soon found that returning upriver as a steamboat decker was far easier than the arduous task of working their way back as a hand, and their numbers swelled the ranks and filled the coffers of boat owners. As more and more of the western lands were opened up, families hoping to make a new start added their prodigious numbers. Then came the foreigners, driven to America by harsh economic conditions, the Irish potato famine, or the dream of finding some level of equality in the new land. Those lacking money discovered that hard to come by.

Frances Trollope made this sarcastic comment about the company she did not keep, when traveling upriver from New Orleans in 1827:

> The deck, as is usual, was occupied by the Kentucky flat-boat men, returning from New Orleans, after having disposed of the boat and cargo which they have conveyed hither, with no other labor than that of steering her, the current bringing her down at the rate of four miles an hour. We had about two hundred of these men on board, but the part of the vessel occupied by them is so distinct from the cabins, that we never saw them, except when we stopped to take in wood; and then they ran, or rather sprung and vaulted over each other's heads to the shore, whence they all assisted in carrying wood to supply the steam engine; the performance of this duty being a stipulated part of the payment of their passage.
>
> From an account given by a man servant we had on board, who shared their quarters, they are a most disorderly set of persons, constantly gambling and wrangling, very seldom sober, and never suffering a night to pass without giving practical proof of the respect in which they hold the doctrines of equality, and community of property.[2]

Charles Dickens, on his return home to England, wrote with more concern.

> We carried in the steerage nearly a hundred passengers: a little world of poverty.[3]

Curious to know the history of the travelers, he got their stories from the ship's carpenter, as first-class passengers were not allowed to openly communicate, much less mingle, with the less fortunate. Some had been in America for several years and others a matter of days. A number had sold their clothes for passage money and, being without food, relied on the rest for charity.

Dickens urged that more be done to protect these unfortunates: the very least to provide medical attention, and to limit the number of bodies stuffed 'tween decks. More to the point, Dickens argues against the "crimping agents," who earned a percentage on all the passengers they inveigled, "tempting the credulous into more misery, by holding out monstrous inducements to emigration which can never be realized."[4]

The author was not alone in his indictment against those who profited by turning on their own: "the German preying on the German — the Irish upon the Irish — the English upon the English."[5] When the immigrants arrived in New York, runners, or "wolves," were hired to cheat the newcomers out of whatever money they possessed, either by steering them in the wrong direction or selling them spurious western land titles.

A Cook Stove and a Bucket

The trek from Ward's Island, New York, to the interior of the United States took many water routes, offering newcomers the opportunity to travel north, west or south.

Deck Fares (Eastern Route)

Year/Decade	Trade	Deck Passage
1830s	NYC to Albany	as low as 25 cents
	Albany to the Erie Canal	2½ cents per mile
	Buffalo to Detroit	$4
	Buffalo to Chicago	$9–12[6]

Deck Fares (Lower Mississippi)

Year/Decade	Trade	Deck Passage
Late 1820s	New Orleans to Louisville	$8–10
Early 1830s	" "	$6 upstream/$4 downstream
	New Orleans to St. Louis/Cinn	$5
1850	New Orleans to St. Louis/Louisville	$3
	New Orleans to Pittsburg	$6
	Pittsburg to Cinn	$1
	Pittsburg to St. Louis	$2
1855	St. Louis to St. Joseph	$2[7]

Fares averaging one-quarter cent per mile were the norm in the 1840s and 1850s, but what steamboats offered by way of cheap transportation was more than made up for by hardship.

Those residing on the lower between deck were subjected to conditions of almost inhuman overcrowding. They suffered extreme heat from their close proximity to the open furnaces and boilers and had little shelter, for the lower deck was left open, permitting wind and rain to penetrate nearly the entire area. When a disaster occurred, deckers were the ones most likely to lose their lives by scalding, being burned alive, or drowning.

A sadly typical account comes from the Newport *Daily News,* December 3, 1847:

> STEAMER TALISMAN.—We published last week an account of the sinking of the steamboat Talisman, by collision with the steamer Tempest, by which at least fifty lives were lost. The only additional information is in the following paragraph from the Louisville Journal:—
> From the clerk of the Anglo Saxon we learn that this boat rounded to at the wreck, and found two deck passengers on a wood-boat alongside of it, who said that of ninety-six deck passengers only ten were saved, and one cabin passenger, the engineer and two deckhands were drowned.

Deck passengers were obliged to bring their own food aboard, for none was offered by management free of charge. While this practice was not uncommon, Dickens complains that at the very least there ought to be government regulations requiring no man be taken aboard without an adequate stock of provisions to tide him over the long trip, for it often proved a tragedy for those who consumed their food before the trip ended or found that which they packed molded or rotted in the steamy temperatures. The suggested fare that best stood up to such conditions were bologna sausages, dried herring, water crackers, cheese and the invariable bottle of whiskey.[8]

Rarely, captains permitted deckers to buy leftovers from the main cabin, but for those already strapped for cash, this was a luxury they could hardly afford.

A river passenger left without food had but two other alternatives: rely on the mercy of others sharing their accommodations, or go ashore when the boat landed and attempt to buy inexpensive fare. This, however, was often impossible, for shopkeepers along the routes were well aware of the need and those closest to shore charged exorbitant rates.

Typically, steamboat owners provided one cook stove for the entire lower deck and passengers were required to stand in line to use it. The stove also represented the only means of heat, and those too far from the machinery suffered greatly during winter and cold spring nights.

Little if any accommodation for modesty was offered the deck passenger. Unlike those living above, where the sexes were separated, men, women and children were packed in together. The boat provided a bucket filled with river water for drinking and presumably another for toilet facilities, but it was more likely the animal stalls would be cleaned than any attempt by management made for the health and sanitation of their two-legged freight. There were, of course, no medical facilities, and disease ran rampant through the dirty, smelly, cramped quarters, resulting in a harrowing loss of life. (See chapter 21, "Potions, Purging and Practitioners")

Occasionally, efforts were made to improve conditions for those paying the lowest fares as revealed in this advertisement for the *Great Western*, a "brag boat," taken from the *Racine Advocate,* June 21, 1843:

> The traveling public are informed that the Great Western will be punctual in her departures, as above specified. Her State-rooms have been enlarged, and she has been much improved in all her apartments, particularly in her arrangements for steerage passengers. Emigrants will find her steerage accommodations superior to any other boat on the Lakes.

This, however, was the exception rather than the rule, and pertained to a boat working the Great Lakes trade. But it was a step in the right direction.

"I hate the Mississippi"

Perhaps Captain Marryat was thinking of the harsh realities faced by innumerable emigrants when he wrote:

> I hate the Mississippi, and as I look down upon its wild and filthy waters, boiling and eddying, and reflect how uncertain is traveling in this region of high-pressure, and disregard for social rights, I cannot help feeling a disgust at the idea of perishing in such a vile sewer, to be buried in mud, and perhaps to be rooted out again by some pig-nosed alligator.[9]

The lower deck lacked protection, and many deckers fell over the sides or were washed overboard. Tempers flared in cramped quarters, and injuries were sustained from fighting and roughhousing, at times with the crew, who had as little regard for them as those living in luxury above. In 1837, the *Dubuque* was traveling upstream from St. Louis when its larboard boiler exploded at Muscatine Bar, eight miles below Bloomington. The deck passengers were scalded and twenty-two died in one of the earliest and most devastating accidents on the river to that date.

On October 4, 1844,* the *Potosi* suffered a collapsed flue after a boiler burst, killing two deck passengers and badly scalding twenty others. Wind blew down the chimneys

---

* Note the discrepancy in the dates. The first came from George Merrick's list of Upper Mississippi steamboats and the second came from a passenger on the boat. The passenger's account made it into the newspaper six days later and may be considered the more accurate of the two. In compiling lists, authors like Merrick, who wrote decades after the fact, often relied on word of mouth or personal reminiscences to supply dates and casualty figures.

of the *Michigan*, throwing thirty crew and deck passengers into the river; four were drowned.[10]

The *Hawk Eye* (Iowa) reported on the sinking of the *Potosi*:

> An explosion took place last Friday morning [September 27th*] between two and three o'clock on the steamer Potosi while at Quincy on her passage from St. Louis to Galena. She was about starting, the wheels having made one revolution, when one of the flues collapsed. All of the cabin passengers [except two] had retired to their state rooms and not one was injured. A Mr. Perrin, of one of the up river towns, a deck passenger, was mortally injured. He died on Friday evening. A deck hand named Miller was supposed to have been blown overboard as he was missing. About fifteen others were more or less scalded. Quite a number of the passengers jumped off the boat and arrived safely on shore. The boat floated down about half a mile below Quincy, where she was moored to the shore. The citizens of the town visited the scene and were very active in rendering assistance and relief to the distressed and the needy.

One other account, typical of the era, gives a good idea of deck passenger deaths. It comes from the *Republican Compiler*, May 1, 1838:

> Little Rock (Ar.) April 2: STEAM BOAT ACCIDENT. The steamboat Liverpool, which left this place about two weeks since, burst her boilers on the evening of the 4th instant, on the Mississippi river, about a mile and a half above the mouth of the White river, scalding eleven persons, deck passengers, four of whom are not expected to recover, and 4 were seen to jump overboard, and were drowned. She was towed down by the Paul Jones, had repaired, and proceeded up to Memphis, her destination.

Working One's Way

If economic necessity (or a quest for adventure) prompted a passenger to pay even less for his transportation than basic deck fare, he was afforded the opportunity to work off a portion of the trip by serving as a wooder. Many no doubt came to regret the decision, for the task was arduous and required a man to be called at all hours of the day and night, in fair and foul weather. (Women were not allowed this privilege.) For a reduction of $1 in fare, the decker was called with the deckhands to race ashore, grab heavy logs and bring them back to the boat, all in 30 minutes to an hour. If the steamer was wooding up from a wood scow, the hands jumped aboard that vessel and tossed the wood up on deck.

Not a few of the volunteers withered under the physical exertion, and opted to hide when the call for wooders came. In such cases, the mate went through the boat, ferreting the reluctant deckers out of their hiding places with cursing and threats.

A Touch of Humor, Deck-Passenger Style

If deck passengers were no more than walking cargo, occasionally a smile might be elicited from the experience. The *Adams Sentinel*, June 4, 1838, recounts one incident with a nod and a wink:

> *Going as Freight.*—An Irishman, whose funds were rather low, had footed it all the way to Wheeling, and was still desirous to get as far as Portsmouth, thence to proceed by canal to a point not far distant from the latter place, where work was to be obtained. Having worn his toes through his boots, and the heels of a pair of old shoes quite low, he gave up the idea of using "shank's mare" any longer. There were plenty of steamboats

puffing and blowing at the landing and he became quite fascinated at the idea of such an easy mode of conveyance.

"Captain, dear," said he, stepping on board a beautiful craft — "Captain, dear, an what'll you charge to take me to Portsmouth?"

"Siven dollars! arragh! siven dollars!"

"Why, Captain, dear, I hav'nt the half of that sum."

"Oh, never mind that, Pat, I'll take you as a deck passenger for three dollars, if you'll work half your passage, that is, help the hands to wood the boat."

Pat mused some minutes on this proposition, and then put another question.

"And Captain, dear, what'll you take about a hundred and sixty pounds of freight for?"

"I'll charge you seventy-five cents for that."

"Then, Captain, you see, I'm jist the boy that weighs that — so you can enter me as freight, and I'll stow away snug enough some where below stairs."

A proposition so novel pleased the Captain highly, and calling one of the hands, he gave directions to have Pat stowed carefully away in the hold — and ordered the clerk to enter on the freight list — *"One Irishman, weighing 160 pounds."*

Pat kept snug until he reached Portsmouth, a distance of 300 miles — having shown himself but twice, and for only a few minutes at a time during the whole passage. There he paid his freight, of seventy-five cents, honorably, and was next seen with his bundle, tramping it along the two paths of the canal for his desired destination. *(Baltimore Atheneum)*

Life was hard, conditions were extreme, and segregation proved that even in the New World, rank had its privileges. But most accepted their lot and some, as indicated above, managed to elicit a few smiles along the way.

20

An International Incident

In 1830, Americans did not have far to look for their entertainment. The publication of Captain Basil Hall's *Travels in North America* created an immediate sensation, that for a time eclipsed discussions of Andrew Jackson's populist movement and his pending fight with the Bank of the United States. Frances Trollope noted, "the book was read in city, town, village, and hamlet, steamboat, and stage-coach, and a sort of war-whoop was sent forth."[1]

Everyone wanted to know who this Basil Hall was, and why he had taken advantage of his status as a foreign dignitary to write such atrocious lies about the country. His background and reputation were investigated, and collective fists were raised whenever his name was mentioned, uniting the inhabitants from New York clerks to Mississippi captains.

"For my part, I consider that no thoroughly independent man is worth a fig."

If Basil Hall, who, if he did not start this war of letters, contributed greatly to its continuance. Born December 31, 1778, the second son of James Hall, 4th Baronet, a noted man of science, Basil was educated at Edinburgh High School and entered the Royal Navy. Commissioned a lieutenant in 1808, he rose through the ranks to captain. Exploring Java in 1813 and 1817, he conducted many scientific experiments and even interviewed Napoleon (who was a friend of his father) at St. Helena. The author of several travel books, he toured the United States with an eye toward an additional work. The resulting three volumes, entitled *Travels in North America in the Years 1827 and 1828* (1829), ignited bitter recriminations on one side of the Atlantic and wild accolades on the other.[2] Hall reportedly received $7,125 for the copyright of his travels in North America from his publisher.[3]

Armed with his reputation and numerous letters of introduction, Basil Hall visited the major eastern cities, toured Washington, traveled down the Mississippi and eventually worked his way north to Canada. Writing on the premise that a spirit of "mutual animosity" existed between the Mother Country and the United States, he added that amending these "unkindly feelings" was not "either practicable or desirable."[4]

Hall took considerable pains to describe the failings of U.S. legislators, which

> come straight from the plough, or from behind the counter, from chopping down trees, or from behind the bar.

He took particular exception with the lack of amenities provided to travelers, describing one scene:

> One day I was rather late for breakfast, and as there was no water in my jug, or pitcher as they call it. I set off post haste, half-shaved, half-dressed, and more than half vexed, in quest of water, like a seaman on short allowance, hunting for rivulets, on some unknown coast. I went upstairs and down stairs, and in the course of my researches, might have stumbled on some lady's chamber, as the song says, which, considering the plight I was in, would have been awkward enough.

Basil Hall was the not the first to condemn the new nation. Prior to 1800, travelers were generally positive in their praise, but the turn of the 19th century marked a more vitriolic period. Weld writes that Pennsylvania farmers "live in a penurious state," and are "greatly inferior to the English," and bemoaned the fact he did not "meet with a man of decent literature." The *Edinburgh Journal* published an article by Sydney Smith, in January 1820, in which he remarked, "In the four quarters of the globe, who reads an American book? or goes to an American play? or looks at an American picture or statue?" The *Quarterly Review* of 1829 went so far as to sneer, "The memory of Washington will probably be nearly extinct before the present century expires."

A Moral Earthquake

Reactions to Captain Hall's adventure stories in the New World were swift. It was the subject of vocal derision, and writers from all walks of life hurried to put in their two cents' worth. Writer Frances Trollope came across a "little anonymous book," critiquing *Travels in North America*, "written to show that Capt. Basil Hall was in no way to be depended on."[5] This text, actually written by Richard Biddle, a Philadelphia lawyer, was printed in Philadelphia in 1830, and reprinted by R. J. Kennett, 59 Great Queen Street, Lincoln's-Inn-Fields, London, the same year. Titled *A Review of Captain Basil Hall's Travels in North America in the Years 1827 and 1828,* it was credited to "An American," and came in at an impressive 148 pages. An erudite and witty work, the author listed a number of reasons why Hall published his book. (Many Americans suspected Hall of being employed by the British government to counter growing admiration of the United States).

Biddle begins by stating that Hall's troubles began when he attempted to assemble his mass of notes "through a bilious medium." (Hall often referred to digestive ailments.) He goes on to suggest that it was essential for Hall to imbue the work with dignity so that important political reflections might be drawn from it: "This is the trying crisis when anxious thoughts throng upon a weak, and a vain man, looking over his discordant notes and calculating the chances of success," wrote Biddle.

He adds that in order to take out a copyright, an author must be a resident of the United States, and as that did not suit Hall's views, he wrote exclusively for his countrymen's tastes. This went hand in hand with Hall's stated observation that he would rather find "his original and preconceived conceptions right, than to discover their injustice had previously been done to the people." Thus, he acknowledges the book would not be well received by those who had befriended him on his journey.

Hall, as a partisan and a Tory, likely supported the view expounded in the *Edinburgh Review* (Volume XXXIII):

> ... the peace and well being of a society in no danger from anything but popular encroachments, and holds the only safe or desirable government to be that of a pretty pure and unencumbered monarchy, supported by a vast revenue, and a powerful army, and obeyed by a people just enlightened enough to be orderly and industrious, but no way curious as to question of right, and never presuming to judge of the conduct of their superiors.

Noting that Hall's only expressions of enthusiasm were for his meals ("A thousand years would not wipe out the recollection of our first breakfast at New York," and "The glorious breakfast ... as lively a picture of Mahomet's sensual paradise, as could be imagined"), Biddle takes many good-natured and not-so-good-natured swipes at the captain, the most amusing of which is when he counters Hall's panic to find water.

> Then it is evident that Captain Hall is himself to blame, for lying in bed until she (the chambermaid) was called off to wait on the breakfast table. That he is rather indolent and aristocratic in his habits, he has obligingly informed us.... Why not put on his clothes? But for his own comparative sluggishness, Captain Hall would have found in these chambers ladies, he knew not, and he cared not whom.... Here, then, we find a gentleman, going about the rooms of a house, expecting any moment to meet females, and conscious that his person was indecently exposed.

Another author, quoted in the *Monthly Commentary*, under the heading "America," writes in response to a query about Captain Hall's book:

> "There is a strong disposition in most of the European governments to represent the condition and the character of the Americas in a false light. One is jealous of the example of their institutions; (I might have said, most have this jealousy;) another is apprehensive that her artisans may carry the knowledge of their crafts into the New World, and change a customer into a rival; while a third is desirous of making its own subjects believe there is no nation freer or happier than their own." He finally drew the conclusion, "It would require a book as large as his own properly to dissect the three volumes of Mr. Hall. I must repeat, that so far as facts are concerned, he is constantly in error."

From "Thin Skinned" to "No Skinned"

Seizing on the controversy, Frances Trollope, an Englishwoman in reduced circumstances, decided to pen a book about her own travels (gained while she attempted various money-making schemes in the United States, including a failed attempt at starting a theater/coffeehouse/lecture hall that was dubbed "Trollope's Folly"[6] Having read Hall's book while in America, she placed herself squarely in his corner. *Domestic Manners of the Americans* (1832) was published by the same house that had released Hall's book.

Trollope's work made her a favorite in England and a pariah in America. Trollope fueled the flames by writing of Hall's *Travels in North America*, "To say that I found not one exaggerated statement throughout the work is by no means saying enough." She then added one of the most often quoted observations, "Other nations have been called thin-skinned, but the citizens of the Union have, apparently, no skins at all."

In an impressive feat of literary log-rolling, Basil Hall wrote an anonymous article in the Tory *Quarterly Review*, announcing Mrs. Trollope to be "an English lady of sense and acuteness," and her work a necessary treatise, "when so much trash and falsehood pass current respecting that 'terrestrial paradise of the west.'"

In the 1832 introduction to *Domestic Manners*, Trollope writes of herself (in the third person), "The chief object she has in view is to encourage her countrymen to hold fast by a constitution that ensures all the blessings which flow from established habits and solid principles. If they forego these, they will incur the fearful risk of breaking up their repose by introducing the jarring tumult and universal degradation which invariably follow the wild scheme of placing all the power of the state in the hands of the populace."

Trollope described the day-to-day existence of everyday Americans. Of steamboat travelers, she noted:

> On the 1st of February, 1828, we embarked on board the Criterion, and once more began to float on the "father of waters," as the poor banished Indians were wont to call the Mississippi. The company on board was wonderfully like what we had met in coming from New Orleans. I think they must have all been first cousins; and what was singular, they too had all arrived at high rank in the army.[7]*

Reiterating a common European theme, Mrs. Trollope summarized America:

> The want of warmth, of interest, of feeling, upon all subjects which do not immediately touch their own concerns, is universal, and has a most paralyzing effect upon conversation. All the enthusiasm of America is concentrated to the one point of her own emancipation and independence; on this point, nothing can exceed the warmth of her feelings.[8]

In many respects, Trollope's text followed Hall's. He wrote:

> The perfect difficulty which men who have become wealthy have to encounter in America is the total absence of a permanent money-spending class in the society, ready

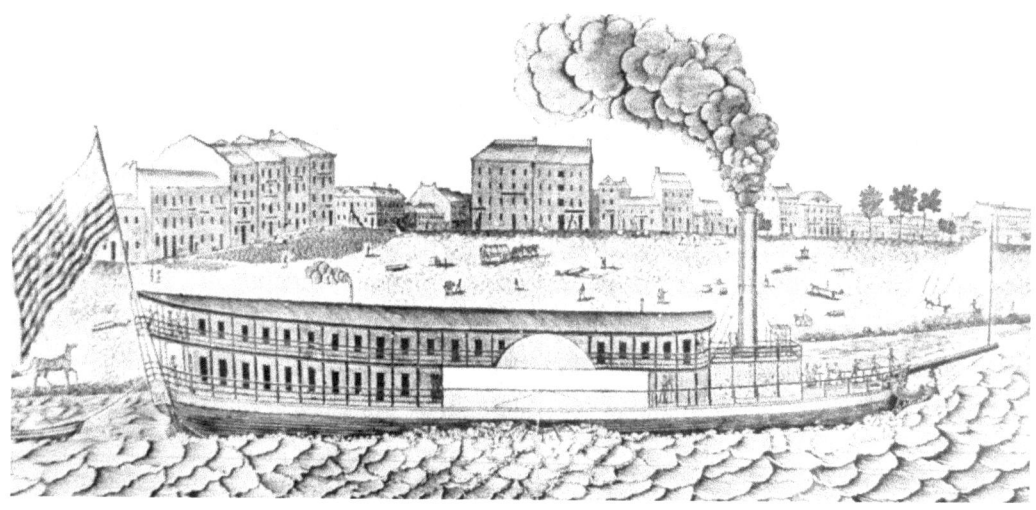

A very early sketch of a steamboat with a major city in the background. The place for the name is conspicuously blank, perhaps so it could be used in all manner of advertisements.

---

*Mark Twain observed, "It was Sir Walter (Scott) that made every gentleman in the South a Major or a Colonel, or a General or a Judge, before the war; and it was he, also, that made these gentlemen value these bogus decorations."[9]

not only to sympathize with them, but to serve as models in this difficult art ... (those) who, having derived their riches by inheritance, are exampled from all that personal experience, in the science of accumulation, which has a tendency to augment the difficulty of spending it well.[10]

And she echoes:

> ... where every man is engaged in driving hard bargains with his fellows, where is the honored class to be found, into which gentlemanlike feelings, principles, and practice, are necessary as an introduction?[11]

Messieurs de Trois-Idées-Européennes

Reaction to Trollope's work was as swift as it had been to Hall's. Among the many who critiqued (and mocked) her was James Fenimore Cooper, then living in Paris. In 1832, he wrote and published, in French, a satire entitled, "Point de Bateaux á Vapeur" ("No Steamboats: A Vision"). In the tale, he is visited in a dream by three wise men ("their footmen call them abstractions") who wish to explain their ideas to him on how America (her government and her manners) are inferior to those of Europe. The trio open by explaining:

> ... we are touched to the heart by the danger of a people possessing but *one idea*; an idea so selfish that it confounds an entire nation with itself. We see your perils, moral, social, and pecuniary, and have resolved not to abandon you to your own movements without one effort to show you the gulf into which you about to fall.

The Men of Three Ideas present Basil Hall, the *British Review,* the *Quarterly Review* and the work of Mrs. Trollope as proof of their assertions. Among them is the proper name of the country ("The United States of North America"), noting, "It is clear the Constitution is wrong" on that point. They assert, "The impartiality of Madame Trollope is beyond approach," and add, "You are found to be plain speakers," to which Cooper replies, "We shall be feared the more."

After waking from his dream, Cooper holds out his hands to "grasp the documents from Messieurs the 'Three Ideas,' as precious relics. They had disappeared. Nothing of them remained." His character then declares in a note of finality, "We are leaving for America!"

Dickens Trollopizes

Ill feelings over Hall's and Trollope's books had hardly faded away when Charles Dickens made his celebrated voyage to America in 1842. In his very readable and intricately detailed *American Notes*, he describes his journey through the United States and Canada.

As Trollope had before him, Dickens, too, remarks on the American custom of spitting, calling Washington "the headquarters of the tobacco-tinctured saliva." He goes on to say:

> In all the public places of America, this filthy custom is recognized. In the courts of law, the judge has his spittoon, the crier his, and the prisoner his; while the jurymen and spectators are provided for, as so many men who in the course of nature must desire to

spit incessantly. In the hospitals, the students of medicine are requested, by notices upon the wall, to eject their tobacco juice into the boxes provided for that purpose, and not to discolor the stairs. In public buildings, visitors are implored, through the same agency, to squirt the essence of their quids or "plugs," as I have heard them called ... into the national spittoons, and not about the bases of marble statues.[12]

In his concluding remarks, Dickens summarizes his impressions:

It would be well, there can be no doubt, for the American people as a whole, if they loved the Real less, and the Ideal somewhat more. It would be well, if there were greater encouragement to lightness of heart and gaiety, and a wider cultivation of what is beautiful, without being eminently and directly useful.[13]

In a private letter to his friend, William Charles Macready, Dickens put the finishing touches on "Trollopizing" (a verb in vogue after the publication of *Domestic Manners*, meaning, "to abuse the American nation"):

I *am* disappointed. This is not the Republic I came to see. This is not the Republic of my imagination.... I believe the heaviest blow ever dealt at Liberty's Head, will be dealt by this nation in the ultimate failure of its example to the Earth.)[14]

Sad closing comments, but one rejected by the bulk of Americans. They persisted with their indomitable will, their peculiar "follies and vices," casting those of the Mother Country, it may be said, by the wayside. The Sandusky *Clarion* (Ohio), February 3, 1846, reprinted an editorial that best sums up the situation:

Dickens, in his Notes on America, Mrs. Trollope, and other English writers, have spoken in a very satirical manner, of the braggadocio which characterizes Americans. It is true we may have too much of the boasting spirit; but have we not reason for it? The Philadelphia North American very truly remarks that —
The greatest man, "take him for all in all," of the last hundred years, was George Washington, an American.
The greatest natural philosopher was Benjamin Franklin, an American.
The greatest of living sculptors is Hiram Powers, an American.
The greatest of living poets is William Cullen Bryant, an American.
The greatest of living historians is William H. Prescott, an American.
The greatest living painter, in portraiture, is Henry Leman, an American.
There has been no English writer to the present age, whose works have been marked with more humor, more refinement, or more grace than those of Washington Irving, an American.
The great lexicographer and philologist, since the time of Johnson, was Noah Webster, an American.
The inventors, whose works have been productive of the greatest amount of happiness to mankind, in the last century, were Godfrey, Fitch, Fulton and Whitney — all Americans.
If one of these facts or estimates is doubted, we can prove them by *foreign* authorities, and so prevent all controversy.

Two final facts and one anecdote: Basil Hall, who set off the "international incident," suffered from mental illness and was held at the Royal Hospital Haslar, Portsmouth, where he died in September 1844. His passing warranted a small obituary in the Sandusky *Clarion*, October 12, 1844:

We notice the death of Capt. Basil Hall, R.N., known to our readers by his travels in this country and his many literary works. He had for some years been an inmate of one of the insane asylums near London.

Frances Trollope eventually earned about 600 pounds from *Domestic Manners,* but it was not enough to save her husband from his creditors. After his death, she and her son Tom moved to Florence, where she continued her literary output. At the time of her death, in 1863, she had penned six travel books and thirty-five novels.

From the *Democratic Banner* (Iowa) of November 25, 1853, comes what is, possibly, the final word:

> A Sensitive Captain — The captain — that is to say, him of the sailing vessel — when he learned the name of the lady passenger who wished to sail in his vessel, refused to receive her on board; and when Marcus insisted on knowing his reason why, he replied that he did not wish to have any authors on board of his ship, who would laugh to scorn his accommodations, and who would put him in a book. Marcus laughed, and wanted to persuade him to run the risk, assuring him that I was not dangerous, and so on. But the man was immovable. He would not take me on board; and I have now to wait till the next steamboat goes, which is eight days later. And for this I have to thank Mrs. Trollope and Dickens.

21

Potions, Purging and Practitioners

The early and mid–1800s were a time of radical change. Technological developments in transportation led to the steamboat, and the steam engine was soon put to work in factories, revolutionizing the manufacturing industry. It was a period of wild speculation in land, mass immigration and social unrest, touching every aspect of life. No one was immune, although not everyone embraced these new concepts.

Politically, America slowly broke away from the old, established order of the Revolution, moving toward the idea of greater personal independence. The women's movement, abolitionist movement, temperance societies, religious reawakening and health crazes spread across cities, towns and frontiers. Many of these groups saw mutual points of agreement and gathered together, forming even larger, stronger organizations. Female enfranchisement, those advocating an end to slavery and wellness reformers formed societies, spoke at large gatherings, toured the southern states and demanded to be heard. Often they were heard; more often they were ridiculed; but once begun, nothing swayed these men and women from their appointed courses. Developing primarily on the East Coast, these movements encouraged independent thought and the breaking down of barriers. While most of their ideas faded the farther they got from the principal cities of New York, Philadelphia and Boston, they affected in some manner even the rough-and-tumble men of the frontier.

The medical profession underwent radical changes. Health, diet and doctoring enjoyed general interest, and medical improvement competed with outright quackery.

A concern for human life prompted legislation forcing the improvement of steamboat safety; portside chapels were constructed for use of sailors; federal maritime medical protection was extended to boat crews, and those committed to the abandonment of alcoholic spirits made their appearance, if not their mark, on western cities. Even the recognition, treatment and fear of epidemics, rampant among the warmer states and along the coastlines, affected those who worked the steamboats. How they and their contemporaries dealt with such concerns is a significant topic of discussion.

"Sew people up like garments"

One of the greatest destructive forces in nature is the possibility that disease may strike anywhere, at any time. Before the development of antibiotics, the generalized use of the stethoscope and thermometer, or any full appreciation of the microscope and the hidden world it revealed, rendered most medical treatments a combination of the barbaric and the wistful.

Surgery was a haphazard procedure, resorted to as a last option. In fact, few medical colleges of the time taught the techniques of operation, relying instead on tonics, potions and purging. It was not until the Civil War that physicians were forced to hone their surgical skills. Incredible numbers of arms and legs were amputated, and mortality rates were horrendous. Soldiers who survived the operating theater with their limbs "sewn up like garments" frequently died from gangrene and infection, for basic sanitary precautions such as hand-washing were not practiced.

Doctors went from patient to patient with blood and gore on their hands, presenting a perfect medium for the transfer of germs. Many physicians would go from one delivery to another without so much as wiping their hands on a towel. If a routine birth happened to follow one with complications, the unlucky woman and her offspring were likely to die from the unknown danger of cross-contamination. (During this period, there was no such thing as specialized medicine. Women's heath was considered synonymous with that of children's.)

In the 1800s, cholera, yellow fever, typhoid, smallpox and malaria were common. In many cases, the first appearance of fever or a blister sounded the knell of impending death. People were afraid and they had a right to be.

Epidemics spread across continents, states, boundary lines and from neighborhood to neighborhood without rational explanation. In fact, many diseases were considered part of the natural order. Such an attitude did not preclude either the men of science or the family apothecary from attempting to prevent, if not cure, disease. In more instances than not, their wild speculations gave rise to treatments that hastened or even caused death.

The Centers for Disease Control

In the 18th century, mainstream doctors were referred to as "regulars," and until 1745, British physicians belonged to the same guild as barbers.[1] The first medical school in the United States was opened in Philadelphia in 1762 by William Shippen, Jr. His lectured solely on midwifery. The following year he expanded the curriculum to include the study of "physic" (medicine), "and also for the entertainment of any gentleman who may have the curiosity to understand the anatomy of the Human Frame."

Morality also played a role in health care. In the 1700s, hospitals refused entry to women in the "family way" who were not wearing wedding rings, those suffering from venereal disease, and patients with contagious diseases. Disfigurement was often considered a punishment by God, and only those deemed "morally worthy" were admitted into clinics. Prejudice also played its ugly role, and neither slaves nor freed blacks working the steamboats were permitted free government health care granted "regular seamen" on the inland rivers through the expanded Maritime Act.

The study of anatomy meant dissection, and dissection required cadavers. In the 18th and 19th centuries, it was an open secret that medical schools often obtained corpses from grave robbers (otherwise knowon as "resurrectionists"). In New York City in the late 18th century, not even the cemetery at Trinity Church was safe from grave robbers; and in the early 19th century, Great Britain was shocked by the case of William Burke and William Hare, who allegedly murdered 17 people and sold the corpses to Dr. Robert Knox for use in his anatomy lab. Later (1832 in Britain and 1854 in New York), legislation was passed that gave medical schools legal rights to human remains under certain circumstances.[2]

Formal education, however, was not considered necessary to practice the healing arts. On-the-job training with an established physician, called a "preceptor," was considered the only prerequisite before hanging out a shingle. Those who did opt to enter a medical college were not required to meet any academic standards. If they had the requisite fee, they were admitted. A standard course of study lasted from sixteen to twenty weeks.

In 1828, Andrew Jackson was swept to the presidency under the banner of anti-elitism. The people of the United States did not believe the advantaged, the moneyed or the highly educated had the sole right to dictate politics or maintain a stranglehold on any profession. In the Era of the Common Man, it was believed anyone had a right to choose his profession without the approval of any governing board. Such an attitude worked well in the western portion of the country where the rogue investor, with only a few dollars in his pocket, could buy a steamboat and declare himself a captain.

The same held true for the practice of medicine. During Jackson's presidency, nearly all government regulations of the healing arts were discarded in favor of a belief in the innate goodness and wisdom of the individual practitioners. By 1845, only three states still licensed medical doctors.

In *The American Phrenological Journal*, August 1845, an unsigned author writes, "Another recommendation of this new panacea [the water cure], consists in allowing all to become their own doctors, or, at least, allows families to do their doctoring *within themselves* ... to render the heads of every family the doctors of their own families. This is the true system. It is *not* the order of nature that one class should fatten on the miseries of another, and that the larger; for then, it is to the interest of the few that disease should multiply its horrors, and that the masses should not inform themselves as to the way to preserve health."

Cleanliness was another aspect of 19th-century life sorely neglected by Americans. Hygiene was considered a personal matter and left to individual tastes. Bathing was an uncommon practice even among the well-to-do. Irregular sponge baths usually sufficed. To somewhat disguise bodily odor, perfumes were preferred, not only on the face and under the arms, but splashed on clothing. Expensive wardrobes of lace and crinoline were difficult to launder and scents made from ambergris were often used, although they left oily stains.

Public Sanitation

Early towns were laid out without a thought to the removal of waste and squalid, densely populated cities encouraged the spread of disease. Insects and other vermin infested bedclothes. Municipal water systems were late coming to the United States. The first appeared in Philadelphia in 1830. New York City got its Croton Water Works in 1842, Boston finally got theirs in 1848.

Europeans traveling across the United States were appalled at what they considered a barbaric lack of sanitation and took pains to describe conditions as they found them. Describing Cincinnati in 1828, Frances Trollope says:

We were soon settled in our own dwelling which looked neat and comfortable enough, but we speedily found that it was devoid of nearly all the accommodations that Europeans conceive necessary to decency and comfort. No pump, no cistern, no drain of any kind, no dustman's cart, or any other visible means of getting rid of the rubbish, which vanished with such celerity in London, that one has no time to think of its existence; but which accumulated so rapidly at Cincinnati, that I sent for my landlord to know in what manner refuse of all kinds was to be disposed of.

Your Help will just have to fix them all into the middle of the street, but you must mind, old woman, that it is the middle. I expect you don't know as we have got a law what forbids throwing such things at the sides of the streets; they must just all be cast right into the middle, and the pigs soon takes them off.

In truth the pigs are constantly seen doing Herculean service in this way in every quarter of the city; and though it is not very agreeable to live surrounded by herds of these unsavory animals, it is well they are so numerous, and so active in their capacity of scavengers, for without them the streets would soon be choked up with all sorts of substances in every stage of decomposition.[3]

Charles Dickens noted the same phenomenon when he wrote in *American Notes*, "Here, as everywhere else in these parts, the road was perfectly alive with pigs of all ages, lying about in every direction, fast asleep; or grunting along in quest of hidden dainties."[4]

Night air was considered to be polluted by poisoned particles dissolved in the vapor, prompting people to keep their doors and windows closed. Sickrooms, especially, were shuttered and darkened in the belief that contamination might be carried on the wind. This was particularly true of birthing chambers, where postpartum women were kept inside dark, stifling chambers for weeks after the birth, forbidden sunlight, exercise or nourishing foods.

People had no any idea of how disease was actually spread, so outhouses were often constructed close to the family garden or near the well. Lime was used to lessen the odor and eat away the refuse, but lime was expensive and most rural families did not go to the trouble. In cities, the somewhat tongue-in-cheek expression "being naturalized" referred to having the contents of a chamber pot thrown on one's head.[5]

Vegetables, Meat, Knives and Liquor

Vitamins and minerals necessary for a balanced diet were unknown in the 19th century. Although early seamen learned that fruits were necessary to prevent scurvy (they brought aboard apples or cider, onions and limes, hence the description of a British seaman as a "limey"), Americans ate few green vegetables, and people even believed that tomatoes were poisonous.

Meals consisted primarily of various cuts of beef, beef by-products and fatty pork, along with potatoes and corn (primarily eaten as cornbread). During winter months, when the last of the spuds, cabbages and onions from the root cellar were consumed, families subsisted on dried meat, what they hunted in the woods and whatever chickens they cared to sacrifice.

Food was often eaten in a hurry, with knives used to bring food to the mouth. The primary use of the fork was to hold food while being cut. In her famous description of a western meal, Frances Trollope remarks:

> The total want of all the courtesies of the table, the voracious rapidity with which the viands were seized and devoured, the strange uncouth phrases and pronunciation; the

loathsome spitting, from the contamination of which it was absolutely impossible to protect our dresses; the frightful manner of feeding with their knives, till the whole blade seemed to enter the mouth; and the still more frightful manner of cleaning the teeth afterwards with a pocket knife soon forced us to feel that we were not surrounded by the generals, colonels, and majors of the old world; and that the dinner hour was to be any thing rather than an hour of enjoyment.[6]

Alcohol was also a staple of the American meal, served at all times, including breakfast. It was also considered hospitable to serve while entertaining; the well-to-do host would offer sherry, rye and Madeira. Americans had a long history with intoxicating beverages, and in the early years whisky often substituted for money. Aboard ships of all nations, rum was standard issue to sailors, and on an occasional "dry ship," when tea was substituted, the owner often did so more as a cost-savings measure than for any concern for his laborers.

Tonics and Pitchmen

Perhaps nothing recalls the 19th-century heath fads better than patent medicines and the men who sold them. Marching together with the idea that every man was his own doctor, chemists, apothecaries and "sure–cure schemes" dominated the era. Because there was no government or state regulation of what could be sold or how it could be marketed, outrageous claims were made to a gullible and often desperate public. The laying on of hands (faith cure), branch water tonics (allopathic cure), mesmerism, or "animal magnetism" (hypnosis), and good, old-fashioned alcohol, cocaine and opium all played their part.

Primary among these miraculous treatments were the little brown, corked bottles, covered in paper and tied at the neck with string, and the equally disarming small clear jars filled with white powders. In order to give these "medicines" dignity, they all came with important-sounding names and were advertised in all the newspapers sold along the riverfronts:

> Dr. Bateman's Pectoral Drops
> John Hooper's Female Pills
> Daffy's Elixir Salutis for "colic and gripping"
> Aqua anti-terminales
> Bilious Pills (to combat yellow fever, jaundice, dysentery, dropsy, worms and "female complaints")
> Swan's Panacea (America's first sugar-coated pill)
> Vermifuge (a tapeworm medication for children)[7]

The original idea of patent medicine came from England, where "pitch doctors" actually performed on stages. In this, they were much like the European mountebanks, who put on exotic-animal performances, pulled teeth and performed magic tricks to augment the wonders of their medicines. Clowns (called *zanni* in Italian) also accompanied this traveling show, and gave rise to the English word "zany."

In the United States, there were two forms of inducement: the high pitch and the low pitch. The high pitch consisted of a "doctor," surrounded by brightly colored anatomical charts and occasionally a full skeleton, selling his wares off a wagon or stage. The low pitch was a toned-down version, wherein the "healer" sold his goods from a satchel mounted on a tripod. Usually well-dressed in long black coats and broad top hats, they offered proof of their products by employing shills to drink the tonic and profess a complete cure.

Other tricks of the trade included:

(1) Rubbing the affected limbs of arthritis and rheumatism patients with their liniment. The rubbing and pressure dulled the pain and made the patient believe he had been healed.
(2) For those suffering from deafness, the pitchman would place a few drops of his medicine in the patient's ear canal and clean it with a swab. If the actual cause were no more than an accumulation of ear wax, a cure was achieved.
(3) To cure tapeworms, the practitioner would sell white strings coated with sugar. After swallowing, the sugar would dissolve and the patient evacuated the "tapeworm," which was actually the string.
(4) For people suffering from liver disease, "liver pads" containing red pepper and glue were externally applied to the affected area. Melted by body heat, the combination provided a comforting, temporary warmth, which seemed to alleviate symptoms.[8]

In addition, hypnosis could reduce violent symptoms for as long as the spell lasted. Some "doctors" purveyed tonics heavily laced with alcohol or derivatives of the poppy. These were sometimes darkened with brown sugar, and more often than not they relieved pain. However, of course, their use often brought serious consequences, particularly because the medicines were overused or taken "straight," despite instructions to dilute small quantities in water.

Dubious Medicines

"Regular" physicians were not immune from the wonders of tonics, but they also had other medicines in their little black bags, some of which had been handed down from the very first practitioners of the healing arts, and were effective for some diseases. Perhaps the most effective (and of most concern to southerners) was quinine, extracted from the bark of the Peruvian cinchona tree. It was used to combat the effects of malaria, very prevalent in warm climates. During the Civil War, when the southern ports were blockaded, high prices for quinine were paid to blockade-runners bringing it in from Europe.

In the 1800s people were obsessed by two health concerns: bilious conditions (acid indigestion) and moving their bowels. The first is easy to explain: people did not eat what we now know as a well-balanced diet. That led to stomach complaints and bowel disorders. Purgatives, emetics and enemas were liberally prescribed, frequently dehydrating or poisoning the sufferer.

Laxative salts, opium and castor oil were often used to "cleanse" the system. When used to excess (under the philosophy "more is better"), these often caused patients to salivate uncontrollably, bleed from the gums, lose teeth and develop sores in the mouth. They also led to acute diarrhea, bloody stools and internal bleeding.

The Terror of Cholera

If there was anything more dreaded on steamboats than fire, explosions and fog-enshrouded obstructions, it was without question the appearance of disease. Unfortunately, the boat decks of the era were breeding places for the most feared of all diseases: cholera.

Steamboats were natural carriers of the disease. Some immigrants arriving in New York were already infected with cholera, and their passage westward spread it along the inland waterways. The squalid conditions aboard steamers augmented the spread with horrific overcrowding, unwholesome food and the commingling of close air with noxious exhalations and ash-laden pollutants.

In the 19th century, it was held that cholera was caused by a shortage of ozone in the atmosphere. The way to combat it was by supplementing the body with sulphurous medicines. Sulphur pills were marketed as the cure, and druggists could not keep them in stock. To aid in clearing the air, sulphur candles were burned in homes and businesses. In the 1840s, when the disease killed hundreds in England, one theory held that the new government-issued gummed postage stamps might transmit the disease to anyone licking the back. Not surprisingly, fundamentalists saw the disease as God's punishment of a sinful, slothful people.[9]

Cholera is actually caused by the cholera bacillus (*Vibrio comma*) and spread through contaminated milk, water or other foods contaminated by the excreta of patients or carriers, but this was not known during the great epidemic years.

The Four Stages of Cholera

(1) Invasion: Malaise, headache, diarrhea and anorexia. In some, but not all cases, there may be a slight fever, lasting several days before subsiding
(2) Evacuation: Purging, violent vomiting, muscular cramps, particularly in the calves, copious and watery stools presenting the characteristic rice-water appearance. Unquenchable thirst, hiccoughing and signs of overall depression, ending in collapse. The duration of this stage is usually no more than 2–12 hours.
(3) State of collapse: eyes sunken, cheeks hollow, skin dry and wrinkled, body surface cold and covered in clammy sweat. The mind remains clear until coma develops in hours to days, after which death occurs.[10]

Death resulted from dehydration. The 20th-century treatment consisted of vigorous replacement of fluids; sodium lactate or sodium bicarbonate for acidosis.

At mid–century in the 1800s, physicians classified cholera into two types:

(1) Cholera spontanea: Which happens in hot seasons without any manifest cause
(2) Cholera accidentalis: Which occurs after the use of food that digests slowly and irritates

The medical community followed the following regime:

(1) In the early stages of the disease when the strength is not much exhausted, the object is to lessen the irritation and facilitate the discharge of bile by tepid demulcent liquids and compresses to the abdomen
(2) When the symptoms are urgent it is necessary to give opium freely, but in small bulk. A blister applied over the stomach may lessen irritability
(3) Strength must be restored by gentle tonics as the aromatic bitters, calumba, with a light, nutritious diet. Strong toast and water is the best drink, or a little burnt brandy if there is much languor
(4) Exposure to cold must be avoided and great attention paid to the bowels.

> Inflammation in the abdomen should be treated with leeches and blistering the skin: to draw off fetid liquids Poultices of wheat bran: to ease distress
> Phlebotomy: Bleeding, or drawing off blood by piercing the skin with a sharp lancet.[11]

Homemade remedies called for nutmeg or the essence of peppermint and water added to a portion of burnt cork, then mixed with sugared brandy. Commercially available mixtures included Wright's Indian Vegetable Pills. Four or five taken at bedtime would quickly "rid the body of every description of suffering."[12]

Transmission of Cholera along the Western Waterways

None of these remedies were available to steamboat passengers unless there happened to be a doctor aboard, and since cholera occurred almost exclusively among the deck passengers (due to overcrowding and the generalized filth), even the most dedicated healer would have found it impossible to stem the tide.

The first U.S. epidemic of cholera (known as Asiatic cholera because of its supposed place of origin) occurred in 1832. It came in via infected passengers alighting on the Atlantic seaports and spread from the St. Lawrence, through the Great Lakes, down the Ohio River through Cleveland, and then down the Mississippi to New Orleans.

A note in the *Star and Adams County Republican Banner,* February 21, 1832, recounts:

> Letters have been received in Boston from Smyrna [Greece] dated 4th November [1831] which states that the Cholera has considerably subsided. The whole number of deaths during its progress was from 6–8000.

The steamer *Autocrat,* built in St. Louis in 1847.

In July 1832, a cholera epidemic struck New York City, but the calamities there "were but a tythe [sic] of those which have fallen upon New Orleans."[13]

The *Star and Adams* details a story from New Orleans, from November 5, 1832:

> The Cholera or Cold Plague, together with the Yellow Fever, is raging to so great an extent, that coffins cannot be made fast enough to put the dead into. The Yellow Fever is very bad, and persons are taken off with Cholera in *two hours*—very few over that time. Business is completely prostrated, stores shut up, and one half of the people have fled from town. Last night, upwards of seventy coffins were at the grave yard, and none to bury them, and in consequence had to remain overnight. The grave yards are now full, and they are burying them outside of the yards. Last week there were 1070 interments—yesterday 176. Almost every hour you can see hearses with *six coffins* in them at once. All the Irish on the canal are killed, and some of our most respectable citizens have fallen victim. All our passengers left town immediately.... Ten thousand pounds of powder were shot off on Saturday to purify the air, and tar burned in different parts of the city.

The New York *Enquirer* (reprinted in the *Republican Compiler,* November 27, 1832) wrote that "The public authorities therefore had been compelled to resort to the revolting measure of sinking the unburied bodies in the river to avoid the consequences that might result from putrefaction."

The New Orleans *Bee* on November 8 continued the story from their city:

> The epidemic is evidently abating, both in malignancy and in the number of attacks. The Weather is now colder; and were it but only clear, the stiff north wind would blow it away altogether. "Courage? friends, courage!"

All along the interior river system, the disease passed from one community to another. An epidemic in Frankfort, Kentucky (on the Kentucky River), was reported to rage "to a greater extent than at any other point in that country. Containing about 2,000 inhabitants, 22 cases in 27 hours preceding the 7th instant, resulted in eleven deaths."

Louisville and Lexington, Kentucky, and Vicksburg, Mississippi, also saw new cases break out during the same time period.

The second major epidemic, in 1833, was believed to have originated in Cuba. The Huron *Reflector,* May 7, 1833, reported that cholera had made its appearance among the United States troops stationed in Key West, Florida. This time, the disease traced the reverse route: from New Orleans to Vicksburg, Memphis, St. Louis, Louisville, Cincinnati, Maysville, Wheeling and Pittsburg. The disease reappeared again in 1834.

Minor outbreaks in 1835 struck Columbia, Kentucky, "with great violence," as well as in Madison, Indiana, in Adams and Crawford County, Ohio, in Palmyra and Nashville. St. Louis, New Orleans and Lexington reported no cases that year.[14]

In 1840, one treatment for cholera was Evans' Family Aperient Pills, described as "purely vegetable, composed with the strictest precision of science and of art." They were "warranted to cure," among other diseases, apoplexy, scarlet fever, cholera, "Affections peculiar to Females," and "all diseases of whatever kind, to which human nature is subject, where the stomach is affected."[15]

The third major cholera epidemic came in 1848, again moving upriver from New Orleans and along the Mississippi and Ohio river cities and as far up the Missouri as St. Joseph. It was not contained, and recurred in 1849 and 1850, Nearly 1,400 people died in Cincinnati, "chiefly among the floating population and emigrants."[16]

A particularly telling article was printed in the Sandusky *Clarion,* January 1, 1849:

The Cincinnati Gazette says that the cholera has reached that city, having come up the river in over-crowded steamboats from New Orleans. It appeared in New Orleans a few days after the arrival of a couple of emigrant ships from Havre. The mayor and board of health published a proclamation in the Evening Mercury stating that the cases were not Asiatic cholera, but cholera-morbus, to quiet the alarm, when it *was* in fact, the dread scourge. They also said that most of the "emigrants had gone up the river." These emigrants have reached Cincinnati and they have mixed with the citizens of the Quoon City. Several deaths occurred on the passage, but all amongst the steerage passengers.

The editorial goes on to state the known facts of cholera:

1st. It is clearly proven that cholera advances with the tide of emigration — with the tide of travel along all our great thoroughfares 2d. It is already in three of our great cities 3d. It is communicated from one person to another.

It concludes by asking:

Who will stop the tide of emigration? Who can prevent the onward march of the fell destroyer? Who can prevent its arrival here? It is at New Orleans — it is at Cincinnati — and must we wait *until it is in Sandusky,* before we act?

In 1849, cholera reappeared in Natchitoches, where the Parish of St. Mary reported more than 500 cases and about 130 deaths since the end of December. An editorial in the *Adams Sentinel,* April 9, 1849 (reprinted from the *York Republican),* offered what little advice it could:

Cleanliness ... is the best preventative in towns and cities, as is the same virtue, unaccompanied with terror of the disease, and conjoined with caution and temperance in diets and drink, in individuals.... Our Borough Authorities should begin in time to search after and abate all nuisances, and citizens should purify and keep their premises clean. By this means the plague may be entirely averted.

Isolated cases reappeared in the western United States in 1851, 1852 and 1853. In 1854, cholera reached a fourth epidemic stage, with St. Louis suffering the most casualties.

In 1853, Mark Twain wrote about the epidemic in New Orleans, where mortality was 452 per thousand in the Fourth District. In ordinary times, he said, the corpse was covered in ice and preserved for a matter of several days while friends and family gathered. If the body was to be embalmed, the cost for such labor was $400. Because of the high water table, the moneyed dead were not buried, but laid to rest in aboveground vaults.

The appearance of cholera on a steamboat not only rang the death knell for deckers, it presented intractable problems for the captain. News of the disease on board spread as rapidly as the sickness. Often, hundreds of deck passengers would perish within a matter of hours, and dead bodies were freqently seen floating in the water as officers tried to rid the boat of those fatally stricken. A traveler on the *Constitution* in 1832 wrote, "I saw men perishing every minute about me, and thrown into the river like so many dead hogs."[17]

If not disposed to heave bodies overboard, captains would quietly bring the boat to shore in the early morning hours and arrange for quick burials. It was not uncommon for passengers to report seeing mile after mile of hastily dug graves marring the shoreline. Disposal of the dead was important for the profitability of the voyage. No passenger would willingly embark on a boat known to be contaminated, and quarantine regulations often kept a boat in dock until it was thoroughly cleansed. In order to get around this, captains frequently "doctored the books" as to how many (if any) persons perished on the trip.

Crews were not immune from fear and many deserted cholera-infested boats. Slave owners also tended to pull their hired property off such vessels known to have a history of disease for fear of losing their investments.

On May 21, 1849, an editor of the Louisville *Journal* wrote:

> Steamboat officers are morally responsible for the comfort and health of their passengers. They know how many persons can be safely and comfortably accommodated on their boats, and when they, for the sake of a few dollars, take on board two or three times as many persons ... they willfully peril the health and lives of their passengers.

In 1851, Cincinnati passed health regulations requiring all persons to undergo physical inspections before docking to keep out infected persons. Large grounds, set off by yellow flags, indicated quarantine areas. In St. Louis, all immigrants desiring to disembark in that city were held in quarantine for five days.

Precautions aside, nothing stemmed the tide, particularly at those cities along the major steamboat trades. The *Daily Globe* (Washington, D.C.), on September 23, 1854, reported seventy-four deaths by cholera at Pittsburg on September 22 and up to twenty-two deaths by noon on September 23. The total number for eight days since the commencement of the epidemic is listed at 500—terrifying numbers for any time.

Yellow Fever—Red Alert

The appearance of yellow fever in the southern states was almost always a yearly occurrence, beginning when spring weather turned warm and continuing until the colder temperatures of fall. The well-off traveled to cooler northern climates; plantation owners unable to do so left their swamp-infested grounds and moved into the city, where the fever was somewhat less prevalent. New Orleans was usually hardhit with yellow fever, which, like cholera, traveled by steamboat as far along the line as temperatures permitted.

Yellow fever struck without warning, inducing chills and fever, dehydration, repeated vomiting and eventually jaundice, giving victims the characteristic yellow tint that gave this disease its name. The disease is caused by the bite of infected mosquitoes, a fact not discovered until 1900. Stagnant ponds, open areas of water, and rain barrels were the breeding grounds.

According to Hooper's 1842 *Lexicon Medicum*, the 19th century medical name for the condition was *Typhus icterodes*. It was described as a "putrid-tending fever which is contagious, and is characterized by moderate heat; quick, weak and small pulse, senses much impaired and great prostration of strength." When fevers arose in the spring of the year, they were called "vernal." The lexicon continues, "It seems to be pretty generally acknowledged, that marsh miasma, or the effluvia arising from stagnant water, or marshy ground, when acted upon by heat, are the most frequent exciting causes of this fever. In marshes, the putrefaction of both vegetable and animal matter is always going forward, it is to be presumed; and hence it has been generally conjectured, that vegetable and animal putrefaction imparted a peculiar quality to the effluvia arising from thence."

The text goes on to describe other probable causes: "A watery poor diet, great fatigue, long watching, grief, much anxiety, exposure to cold, lying in damp rooms or beds, wearing damp linen, the suppression of some long-accustomed evacuation, or the recession of eruptions have been ranked among the existing causes.... One peculiarity of this fever is, its great susceptibility of a renewal from very slight causes, as from the prevalence of an easterly wind...."[18]

Treatment consisted of an emetic or a dose of opium. A cold bath, violent exercise and "strong impressions on the mind" were occasionally employed "with effect." Medication included cinchona, which, when taken in large doses, "will seldom fail to cure the disease." If cinchona failed, "other vegetable tonics may be tried, as the salix (willow), gentian (the root of the plant), calumba and other bitters. When there is obstruction of the liver, mercury will be most likely to prevail."

Less medically approved treatments included firing cannon and muskets to propel gunpowder particles into the air, soaking cloths in camphor and vinegar and holding them over the face, or placing pieces of tar or cloves of garlic in shoes.

Mark Twain described a visitation of the yellow fever in Memphis. Elsewhere he noted that the sewerage system of the city in the decades prior to the 1880s (a reform resulting from an earlier epidemic) had remarkably improved.

> In those awful days the people were swept off by hundreds, by thousands; and so great was the reduction caused by flight and by death together, that the population was diminished three-fourths, and so remained for a time. Businesses stood nearly still, and the streets bore an empty Sunday aspect.

He goes on to quote a German tourist who had been present during one of the terrible outbreaks.

> In August, the yellow-fever had reached its extremest height. Daily, hundreds fell a sacrifice to the terrible epidemic. The city was become a mighty graveyard, two-thirds of the population had deserted the place, and only the poor, the aged and the sick remained behind, a sure prey for the insidious enemy. The houses were closed; little lamps burned in front of many — a sign that here death had entered. Often, several lay dead in a single house; from the windows hung black crepe. The stores were shut up, for their owners were gone away, or dead.[19]

The *Daily Globe* noted the following information:

> Columbia, (S.C.) September 23rd — Intelligence has been received here that nearly the whole population of Augusta, Georgia, have left the city. The country for miles around is covered with tents, and the fever is increasing. The post office is closed, and the mails, in consequence, delayed, there being no one to distribute them.
>
> (The Augusta Constitutionalist, of Wednesday morning, says there were three deaths the day previous, but says nothing about the people leaving the city. It contains the following paragraph about the post office:
>
> "We are requested by Mr. Smythe, post master, to state, that in consequence of his clerks having quit the office, it will be impossible for him to distribute papers."
>
> *The people of Jacksonville fired into the steamer Weloka, from Savannah, on the 17th, while she was passing up the river. The cause of this conduct is said to be the fear of yellow fever.**

An additional note in the same newspaper reported that the number of interments at Savannah on Thursday, September 21, were sixteen, including twelve of fever — a great decrease. "The number of deaths during the week ending on Wednesday [September 20]

---

*Italics added.

was one hundred and eighty-nine, of which one hundred and thirty were of fever. The number of deaths by fever in Charleston on Thursday and Friday was thirty-seven."

This led into a note from Philadelphia, of September 22:

> The steamship State of Georgia, from Savannah, has been at quarantine for nearly two weeks, in consequence of the fever among the crew and passengers. Eight deaths are said to have occurred aboard.

A report from New Orleans of September 19 in the same paper gave the following grave report:

> The deaths by yellow fever this week have been three hundred and forty. *The riots are all over, and the city is now quiet.*

With an eye toward emigrants, a report from Brooklyn, New York, on September 8, 1856, printed in The Oshkosh *Courier*, gave the following notice:

> The Brooklyn Board of Health report no new cases of yellow fever. One death reported since last meeting, that of John Bergen, who died at Gowannus on Saturday — Mayor Hall stated to the Board that he examined the city yesterday and found no cases of yellow fever existing in the city proper. No new cases on Governor's Island and no additional deaths. All are doing well.

Even today, when yellow fever strikes, the prognosis is grave, and mortality of 5 percent may occur in an infected area. There is no cure and the only prophylaxis is the destruction of mosquito breeding grounds.[20]

Malaria

The word "Malaria" comes from the Italian, meaning "bad air." Like yellow fever, it was thought to be caused by miasma. It is also similar to yellow fever in that it is a chronic, infectious disease caused by the parasites within the red blood cells introduced into the body by the bite of an infected mosquito. The destruction of corpuscles brings on the characteristic paroxysms of chills and fever, which occur in either 48- or 72-hour intervals. Symptoms also include derangements of the digestive and nervous system and chills. Quinine is also used to treat this disease."[21]

Because physicians in the 1800s had no cure, all sorts of remedies were tried, including pills marketed by apothecaries and health stores. Among the most famous, were Wright's "Indian Vegetable Pills" and Moffat's "Vegetable Life Pills and Phoenix Bitters." Both promised relief from a varied assortment of ills. The latter was purported to "Purify the Blood," and thus aid in the treatment of bilious fever and liver complaints, dyspepsia, fever and ague (as Malaria was then known), general debility and piles.

Pernicious malaria has a sudden onset resembling apoplexy, but coma came after obvious, severe and intense symptoms, including hot skin, contracted pupils, coated tongue, loss of sphincter control, rapid, irregular and weak pulse and elevated temperature. The prognosis was often grim, particularly when no immediate treatment was administered.

An 1842 medical dictionary describes malaria as a "seasonable disease," coming during the height of the warm season and being destroyed by the approach of winter. The cause was attributed to an infected atmosphere arising from "marsh-miasma," or the rotting of animal and vegetable matter that polluted the air.[22]

Smallpox

Referred to as "Variola" by 19th-century physicians because it changed the color of the skin, smallpox was a feared enemy. It is characterized by eruption of red pimples which may contain pus that dry and then fall off in crusts by the eighth day.

A contagious disease "supposed to have been introduced to Europe from Arabia," smallpox produces fever and then skin lesions, by which it is spread from person to person. It is seen most often in the spring and summer and was considered more common in young persons than those of advanced age. Smallpox can be transmitted by the effluvia arising from the bodies of those who labor under the disease or by inoculation.

The eruptions are preceded by redness in the eye, soreness in the throat, pain in the head, back and loins, weariness and faintness, nausea, the inclination to vomit and a quick pulse. When the pustules are very thick on the face, the eyelids swell shut, the voice becomes hoarse, and it is difficult to swallow.

Treatment in the 1800s included bloodletting, the administration of an emetic, and keeping the bowels clear. Sometimes opium was administered so the patient could rest. Cooling lotions were applied to swollen areas, blisters behind the ears, and leeches to the temples.[23] Vaccination was used to prevent epidemics. With that in mind, on an expedition to broker and maintain the peace with the Chippewa and Sioux nations in the Upper Mississippi in 1832, Henry Schoolcraft was assigned a surgeon to vaccinate the Indians against smallpox. His research indicated the disease was unknown among Native Americans until 1750, when a war party of Chippewa had gone to Montreal to aid the French against the British in the French and Indian War.[24]

On December 11, 1840, the Hagerstown *Mail* reported that smallpox desolated the city of Panama, the population "having been reduced from upwards of 20,000 to less than one half." The article adds that almost every family suffered from the disease, the inhabitants having no knowledge of any means of stopping its progress.

On occasion, smallpox was carried upriver on steamboats, causing panics and waves of deaths. A writer in the Louisville *Public Advertiser*, on October 10, 1827, wrote, "We have seen the whole town thrown into a state of confusion and alarm, by the arrival of a steamboat from New Orleans, with the small pox on board, and our most ingenious lawyers could find no authority in the laws or customs of the country to prevent their landing."

Smallpox continued to plague Americans throughout the century. The disease raged to a considerable extent in Massachusetts, Maine and Vermont in 1840,[25] and even Abraham Lincoln was said to have suffered a light case of the disease.

Diphtheria

This is an acute infectious disease characterized by the formation of a false membrane on any mucus surface and occasionally on the skin. Diphtheria is caused by gram negative bacillus which was not understood in the 1800s. It is rare in babies under one year, most prevalent in children before the age of ten, and rarely seen in adults. It is most common in the fall and winter months. Transmission is through direct contact with the human carrier or by exposure through articles that have been contaminated.

The incubation period is two to five days. There is a gradual onset of headache and malaise, low-grade fever, and sore throat with a yellowish-white membrane adhering to the tonsils. This disease is difficult to diagnose, in that it mimics tonsillitis, scarlet fever, acute pharyngitis, strep throat, infectious mononucleosis and staph infections. Death is more frequent in the very young and the elderly, although not everyone is susceptible to the disease.

Treatment consists of strict bed rest and a liquid diet. Stimulants are considered counterproductive. In cases of laryngeal diphtheria, a tracheotomy (opening of the throat to facilitate breathing) may be indicated. Cyanosis may result, and if so, death is the likely outcome.[26]

The Marine Hospital Service

In 1798 the federal government established the Marine Hospital System, which provided medical attention and hospitalization for seamen. Based on the British system, it required the master of every American ship in the foreign and coasting trades to collect twenty cents a month from the wages of each sailor. This money went into a general fund, and maritime hospitals were erected to provide free care for these men. The Marine Hospital Act was extended in 1802, requiring captains of rafts and barges descending the Mississippi to pay dues in the same manner as seamen. The failure to collect these fees, however, rendered the act null, and those injured while performing their duty were left to their own devices.

In 1830, the Act of 1798 was extended to the western waters, and a more serious attempt was made to collect dues from steamboat captains. Seven years later, in 1837, Congress provided funds for seamen's hospitals to be constructed at Pittsburg, Cincinnati, Paducah and Evansville on the Ohio, and New Orleans, Natchez, Vicksburg, St. Louis and Galena on the Mississippi. None, however, were opened before 1850.

The benefits were only available to crews of registered ships, which generally left out those working the barges and flatboats. Slaves hired out to work the steamboats were also ineligible for this free medical attention, for neither their owners nor the captains employing them paid their dues.

Often, the monetary allotment to hospitals did not cover costs, or beds were unavailable. In those cases, the seaman was left to provide for his own medical needs, in spite of the fact he had paid his dues. Situations like this were frequently encountered in Cincinnati and Louisville. Despite the shortcomings, the medical system on the inland waterways worked as well for steamboat crews as it did for those sailing the oceans, and for the most part, maritime hospitals provided needed relief.

Illness from disease, infection, lack of sanitation, malnutrition and occasionally the practice of 19th century doctors, whether they be of the degreed or the "everyman" variety, was an accepted aspect of frontier life. People survived as best they could. For most of the 1800s, profit vied with charity, and science with fakery. Occasionally, it was difficult to separate the extremes. One fact, however, remained a constant. Whether a captain or crewman, a passenger or a merchant, no one was immune. As always, death was the great equalizer.

22

(In)Famous Steamboat Cities

Looking back from a historical perspective tinged with romanticism, the city most associated with the heyday of the steamboat era — when brightly painted, multi-tiered boats roamed the waters of the Mighty Mississippi, flags flying, gilt sparkling in the sunshine and huge, well-dressed crowds eagerly swarming the wharves in anticipation of arrivals and departures — is certainly New Orleans.

Then there was Natchez, the upriver city more famous for its "below" underground of lawless brawlers and whisky swillers than the refined southerners inhabiting the "upper" regions. Continuing up the Mississippi, moss-covered, low-hanging willows crowd the shorelines. Innumerable channel islands dot the river, which leads to Cairo, at the mouth of the blue, sparkling Ohio. A heave of the great wheel to starboard takes the tourist to the falls, through the imposing man-made canals, and eventually past such smoke-shrouded industrial cities as Louisville, Cincinnati and Pittsburg.

Next, the passenger reaches St. Louis, the connecting point between the Upper and Lower Mississippi. There, amid bustling wharves and waterside buildings, we see Indians, itinerant rivermen, screaming boys, scurrying clerks and impatient merchants. If one maneuvers among the walkways defined by cotton bales and hogsheads, a ticket may be purchased for a trip further north. There, the river is characterized by treacherous rapids, deep, impenetrable forests and towering bluffs. Onward to Galena and the lead mines, or ever upward, into the territories of ice and snow, populated by ragged fur trappers, forts of blue-clad soldiers and patches of corn, planted by newcomers speaking a cacophony of foreign languages.

Any and all of these Catlinesque images have a foundation of truth and the wisp of fantasy. It accentuates the wondrous and ignores the hardships of everyday life. First and foremost, at least in the eyes of those involved, steamboating was a business, and the settlements, towns and cities along the way were places of commerce. Looking at them through a kaleidoscope of work and play gives a somewhat more balanced perspective.

New Orleans

The city of New Orleans (*La Nouvelle-Orléans*) was named after Philippe II, Duc d'Orleans, Regent of France, and founded August 25, 1718, by the French Mississippi Company. It remained under the French flag until the Treaty of Paris in 1763, when it was ceded to Spain. The first of two devastating fires happened in 1788, when 850 structures in the French Quarter (*Vieux Carré*) were destroyed. A second fire occurred in 1794, and 200 more were burned. Rebuilding was done utilizing the Spanish style of architecture, featuring a central courtyard with wrought iron balconies.

Thirty-eight years later France reacquired the territory, but did not keep it long. In 1803, it became part of the United States via the Louisiana Purchase, under President Thomas Jefferson.

The city is located on the high ground of the east and west banks of the Mississippi River, 105 miles from the Gulf of Mexico. The approach to New Orleans offered a view in contrasts. Murky water flowed past muddy banks, revealing the Balize, the cluster of huts inhabited by local fishermen. Beyond that, huge bulrushes emerged from wetlands, and driftwood from upriver clogged the passageway. Occasional residences and sugar plantations dotted the landscape, surrounded by ridges that served as natural levees. Numerous cut-offs and hidden islands, protected by lush, overhanging vegetation and tangled swamps, provided an ideal habitat for pirates and smugglers.

The series of natural and man-made levees stretched from Balize to New Orleans and 100 miles past, protecting the city from floods. At high-river levels, river water reached the top of the levee, permitting a downriver steamboat passenger the unique opportunity of peering into the city from an elevated perspective.

As Twain described it, brick salt-warehouses were jammed along the upper end of the city, and dust and litter clogged the streets. Gutters were full of insect-breeding wastewater, and the sidewalks were blocked by innumerable casks and barrels.

New Orleans was created as a defensive city to protect passage from the Gulf of Mexico to the interior waterways. The French and Spanish who populated the city were soon joined by Americans. Following the Haitian Revolution of 1804, refugees swarmed to New Orleans, including free blacks and huge numbers of slaves. They were welcomed by the French Creoles, who gladly greeted the influx of French-speaking people. Nearly 90 percent of the émigrés stayed in the city. By 1840, New Orleans had become the wealthiest and third most populous city in the United States.[1]

New Orleans had always been the final destination of keelboats and flatboats bringing produce down the Ohio and onto the Mississippi River, as a base of both national and international trade; western goods were shipped to the Atlantic states across the Gulf of Mexico and up the coast, while manufactured goods from the eastern United States and the European continent flowed into New Orleans, where they were distributed upriver to the rapidly developing inland territories. When Robert Fulton first envisioned a steamboat monopoly, his intent was to monopolize the New Orleans-to-Natchez trade as the first step in dominating the entire water route between New Orleans and Pittsburg.

New Orleans also thrived as a slave market. After America abolished the international slave trade in 1808, the city became a trading post for domestic slaves, receiving hundreds of thousands from actively exporting cities such as Baltimore and Charleston. Slaves were sold to sugar planters, and among the eastern slave population, New Orleans earned a reputation as being the "place of no return," for attrition among blacks in the sugar trade was staggering.

The process of turning cane into molasses and sugar was labor intensive and wore out even the youngest, toughest men. Unlike the cotton picking done in the lower southern states where women worked alongside men, and thus had nearly equal value, females were deemed unfit for the sugar-house. Plantation owners were unwilling to keep women merely as wives and mothers, and this created a tremendous male-female imbalance. Slave-owners were thus unable to propagate the enslaved workforce. This made it necessary to import more and more slaves. When the demand could not be satisfied by legal importation, an illegal trade quickly developed. Africans were illicitly brought into the country and hidden in the backwater islands until papers were forged and they could be legally sold at market. Hundreds of local men made a lucrative living in the smuggling and sale of Negroes, either as pirates or dealers. Jean Lafitte, among others, gained a fortune from the business, as well as a substantial reputation and great political influence.

Harriet Beecher Stowe spent time in New Orleans at the home of Judge Francois Xavier-Martin, first chief justice of the Louisiana Supreme Court. It was there she was inspired to write her famous abolitionist novel, *Uncle Tom's Cabin,* after witnessing the activity at nearby slave markets.[2]

The Louisiana Pirate: Terror of the Gulf Coast

The story of New Orleans cannot be told without mention of pirates. These rogues of the sea were an integral part of American lore and none was more famous than Jean Lafitte (alternately spelled "Laffite"). He was well known in New Orleans, and his history was interwoven with subsequent Mississippi pirates.

Jean Lafitte's origins are obscure. Most authorities place his birth around 1776, on the French colony of Saint-Domingue (Haiti). Others place his birth at St. Malo, about 1781. It has also been suggested he was born in Bordeaux, France, in 1780, or Brest, or even Spain. Lafitte, himself, asserted on occasion to have been born at Bayonne. Political circumstances certainly dictated which native country to claim, but as he was later granted American citizenship, it seems fair for the United States to claim him, for better or worse.

Reading Lafitte's history is similar to delving into the exploits of a character in a romance novel. Gaspar Cusachs, in a lengthy dissertation read before the Louisiana Historical Society (1919)[3] catalogued with near adoration the feats of this "corsair, Patriot, pirate, smuggler and warrior." Boldly asserting there were none to compare with Lafitte except Robin Hood, "whom he surpassed in audacity and success," the researcher detailed an early career of murder, mayhem and money.

Touching on the pirate's undeniable charisma, Lafitte's biographer asserted that through suave manners, a gentlemanly disposition, majestic deportment and undoubted courage, Lafitte became esteemed and respected by his crew. Described as tall, finely formed and agreeable (when the mood was on him), he was also known to stand for hours on end with one eye closed, undoubtedly contemplating some lawless scheme. At those times, his appearance was "harsh."

Cusachs elaborated upon many wild adventures of Lafitte's early years: attacking vessels of every nation and accumulating vast hoards of gold and silver, which he divided among his buccaneers, who later squandered the wealth in wine, women and song. After entering the slave trade, the pirate captured an English man-o-war and later, an East Indiaman called the *Pagoda,* which he defeated by pretending to be a Ganges pilot. When they came close enough to board, Lafitte and his men leapt onto the enemy deck, cutlasses in hand, and after a bloody fight, emerged victorious.

More exploits followed. In 1807, he defeated a heavily armed East Indiaman, the *Queen*, by ordering his crew to lie flat on the deck, feigning defeat after a heavy broadside. The captain, believing he had vanquished the marauders, drew too close. Then Lafitte blew a whistle, his men rose up, and those on the yards emptied a devastating volley on the Indiaman, producing "havoc and slaughter."

Turning his attention to the Gulf Coast, Captain Lafitte established his headquarters in Barataria, on the coast of Louisiana. Barataria Bay is situated on eastern Louisiana and Bayou Lafourche on the western side of the state. Several islands populate the area, with Barataria being the largest. Another is Grande Terre, six miles in length and two to three miles wide. Grande Passe, the western entrance, holds from nine to ten feet of water. Situated two leagues from the sea, the bayous and cypress swamps form a nearly impenetrable network of obstructions, the perfect hideout for lawless buccaneers.

In December 1807, amid continuing political turmoil with Europe, Congress passed the Embargo Act, prohibiting foreign vessels from leaving American ports with any cargo, while also denying American ships the right to dock at foreign ports. The intent was to deprive Great Britain and her colonies of raw materials. The act caused acute economic hardship around the world and opened the door for smugglers.

The success of Lafitte's enterprise (which also included the smuggling of slaves into the country), was demonstrated by Governor William C. C. Claiborne placing a $500 bounty on his head. Countering the offer with an audacious one of his own, Lafitte offered a bounty on the governor's head. Such action increased Lafitte's popularity with the citizens of New Orleans, who looked to him as a savior for bringing in needed foodstuffs and luxuries.

Numerous government attempts were made to capture the "gentleman smuggler," to no avail. While Pierre Lafitte, Jean's older brother, handled operations in New Orleans, Jean controlled business from the island. By 1810, the enterprise had swelled into a booming venture, with numerous ships working the privateering trade under the Lafitte's guidance. Two years later, auctions of illegal goods were moved closer to the city at the Temple (a memorial mound halfway between Grande Terre and New Orleans) to facilitate transfers.

The War of 1812 brought both pirates and steamboats together. Henry Shreve dedicated the services of the *Enterprise* to General Jackson, and Jean Lafitte offered the assistance of his crew to Old Hickory. While claiming an army of 3,000 seamen, the actual number was much smaller. Working from the French Quarter, the men under Lafitte's command served as skilled artillerists and reputedly provided valuable assistance in the Battle of New Orleans.

Many stories are told of Jean Lafitte's close relationship to Andrew Jackson, most of which have the ring of tall tales. Whatever the true situation, Jackson did commend the pirates to the secretary of war, the final result being that the state of Louisiana applied to the president for full pardons for the heroes. A presidential proclamation followed on February 6, 1815:

> I [James Madison], President of the United States of America, do issue this proclamation, hereby granting, publishing and declaring, a free and full pardon of all offenses committed in violation of any act or acts of the Congress of the said United States, touching the revenue trade and navigation thereof, or touching the intercourse or com-

merce of the United States with foreign nations at any time before the eight [*sic*] day of January, in the present year, one thousand eight hundred and fifteen, by any person or persons whatever, being inhabitants of New Orleans and the adjacent country, during the invasion thereof as aforesaid.

The good feelings did not last long. Wearing out his welcome in New Orleans, Lafitte departed for Galveston, where he resided in a lavish mansion (once owned by the French pirate Louis-Michel Aury), which Lafitte called Campeche. Mexican governors Longe and Humbert gave him letters of marque, and he proceeded to amass gold and silver in legendary amounts. According to Cusachs, Lafitte commanded the *Jupiter* (the first vessel chartered by the new government) during the War for Texas Independence, and later became governor of Galveston.

Being less than discreet, one of Lafitte's captains captured an American vessel. The crew were subsequently captured, taken to New Orleans, tried and condemned. The *Enterprise* was then sent to remove the mastermind from his seat of operations, and Lafitte agreed to go without a fight. Departing in 1821 or 1822* aboard the *Pride,* an embittered Lafitte burned his fortress and took with him a wealth of treasure.

> A private letter from an officer on the New Orleans station, informs me that the pirates of Galvezton, under the notorious La Fitte, have stipulated to quit their old rendezvous at the place of that name, so that vessels passing the Gulf of Mexico, will be in less jeopardy that heretofore.

Soon Lafitte's fortunes turned sour. Two of his three ships deserted his command and he made for Jao de la Porta, later establishing a base of operations on Isla Mujeres, off the coast of Yucatán. In Casachs' lurid account, Lafitte died after being attacked by a British sloop-of-war. Although he sustained terrible injuries, Lafitte's last act of barbarism was to murder the enemy captain, whose mortally wounded body had been placed next to his. A more likely account is that he died at Teljas of a fever.[4]

To this day, rumors of buried treasure abound. In Louisiana, the banks of the Mississippi upstream from New Orleans, Caillou Island, the LeBlue plantation in Calcasieu Parish, Pecan, Jefferson and Kelso's Island, and Isle Dernier are considered likely hiding places of Lafitte's gold.[5]

The Development of New Orleans

A center of international trade and a gateway to the western and northern heartlands, New Orleans was integral to the development of the United States. To underscore its importance, even before America gained possession of the territory, the Bank of the United States was constructed in 1800 at 343 Royal Street. Designed with a hand-forged, wrought-iron balcony, it was the first building used as a bank in the city. In 1821, the Louisiana State Bank at 403 Royal Street opened for business, and five years later, in 1826, the Bank of Louisiana was built at 334 Royal Street,[6] testimony to the city's growing population (17,242 in 1810 to 116,375 in 1850), and the importance of the income derived from the rapidly developing steamboat trade.

---

*An article published March 29, 1820, in the *Adams Sentinel* provides an earlier date, although it is possible that Lafitte only reluctantly kept his promise and delayed departure for a year or more.

Value of Western Produce Reaching New Orleans

Year	Receipts
1801	$4,000,000
1817–8	$13,500,000
1830–1	$26,000,000
1849–50	$97,000,000[7]

The increase was obvious: as steamboats proliferated and technology advanced, the vessels became faster and were thus able to make more trips per year. They were also capable of carrying far heavier loads.

Steamboat Arrivals in New Orleans

Year	Number
1820	198
1825	502
1830	989
1835	1,005
1840	1,573
1845	2,530
1850	2,784
1855	2,763
1860	3,566[8]

Along with the great swell in commerce, the population grew. The greatest leap (121.8 percent) came in the 1840s, which coincided with a huge increase in river business.

Year	Population
1810	17,242
1820	27,176
1830	46,082
1840	102,193
1850	116,375
1860	168,675[9]

The City with "the worst people in the United States"

Despite the above, which was contemporary chronicler Ole Rynning's wry comment on the citizens of New Orleans (a remark he made when warning émigrés to avoid the southern route when traveling west),[10] the city was justly famous for its rich heritage. It had a mixture of nationalities, an international reputation for cuisine, a fascinating mix of architecture (which, interestingly enough, both Twain and Trollope disparaged), and the erstwhile religious celebration of Mardi Gras on the Tuesday preceding Ash Wednesday.

In stark contrast to the predominantly Protestant bent of the southern states and the midwest, the French and Spanish population of New Orleans was Roman Catholic. The simmering melting pot of cultures, especially from the Caribbean, added two distinct factors to religious practices of the era: those of Voodoo and the supernatural.

The most famous practitioner of the so-called black arts was Marie Laveau. She was born around the turn of the 19th century. Her life spanned the most interesting period of New Orleans' history, encompassing the exploits of Jean Lafitte, the War of 1812, the entire antebellum steamboat era, and the Civil War. How she may have affected any of these is a matter of conjecture, but she was famous, and it is doubtful any river captain working the Lower Mississippi trades was unfamiliar with her prowess.

Research provides few exact facts on her origins: some believe her to have been born in the French Quarter, daughter of a white planter and a free Creole woman; others believe she came to the city in her youth. In 1819 she married a Haitian émigré named Jacques Paris (or Santiago) who had come to the United States after the Haitian Revolution of 1804. He subsequently disappeared under mysterious circumstances, and Marie took up residence with Cristophe Glapion. She bore fifteen children between the two men.

Early in her career, Marie was asked by the father of a young Creole to assist in freeing the youth from murder charges, promising her anything in return for a non-guilty verdict. She asked for his house in payment, and placed various charms throughout the courtroom. When the boy was declared not guilty, she collected her winnings, along with a substantial reputation for magical powers.

It is reported that Laveau assisted in the care of wounded American soldiers during the War of 1812, and later expanded her influence by developing a vast network of spies, slaves owned by the families of the rich and influential. She was said to own a large snake named Zombi, and dance with the reptile wrapped around her.

One of Marie's daughters, also called Marie, bore an uncanny resemblance to her mother. She also practiced Voodoo, and was often mistaken for her mother, adding to the legend of the elder's timelessness.[11]

New Orleans was also famous throughout the steamboat era for its mysterious hauntings. In the mid–1800s, the aristocratic Delphine Lalaurie was renowned for her cruelty. She reportedly tortured slaves whom she kept hidden in her attic. After a conflagration broke out in the house, firemen discovered her secret, and a mob chased her out of town. Rumor has it that Madame Lalaurie's ghost is one of many which haunt the old house.

Another haunting involves a slave woman named Chloe. She was accused of poisoning two of her master's children by putting oleander into a birthday cake. Chloe was hanged and her body thrown into the river, but her spirit remains in the Myrtles Plantation, accompanied by the two children she murdered.[12]

If not actually haunted, the Pirate House bears considerable mystery. It was built in 1802 along the river in Hancock County by a New Orleans businessman, but the owner was said to be Jean Lafitte. The house had white stucco walls, square white frame columns supporting the gallery's iron grillwork, a brick ground story and an outside stairway leading to the first floor. However, circumstantial evidence points to the fact an underground tunnel led directly to a small bayou of the type used by smugglers to hide from the authorities. While such a structure would have been extremely difficult to build and maintain, the legend of the Lobrano House and its mysterious owner persists.[13]

Natchez-Over-the-Hill — and Under It, Too

If ever a city were divided in two, then Natchez must be the prime example. The lower half, where boats docked, belonged to the rivermen — those who, from the earliest days of navigating the Mississippi, gathered in the small, shabby taverns to gamble, tussle, frequent the red light district, and spend their hard-earned wages. Little improvement came about as keelboats gave way to steamboats, and the reputation of the lower half remained as it had been: a haven for the rough and tumble, the outlaws, highwaymen and pirates.

Acts such as reported on April 7, 1834, from Natchez give an interesting state of conditions in that city.

> Atrocious Act — Yesterday evening about 6 o'clock, as the steamboat Splendid was leaving the landing, a shot was fired from a house near the wharf from whence the boat started, and killed one of the passengers. The boat immediately returned, and the corpse was brought on shore. This is the third time, and from the same house, that persons have been shot on board boats leaving this place, in the former cases, two persons were wounded.... The citizens of Natchez, we have no doubt, will take care that no more atrocious acts of the kind be perpetrated, at least from the same house.[14]

A principal component of Natchez-Under-the-Hill was the 440-mile trail leading from that city all the way through to Nashville, known as the Natchez Trace. This path linked the Cumberland, Tennessee and Mississippi rivers, and served as both a getaway and a route for illicit activity. The trace originated with Native Americans, including the Cherokee, Choctaw and Chickasaw, who used it to follow bison herds as the animals made their way into Tennessee and the salt licks indigenous to the area.

During its prime, the trace was the best route between the upper parts of the country and the developing cities of the Mississippi and Louisiana territories. Itinerant preachers were among the earliest to use the trail, but they were quickly followed by bandits, who preyed on the unwary and made good their escape to Natchez-Under-the-Hill, where they knew it was unlikely they would be prosecuted. John Murrell was one of those who made a "killing" along the route, and he was said to have pushed his fair share of victims over the steep edges of the trace, thus disposing of the incriminating evidence. If the legends are even partially correct, he and fellow highwayman Samuel Mason were among the first operators of organized crime in the country.[15]

In 1801, the United States government widened the trace for use as a postal route, and later as a thoroughfare for travelers. By 1809, the trail had been widened enough to permit wagons to pass, and trading posts, or "stands," sprang up along the way, providing lodging and refreshment for travelers. It was at one of these stands that Meriwether Lewis met his fate, in what may have been a suicide or an unsolved murder.

By 1816, more direct overland routes by the Memphis and Jackson's Military Road leading to New Orleans reduced the trace's usefulness, and with travel by steamboat becoming the preferred mode of transportation, it soon fell into oblivion. By 1830, the Natchez Trace was officially abandoned as a road and reverted back to wilderness.[16]

Natchez-Over-the-Hill was the exact opposite of its poorer cousin. Constructed on the high bluffs overlooking the Mississippi, the town was settled by well-to-do southerners. Mark Twain remarked that it was always attractive, and Frances Trollope noted in 1827:

> The town of Natchez is beautifully situated on one of those high spots [bluffs]. The contrast that its bright green hill forms with the dismal line of black forest that stretches on every side, the abundant growth of paw-paw, palmetto and orange, the copious variety of sweet-scented flowers that flourish there, all make it appear like an oasis in the desert. Natchez is the furthest point north at which oranges ripen in the open air, or endure the winter without shelter.[17]

Of principal importance in the New Orleans–Natchez trade was the heavy concentration of settlers, constituting the largest population outside the Ohio Valley in the decade of 1810–20. Beyond Natchez, there were no areas of importance or commerce along the river until Louisville.

In setting up his steamboat the *New Orleans* in the trade between New Orleans and Natchez, Robert Fulton wrote that the Mississippi "was conquered." On a trip in 1812, the boat carried 1,500 barrels (150 tons) from New Orleans, against the current, traveling 313 miles in seven days, working in that time 84 hours. In 1817, the Fulton Line charged 10 cents a mile for through passengers and 12½ cents per mile for way passengers aboard the second *New Orleans*. The captain reported a net profit of $4,000 on a single trip. Until 1817, the New Orleans–Natchez trade saw a heavy concentration of steamboat traffic. Only after tonnage increased did the longer, New Orleans–Louisville route assume dominance.[18]

St. Louis: Gateway to the West

In 1803, at the time of the Louisiana Purchase, the population of St. Louis was a mere 925, fully one-third of whom were slaves. Five years later, this base for trappers and Indian traders had nearly doubled in population. With increased numbers came the *Missouri Gazette,* the first newspaper west of the Mississippi. A post office was also established in 1808, although service was spotty.

By 1816, the newspaper editor described his city as "opulent," having lead, soap and candle factories and $1 million in capital. The city was then, as she would remain, a prime consumer of alcohol, purchasing in that year more than five thousand barrels of whiskey. The following year, even before the first steamboat made an appearance, St. Louis had become the trading center of the upper northwest. It had a Presbyterian church, the Bank of Missouri was incorporated, a public school system was begun and a courthouse erected. Early French settlers left a legacy of Roman Catholicism, and the city had the honor of housing the Jesuit St. Louis University, the first institute of higher learning west of the Mississippi. St. Louis was nearly as famous, at least among rivermen, for its billiard saloons. Among the most famous was Bogart's, where the navigators freely spent their wages while discussing river conditions.[19]

A note to that effect was run in the *Adams Sentinel,* December 7, 1840:

> GAMBLING — they play billiards at St. Louis for $500 a side, and give public notice of the games in the newspapers.

The arrival of the first steamboat in 1817 started a trend that quickly developed. In 1823, two years after Missouri's admission as a state, the population had reached 5,000, and the city had become a hub of river traffic, being the point of departure for all expeditions up the Mississippi, Missouri and Illinois rivers and the transition point for those arriving from the Ohio River.[20]

In the early part of the 19th century, great things were predicted for St. Louis. The following, from the *Adams Sentinel,* January 12, 1820, beautifully describes the city and her expectations.

ST. LOUIS

> Occupies one of the best situations on the Mississippi, both as to cite and geographical position. It is perhaps not saying too much that it bids fair to be second to New Orleans

in importance to this river. From the opposite it presents now the aspect of a large and elegant town;* the buildings extend a mile and a half along the shore, and rising above one another from the waters edge like the seats of an amphitheatre. Back of the town is a prairie six miles wide, composed of waving grounds, beautifully swelling and rising into successive heights and eminences, which will eventually be crowned with houses and gardens. In front of the town, on the left bank of the Mississippi, is the *American Bottom*, eight miles wide, and finely diversified with prairie and woodland, and when the curtain of wood on the riverbank is removed, a delightful prospect will be opened into that rich and elegant tract.

St. Louis will probably become one of those great reservoirs of the valley between the Rocky Mountains and the Allegheny from whence merchandise will be distributed to an extensive country above. It united the advantages of the three noble rivers Missouri, Mississippi and Illinois. When their banks shall become the residences of millions, when flourishing towns shall rise, can we suppose that every vendor of merchandise will look to New Orleans or the Atlantic cities for a supply?—There must be a place of distribution somewhere between the mouth of the Ohio and Missouri, and St. Louis occupies the best situation and has taken the start.—Besides, trade to Santa Fe and New Mexico will be opened (when that fine country is freed from the dominion of Spain;) and a *direct communication to the East Indies by the Missouri river may be more than dreamt;* in this St. Louis will become the Memphis of the American Nile.—[*Brackenridge, p. 124.*]

It (St. Louis) is in a state of rapid improvement. Occupying a point where so many vast rivers mingle their streams, and increasingly, rapid and lasting prosperity is promised to this town. Including Louisiana, it is the most certain town yet built in the American union. It may, in the course of human events, be the seat of the empire, and no position can be more favorable for the accumulation of all that comprises wealth and power.—[*Darby, p. 143*]

When we view the central situation—the great confluence of the waters—the extent of the prairies—the salubrity of the climate—and the advantages that will result from the mines in the neighborhood—the mind instinctively looks forward to this place (St. Louis) as one of the first consequence in the United States: probably as the future capital of the greatest country that the world ever saw.—[*Metisch, p. 40*]

St. Louis never became the capital of the nation, and it certainly lost out by the failure of the Missouri to connect with the East Indies, but it had other advantages.

From its inception, St. Louis garnered the greatest portion of the Indian trade, and by 1829, the Indian Department had spent millions of dollars in the city, purchasing goods and paying for their transportation to the various tribes. The rest of the federal money was spent buying goods from Europe and the eastern markets, and transporting these by steamboat across the upper northwest.

These northern routes proved somewhat of a difficulty, as the American Fur Company discovered. Prior to 1840, trading posts below Prairie du Chien shipped their pelts to St. Louis. The agents for the American Fur Company broke ranks, determined to transport some of their furs and hides to New York by way of the Wisconsin, Fox and Green Bay rivers. The passage proved difficult, time consuming, dangerous, and worst of all, unprofitable. Managers eventually discontinued the policy and by 1841, had resumed all shipments to St. Louis.

The fact that they continued to keep a foothold in St. Louis is revealed by a notice in the Hagerstown *Mail,* July 31, 1840, which reports:

---

*Double in population and number of houses since this was written.

> *Riches of the Far West.*—The St. Louis Era says, twelve Mackinaw boats arrived at this port yesterday from the Upper Missouri river. They belong to the American Fur Company, and are freighted with Buffalo robes. We have not been able to ascertain the correct number, but the boats probably contain twenty thousand robes. These returns constitute, we believe, only a part of the proceeds of the year's lucrative business, having been collected at the establishments along the Missouri river.
>
> The boats, we have understood, met the Company's steamer, the Antelope two hundred miles below the mouth of the Yellow Stone, her point of destination. The water was very low, and there was much apprehension that she would not be able to return this season — certainly not with any freight.

Numerous and well-regarded hotels sprang up, warehouses were constructed, and wharves were laid out. Jefferson Barracks, a United States fort, contributed not only to the prosperity of the region but the security as well. From the fort, many excursions into Indian Territory and the forts of the upper regions were organized, and the city maintained a strong military presence that was significant to early settlers. Because of the warmer temperatures and islands dotting the river, many early steamboats working the northern regions of the Mississippi, Missouri and Fever rivers were wintered in the vicinity.

By 1839, the Hagerstown *Mail* of August 23 reported:

> *Trade of St. Louis.*—During the month of July, there arrived at this flourishing capital of Missouri, no less than one hundred and sixty-nine steamboats, and departed one hundred and forty-nine!

In 1836, St. Louis gained a famous citizen when Captain Henry Shreve purchased 300 acres of land northwest of the city. In 1840, the population reached 16,489, and it appeared nothing would retard forward progress. However, in 1844, the Mississippi flooded, widening the river to six miles in some places. Steamboat travel was seriously affected, and much of the business district lay underwater. Shreve personally suffered when his farm at Gallatin Place flooded.[21]

St. Louis recovered, and by 1850, as a direct response to the ease of transportation brought about by steamboats and the subsequent influx of émigrés, the population had risen to 77,860. Smokestacks filled the air with black clouds, the wharves were ever busy, and buildings jammed the waterfront. For whatever positive effects the commerce had, however, the overcrowding proved a very serious drawback when disaster struck.

The Great Fire

The date lives in infamy: May 17, 1849. The *White Cloud* lay at the St. Louis wharf between Wash and Cherry streets. The *Endors* lay above her and the *Edward Bates* below. The *White Cloud* caught fire and the crew was unable to quench the flames. One account has it that they attempted to set the boat adrift but a breeze brushed her against other boats, quickly spreading the flames. Another account relates that the *White Cloud* set the *Edward Bates* on fire. An attempt was made to cut the *Edward Bates* free, and the boat was carried downriver in the current. A strong northeast wind brought it back, and every time it came close to another boat, it caught the vessel on fire. A reported twenty-two boats were burned to some degree.

> These in turn ignited the rest, until in a short time the river presented the spectacle of a vast fleet of burning vessels, drifting slowly along the shore. The fire next spread to the

A very early photograph of a steamboat, "Car of Commerce," taken in 1848. Note the wooden businesses and warehouses constructed close to shore. Structures such as these were susceptible to flooding and conflagrations. During the Great Fire of 1849, fifteen city blocks of St. Louis were burned and twenty-two boats damaged.

buildings, and before it could be arrested had destroyed the main business portion of the city.[22]

A contemporary account relates that the boat bells (warning alert) began ringing about 10 o'clock P.M. By this time, the *Edward Bates* was already on fire and was burning pretty well when she broke loose and floated down the wharf, firing the others she passed. The levee caught fire above Locust Street and burned everything to the Old Market, as far back as 2nd Street; all the newspapers but the *Union* were destroyed.

> There is no telling how many lives are lost some Burnt some drounde and some Blown to pieces with Powder there have been several Bodyes dug out of the ruins some with their heads and legs and arms all Blown off.[23]

The White Cloud

Built: 1843 in Pittsburg
Tonnage: 261 tons
Length: 166 feet
Beam: 26 feet
Depth of hold: 6 feet, 4 inches
Trade: Worked the lead trade from Galena to St. Louis to New Orleans

Fifteen city blocks were destroyed by fire and property damage was estimated at $5,500,000. The *White Cloud* was a total loss, but was fully insured. (Captain Chittenden, quoted above in *Old Times,* noted that the insurance covered only $225,000, but was not all paid as the fire broke several of the insurance companies; see below.)

Steamboats Totally Lost

Boat	Value	Cargo	Insurance
American Eagle	$14,000	None	$3,500
Edward Bates	$22,000	None	$15,000
Montauk	$16,000	$8,000	$10,000
Red Wing	$6,000	$3,000	None
St. Peters	$12,000	None	$9,000
General Brooke			
Alexander Hamilton			
Eliza Stewart			
White Cloud[24]			

The Red Wing

Built: 1846
Style: Side-wheel
Beam: 24 feet
Trade: St. Louis and St. Peters

The General Brooke

Built: 1842
Style: Side-wheel
Sold: 1845 for $12,000
Trade: Missouri

Perhaps the greatest testament to western fortitude comes from the *Daily Sanduskian* (Ohio), in its notice of June 18, 1849:

> DAMAGE BY THE FIRE AT ST. LOUIS Speaking of the exaggerated Telegraphic accounts of the late fire at St. Louis, the Republican of that city says:
> "Many of our citizens have suffered heavy losses, but it is idle to say they have been reduced from opulence to beggary, or that the suffering among the poorer classes is heartrending. Of the three millions of property destroyed, it is computed that two millions were insured, and, aside from this, many goods and valuables were saved. The actual loss to the sufferers will not be more than a million of dollars, and much of this loss falls upon those who are able to bear it. Two weeks have passed since the calamity occurred, and those who were *unhoused* have already found comfortable places, and men of business, mechanics and trades-people are making plans to revive their disasters. Never have we seen a community who seemed to care less for their losses, or who felt so little anxiety about their ability to sustain themselves, in despite of this misfortune. With them there is no such word as FAIL. Nowhere is there any gloom or despondency — nowhere is there any doubt about the ability of our people to survive the shock of this most untoward event; but confidence, self-reliance, industry, and energy is every where to be found, exercising a most wholesome influence upon the sufferers, as well as the entire community."

It makes one speculate that the author had aspirations for public office or membership in the Chamber of Commerce. This might be regarded as an example of "puff" reporting.

But perhaps the author was only exaggerating to a slight extent, for St. Louis did, in fact, survive and rebuild, as George Merrick describes it in 1857:

> I have seen boats lying two deep, in places, and one deep in every place where it was possible to stick the nose of a steamboat into the levee — boats from New Orleans, from Pittsburg, from the upper Mississippi, from the Missouri, from the Tennessee and the Cumberland, the Red River and the Illinois, loaded with every conceivable description of freight, and the levee itself piled for miles with incoming or outgoing cargoes.[25]

Everything along the river cities was in place for continued success. Floating Palaces dominated the passenger trade; freight was being hauled in record tonnage; the mosquito fleet had penetrated the narrowest tributaries, bringing cargo out from small but growing towns; mechanical improvements had increased safety. The 1850s roared.

What came next broadsided the industry, and not even the most skilled pilot or the savviest captain saw disaster looming over the horizon.

23

The Roaring '50s: The Railroads Come Calling

It is an irony of life that just when events reach their highest point and the future seems secure, disaster lurks around the next bend. Such was the case as the steamboat industry roared into the 1850s. While few could have predicted the devastating effects the Civil War would have on western transportation, those involved in river commerce should have anticipated the vast changes coming with the arrival of the railroads. Strangely, they did not.

In 1840, only 3,328 miles of track had been laid; by 1850, that number increased to 8,879, most of it in the eastern part of the country. Only one line between Chattanooga and Savannah, completed in 1849, served the trans–Appalachian region. In 1848, the Ohio and Pennsylvania Railroad was chartered to run into central Ohio. By the spring of 1853, the road had expanded nearly 200 miles. In conjunction with a railroad at Cincinnati, passengers were carried from that major steamboat port to Pittsburg in fifteen hours, and freight in a day and a half, greatly cutting into river traffic.[1] By January 1854, Cincinnati was connected to St. Louis, in a circuitous route through Chicago. The arrangement was inconvenient for the shipment of freight, but quickly garnered the passenger trade.[2]

By the 1850s, railroad track nearly quadrupled, reaching 30,626 miles by 1860, much of it in the west, with an eye toward stealing the interior business that then traveled via the waterways. Lines were built north from the Ohio River and south from Lake Erie. The Ohio and Mississippi were connected, and by 1860, Pittsburg, Cincinnati, Louisville, St. Louis, Memphis, Vicksburg and New Orleans all had railroad lines. So much construction occurred in the west that by 1860, Ohio and Illinois were the leading states in railroad mileage, with Indiana a close third.[3]

The Robber Barons

The principal differences between the development of the steamboat and railroad business were size and capital. From the breaking of the Fulton-Livingston monopolies through the antebellum period, steamboats were primarily small operations, owned by four or fewer individuals. The waterways were free to anyone and no great construction was needed for wharves. That meant that a man with a modest investment could buy a vessel, hire a crew, solicit freight and be in business in a fortnight. It was an era of individualism.

Railroads, on the other hand, required a huge investment in capital, well beyond the means of single investors. As with stagecoaches before them, land had to be surveyed and rights purchased; the laying of track involved thousands of laborers and the importation of iron from foreign countries. The construction of steam engines and cars demanded immense factories and skilled craftsmen. Instead of a go-it-alone attitude, cooperation between privately held railroad lines, states and the federal government were needed for success.

Outside of state-financed canal construction, little if any federal government intervention was required for the steamboat trade beyond safety regulations and the scant assistance provided for actual river improvements. Such was not the case with railroads, where aid was received in various and significant means. At least sixty-one railroad surveys were made by army engineers in the period 1824–38, at a cost of $75,000. From 1830 to 1843, Congress approved a decrease in iron tariffs, resulting in a savings to railroad men of nearly $6,000,000. The dispensation was revived in March 1843, to subsidize the domestic manufacture of railroad iron.

In 1850, Congress passed a bill providing grants for a north-south railroad line extending from Illinois to Mobile, Alabama. The bill provided that Illinois, Mississippi and Alabama should receive federal lands within their borders for the construction of a railroad from LaSalle on the Illinois and Michigan Canal to Cairo, Illinois. It would then go across Kentucky and Tennessee and through Mississippi and Alabama to Mobile. The grant provided a 200-foot right-of-way, and alternate even sections of land on each side of the road for a depth of six miles. The subsidy totaled 3,736,005 acres and also provided capital for the rapid construction of railroads. The money came primarily from mortgages secured by railroad properties, chiefly federal lands.

In the years 1852, 1853, 1856 and 1857, more lands were granted to ten states, benefiting approximately forty-five railroads.[4]

The lower Ohio Valley was not far behind. The Louisville and Nashville Railroad was open for business by 1859. A branch ran to Memphis, posing as real a threat to the steamboat businesses in the southern areas as that of the Cincinnati–St. Louis railroad.[5]

Railroad Lines

Line/Points Connected	Year of Completion
Chicago — Rock Island	1854
Nashville and Chattanooga	1854
Cincinnati — Wheeling	1854
Memphis & Charleston	1857
Cincinnati — Cairo — Pittsburg — Louisville — St. Louis	1857
New Orleans — Canton	1858
Cincinnati — New Orleans	1859
Louisville and Nashville	1859
Louisville — Memphis	1861
Nashville — Bowling Green — Memphis	1861
New Orleans — Cairo*	1861
Mobile — Cairo*	1861

*The terminus of the Illinois Central Railroad. Both lines reached near Cairo by the opening days of the Civil War.

To celebrate the completion of the Chicago and Rock Island Railroad on February 22, 1854, by which the Atlantic was connected to the Mississippi via a series of lines, the contractors from the firm of Sheffield and Farnam arranged for a magnificent celebration. They organized a joint railroad-steamboat trip from Chicago to the Falls of St. Anthony.

Leaving Chicago at 8:00 on June 5, a huge party reached Rock Island by 4:00. They eagerly boarded seven waiting steamboats: the *Golden Era, Spar-Hawk, Lady Franklin, Galena, War Eagle, Jenny Lind* and the *Black Hawk.* At least 1,200 people piled aboard (many more were forced to return to Chicago for lack of space), where they were treated to a great feast.

After a side trip up the Fever River to Galena on the following day, the celebrants spent their time sightseeing and touring the shore during frequent wooding stops. They danced and toasted one another with Ohio and French wine. Arriving at St. Paul with bands striking up lively airs, the passengers disembarked and were transported by carriage to the falls. The trip ended all too soon for most, with former president Millard Fillmore awarding a silver pitcher of appreciation to Captain Bersie of the *Golden Era.*[6]

All was not smooth sailing, however, as factions of the steamboat and railroad supporters fought over what the development of railroads would do to their individual commerce. An article from the Alton *Weekly Courier,* May 18, 1854, held profound implications for its readers:

> ANOTHER SPECK OF WAR—THREE STEAMERS SUNK. The inhabitants of villages on Lake Champlain are now engaged in a quarrel about railroad and steamboat matters, which is not likely soon to end. It has already resulted in violence and outrage upon persons and property. It appears that the Plattsburg people are building a railroad from that place to Montreal, a portion of which was completed. The Company owning the railroad from Rouse's Point to Montreal purchased the Montreal end of the Plattsburg route and left the people of the latter place in a bad fix. The Plattsburg people owned a steamboat called the Sultus, which they designed to run in connection with their road this season. The boat wintered at Sherburn Bay, and when the proprietors were about to move her, they found that a part of her machinery had been stolen. They attempted to tow her down to Plattsburg, but the people of Burlington cut the lines and took her back. The following night two old steamers, the Burlington and Whitehall, were drawn beside the Sultus and sunk in such a position that the latter cannot be moved. The Plattsburg people were much exasperated. The captain and owners of the steamer Saranao were supposed to be concerned in the outrage, and when the boat came to their village, 400 persons rushed on board armed, lashed the wheels, and threatened to sink her, arrested her captain and pelted him and others with rotten eggs. So the matter stood at last account. Compile the above statement from the Ogdensburgh *Despatch.*

(Iron) Horse of a Different Color

Travel by railroad was not as easy for freight and passengers. Many local lines stopped short of connecting with those owned by another company, forcing the manual transfer of cargo. Worse, many railroads used gauge of varying widths, which prevented trains and cars of one line from moving across those of another.

In parts of Virginia and North Carolina, a gauge of four feet, eight and one-half inches was used. Lines operating south of the James River on an east-west axis and at least three North Carolina companies used a five-foot width. Georgia, South Carolina, Florida, Tennessee and most of Mississippi also used a five-foot width, while

Alabama used the four foot, eight and one-half inch gauge. Crossing the Mississippi River, the New Orleans, Opelousa & Great Western, the Memphis & Little Rock and the Vicksburg, Shreveport & Texas line used a five-foot, six-inch gauge.[7] There were many reasons for this, including local jealousies, the required payment of transfer fees and the fact that moving freight from one line to another offered lucrative jobs for townsmen.

Railroad cars were flimsy affairs; tracks warped, giving a jolting, uneven ride; copious amounts of black smoke belched, first from wood and then coal-burning engines, seeping in through closed windows and fouling the air; wooden seats were uncomfortable and the inside temperature was either sweltering hot in summer, or excruciatingly cold in winter, when only a solitary stove at the head of each car was provided for warmth. Delays were frequent, stoppages to transfer lines tedious and passage over rickety bridges a tax on even the most steadfast nerves. There were virtually no sleeping arrangements for travel extending through the night, and most damning of all, at least from the perspective of luxury, no meals were served. Passengers were forced to alight at any of the small, ill-kept stations along the way, much as stagecoach passengers did, to purchase what scant, greasy food was offered by indifferent cooks.

No wonder steamboat men, with their Floating Palaces and luxury accommodations had little fear of the encroaching Iron Horse. Or so they thought.

Differences between the two modes of transportation were significant, but a number of factors worked in the railroad's favor: speed, accessibility to inland territories not blessed with innumerable tributaries; safety, adherence to the doctrine of common-law carriers, whereby the shipper assumed liability for freight (as opposed to having cargoes insured by the merchant) and, perhaps equally important, the ability to keep to a posted schedule of departures and arrivals.

Balancing advantages with disadvantages, the first-class passenger trade, so highly sought by steamboat captains, ultimately chose speed over other considerations. From the earliest days, man has always been fascinated by the ability to get from one point to another as quickly as possible, and racing, first on foot, then later by horseback, in wheeled conveyances and finally steamboats, presented the ultimate thrill. Railroads offered an even faster mode of travel, and within four years of the railroad connecting Pittsburg and Cincinnati, the famed steamboat packet line of the same name failed.[8]

Distance in Miles, Steamboat vs. Railroad Travel

Cities	River Distance	Railroad Distance
Pittsburg to		
Cincinnati	470	316
Cairo	979	675
St. Louis	1,164	612
Cincinnati to		
Knoxville	1,200	293
Nashville	644	301
St. Louis	702	339
New Orleans	1,484	922
Cairo to		
New Orleans	973	560

Travel Times, Steamboat vs. Railroad (late 1850s)

Cities	Steamboat	Railroad Passenger	Railroad Freight
Cincinnati to			
Pittsburg	3 days, 6 hrs	15 hrs	36 hrs
St. Louis	2 days, 22 hrs	16 hrs	30 hrs
New Orleans	8 days	60 hrs	...
Louisville to			
Cincinnati	15 hrs	7 hrs	12 hrs
Pittsburg	4 days	22 hrs	48 hrs
St. Louis	2 days, 12 hrs	14 hrs	24 hrs[9]

The numbers speak for themselves. In a contest between the romance of a steamboat and the rough, uneven travel by railroad, time won, even though Charles Dickens, on his travels in America, had this to say about American railroads:

> There are no first and second class carriages as with us; but there is a gentleman's car and a ladies' car: the main distinction between which is that in the first, everyone smokes; and in the second, nobody does. As a black man never travels with a white one, there is also a negro car; which is a great, blundering, clumsy chest.... There is a great deal of jolting, a great deal of noise, a great deal of wall, not much window, a locomotive engine, a shriek and a bell.[10]

Perhaps the greatest advantage railroads had over the steamboat was the fact the Iron Horse was capable of operating during all seasons, whereas river traffic was entirely dependent on river conditions. When the water froze, or rivers fell below acceptable levels during the summer months, travel for the larger class of boats came to a standstill. In the 1850s, late rises occurred in eight out of ten years, preventing merchants from getting out their stock. This forced them to look elsewhere for redress, and where they looked was to the railroads.

The issue of safety was also a major factor in determining how to travel and transport goods. Steamboat disasters were well chronicled and vivid in their descriptions of loss of life. On the other hand, when railroads were involved in accidents, casualties were typically limited to railroad workers. Similar to river disasters when "none but deck passengers were thought to be lost," Americans tended to overlook the deaths of the lower classes and concentrate on "quality" injuries and demises. They therefore developed the (not unsubstantiated, but more emotionally founded) belief that railroads were safer than steamboats.

Cutting One Another's Throats

The inevitable result of competition from railroads fostered substantial, and at times violent, clashes between steamboat operators. Destructive rate wars developed as owners fought for the diminishing passenger and freight trade, with rates often cut below subsistence levels.

In 1847, the authors of the *United States Statistical Directory* noted:

> Nowhere in the world is traveling so cheap as on our western rivers; the reason is, that there is no preconcertion among boat owners as to price, and there are almost as many different owners as boats....

Hundreds of small- and large-class boat owners found themselves on the brink of disaster. Where once the transients and lines had found enough business to satisfy all, transporting hundreds of thousands of tons of cargo and countless passengers, many from distant cities and foreign countries, and a lucrative season might actually pay for the upfront costs of starting a business. Now the numbers worked against them. One by one, tramps and packets went out of operation; those that continued adopted dangerous practices such as overloading, driving the boats beyond their capacity, and racing one another with careless abandon. This merely exacerbated the problems, for they resulted in an increase in accidents and loss of life.

Attempts were made to organize most, if not all, the steamboats into more efficient lines, but the nature of the beast had always been the stark independence of owners. Even when faced with the inevitable, quarrels over routes, landing sites and times of departure made organization impossible. Some new associations, especially those serving the short, highly competitive trades, survived, but the long-distance routes, run by too many boats, were harder to control. Prices continued to drop and so did profits.

In order to survive, some of the major steamboat lines attempted a sort of pact with the railroads, offering a combined steamboat-railroad through service. Typical advertisements of the day read:

> STEAMER PACKET J. MCKEE, S. HEAIGHT, Master Will leave immediately after the arrival of the Cars from Chicago, on Monday, Wednesday, and Friday evenings. Supper on board, and no Extra Charge
>
> War Eagle and Golden Eagle
>
> Leave Keokuk daily except Sunday at 2:20 P.M., connecting with the trains of the T.P. & W. and K. & D.M. Railroads.

* * *

1855   1855
ST. LOUIS AND KEOKUK MAIL LINE STEAMERS

> This popular line of steamers will make regular daily trips as usual throughout the season, connecting at Keokuk with the regular daily mail packets to ROCK ISLAND, GALENA, and ST. PAUL, running in connection with the Railroads to Chicago and the east.[11]

The Panic of 1857 did much to further depreciate the steamboat business as money became scarce, personal bankruptcies mounted and merchant businesses failed. Four years later, a matter of national importance assumed the forefront, and effectively put an end to the steamboat era: the Civil War.

Leaving Mark Twain, riverman, pilot, and author, with the final obsequies:

> Time drifted smoothly and prosperously on, and I supposed — and hoped — that I was going to follow the river the rest of my days, and die at the wheel when my mission was ended. But by and by the war came, commerce was suspended, my occupation was gone.[12]

Appendix A: Glossary

A1: Top rating for a pilot.

Acts of God: Accidents attributable to causes other than obvious pilot error.

Admiralty jurisdiction: Oversight of all marine sea vessels by the United States Department of the Navy. In 1851, this power was extended to tidewaters (along the coasts) and western rivers. All seamen and rivermen were thus considered "wards of the Admiralty" and brought under its protection.

aft: The back of the boat.

ague: Malaria.

all-purpose carriers: Western steamboats which always combined the transportation of freight and passengers.

anchor ice: Ice that formed in the river before the entire body of water froze over. Anchor ice caused severe damage to steamboats.

anthracite coal: Hard, glossy coal that burns without much smoke. After 1825, anthracite coal was used more frequently in eastern steamboats.

ark: A craft that was shallow of hull and square or V-shaped at the ends. Arks usually lacked decking or housing, but could reach lengths of 100 feet. Built only for one trip downstream, they carried families and cargo. A seldom-used term.

armored diving suit and helmet: Used for salvage operations, replacing the diving bell. The helmet had a glass window and was worn by a diver who was making preliminary repairs on a sunken vessel before it was raised by powerful pumps and tackle. The diving gear was also used when attempting to retrieve valuable cargo such as specie.

article of agreement: See "shipping papers."

association of part owners: People holding shares in a steamboat operation. Shares could be bought and sold at will. Steamboat stock ranged in size from halves to hundredths.

bagasse: Great piles of sugarcane stalks; refuse.

barge: A river vessel, 20–70 feet long, 7–12 feet across, usually employing sails and a rudder. Two pairs of oars were used to increase speed, which reached 4–5 miles per hour downstream. Moving upstream with sails and polemen, 2 miles per hour was the average. Barges typically cost $5.00 per foot of length.

bell boats (also called submarines): Designed by James B. Eads after 1842, when he entered the salvage business. Salvage boats were generally employed by insurance companies. Owners typically received 20–75 percent of the recovered property.

bell, pilothouse: Used as a means of communicating with the engine room. Among the wide variety of signals were "slow bells," "backing bells," and "stopping bells." In some parts of the Mississippi, pulling two bells meant to slacken speed. The engine crew responded by ringing its own bell to acknowledge the order. See also "speaking tube."

big bell: Used as a means of communication. Rung by a rope in the pilothouse. Typically, three taps on the bell was the signal to land. No taps on the bell meant the boat was going to bypass a landing. Two bells followed by one bell was the order to put out the lead. Bells varied widely in tone to distinguish one boat from another, and rivermen and settlers living close to shore could recognize boats by the distinctive tone of their bell.

bituminous coal: Came into use early on the western rivers, but was limited due to availability. By the 1850's, coal use was common on the Ohio River, but nearly always in combination with wood. Coal was seldom used on the Lower Mississippi due to availability and cost.

Black Hawk War: In 1832, the Indian Question became a national issue when members of the Sauk and Fox tribes murdered twenty-eight Menominee Indians in the village of Prairie du Chien. Black Hawk, the Indian leader, defeated state militia at Stillman's Run, and Governor John Reynolds quickly called for 2,000 mounted volunteers. Steamboat owners helped transport troops, bringing this new mode of transportation to national attention. Steamboat owners received great remuneration for their service. In subsequent years, transporting troops and military supplies would be a highly profitable business for steamboat owners.

bluffs: High ground or cliffs along a riverbank. The Upper Mississippi was lined by bluffs. Rockslides were common occurrences at bluffs, and created dangerous, unpredictable hazards for steamboats.

bateau: A big skiff, capable of carrying an entire family. It was propelled by several pair of oars.

bourdon bent tube gauge: A steam gauge designed for use in high-pressure engines, invented in 1845.

bow: Front of the boat.

brag boat: See crack boat.

break-up: The period in springtime when river ice begins to melt.

breast-board and head board: Shades in the pilothouse used to admit or limit light.

broadhorn: A type of raft in use before steamboats.

broadside: Pertaining to the side of a boat. If a boat is turned parallel to the flow of current, the water strikes it broadside, often resulting in a loss of control.

broom: On the Upper Mississippi, a broom displayed on the pilothouse signified the fastest boat. If the boat with the broom was bested in a race, the captain was compelled to pull it down.

bush-whacking: Manually pulling a boat up the current by means of a rope affixed to the bushes on the bank. Applied mostly to keelboats. See also "warping (2)."

cabin: The main area of entertainment and dining on a steamboat, otherwise known as the salon or saloon.

cabin crew: On larger vessels, these included cooks, stewards, waiters, cabin boys and chambermaids. Most were hired for the entire season and received wages somewhat less than those of the deckhands.

cabin fare (1): The cost of first-class accommodations.

cabin fare (2): The food served to first-class passengers. The cost of food was included in the price of a ticket, but liquor was extra.

card: A brief printed or published announcement typically run in a newspaper. Steamboat passengers well pleased with their journey often signed and published a card, thanking the captain for his gentlemanly deportment and/or the speed and comfort of his boat. Steamboat owners would also publish cards announcing itineraries and the fine features of their vessel. Merchants or professional men would publish cards announcing a move or the dissolution of a partnership.

cargo box: On very early steamboats, a term meaning shelter for passengers and cargo.

carpenter: Carpenters became regular crew members early in the history of western navigation. Carpenters were responsible for repairing breaches to the hull (when possible), as well as the paddle-wheel buckets and guards.

certificate: The license earned by an engineer or a pilot after demonstrating competence in his profession.

chain: A succession of rock bars or ledges.

channel islands: Any of the land obstructions that constantly formed and reformed in the river. They divided the strength of the current, and steamboats had to move past on the deeper side. They also reduced the depth of water in the channel and gave rise to sandbars at the head and foot.

chute: River areas close to shore; very narrow passageways. They were very deep except at the head. The current was gentle. Under the "points" the water was absolutely still, and the banks so close a boat sheared off willow leaves. Going "chute to chute" meant navigating the river from one chute to the next.

clerks: On large boats the position was divided into two: the steamboat clerk, or chief clerk, and the second clerk, or "mud clerk." Clerks' duties included collecting fares, assigning rooms, collecting freight bills, paying for fuel, keeping track of waybills, and checking on the loading and unloading of cargo on frequently muddy wharves (hence "mud clerk"). The steamboat clerk on larger vessels was third in the hierarchy, below the captain and pilot. Steamboat clerks were responsible to the owner, and handled the business end of the venture, recruiting passengers and cargo, preparing waybills, handling payroll and hiring and firing deckhands.

***Clermont*:** Robert Fulton's first steamboat, named after his associate Robert Livingston's estate on the Hudson. Historically considered the first boat to demonstrate the viability of steam-powered engines.

"Coast, the": The Bayou Sara below New Orleans. The area included sugar and rice plantations.

code of honor: Taken from the maritime code, the rule which states that the captain must be the last off a sinking or burning boat.

cold storage room: An area in the galley well stocked with ice to preserve perishable food.

common-law doctrine: The rule stating that a transporter was responsible for the value of freight it carried in case of loss. This doctrine was never enforced on the rivers, although later in the steamboat era owners wrote waybills protecting themselves against this regulation. Railroads, however, fell under this doctrine, in effect becoming the insurer of their freight. This gave railroads considerable advantage over steamboats when the margin between water and land rates narrowed.

company: In the early years of steamboating, the term loosely referred to several rivermen pooling their resources to buy a boat and dividing profits according to the shares held.

cordelle: A heavy rope which, when attached to a tree or other object, permitted a boat to be hauled forward. Where the bank provided good footing, the cordelle might be used as a towline by the boatsmen.

corporations: Rivermen who merged their funds to build and operate steamboats, dividing among themselves the better-paying positions and the profits. Most corporations did not work out well, as each man believed he knew best how to increase efficiency and financial reward.

crack boat: A steamboat decked out with finery to attract passengers, and usually renowned for speed.

craw-fished: Failed.

crevice: The area between shore and an island, usually narrow. Steamboats occasionally used crevices as a cut-through. They were potentially dangerous and there was no backing out once inside, so if a boat were caught on an obstruction or a falling river, it was stuck there until the river rose. In extreme cases, this might be six months.

cub: An apprentice pilot or engineer.

cut under: To underbid another steamboat operator for freight or passengers.

daylight license: A partial license granted a steamboat pilot, permitting him to work only in the daytime. After he proved his competency in nighttime navigation he was granted a full pilot's license.

dead hours: The hours between midnight and 7:00 A.M. or the hours of darkness.

Deadly, the: A 19th-century reference to the Mississippi River, stemming from the common belief that nothing that sunk in the river (including swimmers) was ever known to rise again.

deck crew: This classification of steamboat workers comprised more than half the crew of larger boats. In the early days of steam, many men were drawn from keelboat or flatboat crews, but as émigrés became more plentiful, they rapidly swelled the ranks. Most of the deck crew were young men under age 30, and were hired by the trip rather than the season. Of those foreign born, the Irish comprised between 40–50 percent of the deck hands in the 1850–60 period, and Germans made up 17 percent. Slave labor was also used on those boats working the Southern trades.

deckers: The common-usage term for deck passengers.

"deckhands' purgatory": Shoal areas along the Ohio River, where boats frequently ran aground, requiring intense manual labor to free them.

deck passengers: Passengers who individuals purchased a tickets at reduced fares. They were lodged on the main or lower deck (hence the name), which provided little or no protection from the elements, and were not provided food. Whatever they ate had to be brought aboard at the start of the trip or purchased ashore, at inflated prices, at wood stops or towns along the way. There might be as many as 200 deck passengers crowded in the confined deck space at any one time. Deck passengers were derisively referred to as "living cargo."

derelict: A boat without a crew, abandoned.

diving bell: Used as early as 1838, but primarily in the mid–1840's for salvage operations.

"doctor, the: A small auxiliary engine, used to drive the water supply pump independently of the main engines. It was also used to drive the bilge pump and the fire engines. The nickname stemmed from the fact it cured many early steamboat ills, Doctors were in general use on larger boats by the 1850's.

dog watch: From 4:00 A.M. to 7:00 A.M. The night shift.

donkey: The hoisting engine.

double-tripping: A term used when cargo was removed from a steamboat and placed on shore in order to lighten its load during periods of extremely low water or when ice made progress difficult. It was so called because when river conditions improved, the boat had to return (and thus make a double trip) to retrieve the freight.

driftwood: Floating logs and branches in the river. Driftwood was responsible for breaking paddle-wheels and causing damage to hulls.

dug-out: Canoes hollowed out of white pine tree trunks.

dull season: A period when freight was not readily available for transport. Boats usually put up in port to await a full load rather than travel at a loss.

engineer: The officer in charge of the engines and boilers aboard a steamboat. He maintained the machinery and regulated the amount of steam required for speed and navigation, receiving orders from the pilot. The engineer also had blacksmithing skills and was required to repair and make parts for the machinery. His salary was less than a pilot's.

Engineer's Association: Established in Cincinnati and St. Louis in 1842, establishing standards of character and training. An engineer meeting the requirements was issued a certificate. Like the Pilot's Association, it had little impact. Eventually, engineer and pilots' qualifications were outlined by Federal legislation.

Evans' Safety Guard: Created by Cadwallader Evans as a tool used to prevent boiler explosions. It came into use in the 1840s and was highly promoted by steamboat captains for its effectiveness. Steamboat lines often cited its use aboard their boat as proof of the inability of its boilers to explode.

falls: A place where the river abruptly dropped down a shelf of rock. The largest and most significant falls on the Ohio River was the Falls of Ohio at Louisville, where the river fell 22 feet over a series of ledges for a distance of two miles.

Fashionable Tour: Term coined in the 1830's by George Catlin, a western illustrator, describing a holiday trip by steamboat from St. Louis to Rock Island, Galena, Dubuque, Prairie du Chien, Lake Pepin, St. Peters and the Falls of St. Anthony. A return trip went from St. Anthony back to Prairie du Chien, and from there to Fort Winnebago, Green Bay, Mackinaw, Sault Ste. Marie, Detroit, Buffalo and Niagara. Pleasure trips on the Upper Mississippi were limited until the end of the Black Hawk War when the threat of Indian attack was lessened.

Father of All Rivers: The Mississippi.

fellow-servant rule: The common-law rule that dictated an employer was not responsible for injuries resulting to an employee through the negligence of a fellow employee — an employee with limited protection under the law.

Fever River (Galena River): The word comes from a corruption of the French "*féve*," meaning "broad bean". The story was that a band of Indians died of smallpox at the river, and the name stuck. "Fevre" was the alternate spelling.

firebox: Where wood (and later coal) was stored aboard steamboats.

firemen: A specialized subgroup of deckhands whose primary task was the wooding (taking aboard wood) and stoking the fires used to create steam. For this back-breaking labor, they were rewarded with slightly higher wages than ordinary deckhands.

flatboat: A creation of the Ohio Valley. This was the most popular craft for transporting families and their household goods in the era before steamboats. The average raft rose 3–6 feet above the water's surface. There were small structures built on top, which were used for bedrooms, a dining room, and a place for cooking. Flatboats cost from $1.00–1.50 per foot of length and ranged from 20–80 feet length by 10–20 feet width.

flat-bottomed skiff: A raft fit for carrying 2–3 people.

floater: Slang term for the body of a drowned person on the river.

Floating Palace: A gaudy, flashy steamboat known for its speed, and luxury accommodations and bounteous food. They were most expensive and largest of all steamboat classes. The owners attracted passengers by the reputation of the boat. Most floating palaces offered service equivalent to a first-class hotel as well as providing transportation for roughly the same fee.

floating teakettle: Early derisive name for a steamboat.

fo'c'sle: Bow of the boat.

forest-bordered crevice: The space on the river between an island and shore.

found: A term typically associated with the west, and used in context with wages. Thus, "Thirty a month and found" meant the employee received a salary of $30 a month and sundries, including meals and shelter.

free list: A list of dignitaries, important persons or steamboat pilots who could travel aboard a steamboat without charge.

freight classifications: "Heavy" and "light." Heavy freight: Products of rolling mills, foundries, or glass works; hardware and implements. Light freight: Dry goods and merchandise of European manufacture having high value in proportion to weight. Rates were determined by the containers in which they were shipped: casks, barrels, bales, sacks and hogsheads.

freshening: A sudden swift current from a rise in the river.

freshes: Influxes of water from tributaries into the main river system, usually arising the water level 12–15 feet.

fruit and furniture: A boatman's derisive term for the goods families brought with them downriver on rafts.

full license: The license allowing a pilot to work both day and night.

Fulton, Robert: Born November 14, 1765; died February 23, 1817. Credited with being the first individual to successfully demonstrate the use of steam power aboard a boat in 1807.

gentlemen's cabin: Assigned to the forward (most dangerous) part of the boat and supplied with berths. In the early days, sleeping accommodations were usually shelves, two or three tiers high. Women were never allowed in the gentlemen's quarters.

get out of the river: An expression meaning to run from New Orleans to Cairo (the length of the Lower Mississippi).

Golden Age of Seamboats: The 1850s, when floating palaces were at their height and public enthusiasm was high. It has also been referred to as the "critical decade," for despite outward appearances of high profits and numerous show boats, the railroads were making headway in siphoning off freight and passenger service, competition between steamboat captains reached a cutthroat level, and serious droughts underscored the problem with water transportation.

Grand Union Association of Steamboats and Steamboat Engineers: A conglomeration of local associations gathered together in 1855.

grass-hoppering: See "sparring."

gravel bars: River obstructions more stable and more dangerous than sandbars, they were categorized as either hard and permanent, or less compact and stable.

graveyard, the: The area of the Mississippi River between St. Louis and Cairo, littered with sunken wrecks. By 1875, there was an average of one hulk per mile, causing great danger to steamboats.

Great Through Route: The steamboat mail service line from Louisville or Cairo to New Orleans in the 1850's.

grounding: Getting stuck on a sandbar.

"guard deep with freight": Heavily loaded; cargo stowed up to the guardrails.

guards: A distinctive feature of western steamboats. The guards were extensions of the main (lower) deck, originally designed to protect the paddle-wheels from injury. The guards were held in place by braces. The extensions ran from either side of the hull, from stem to stern, and provided additional space for cargo and passengers, as well as providing passageways to different parts of the boat.

"hand down": An order to give the engines all the power they had; used to increase speed to get over an obstruction or during a race.

heavy goods: A term of freight classification, indicating iron, foundry cargo, glass and hardware.

high-pressure engine: Developed rapidly in the West, where greater flexibility and reserve power were needed. They were compact and weighed less than low-pressure engines. They also permitted the boilers to use muddy river water. High-pressure engines were easier to operate, were relatively inexpensive, and supplied reserve power for emergencies. As the high-pressure engine was developed, the condenser was discarded.

hog-frame: Longitudinal trusses arched above the superstructure of the steamboat, giving strength to the long hull.

hogging: Bending or breaking the hull out of shape. Used when the boat was being "sparred off" a sandbar.

hold: Traditionally storage area aboard ships. The addition of 'tween decks on a steamboat lowered the headroom in the hold, making it impractical for the storage of cargo, especially in small and medium-tonnage vessels.

home port: The principal port where steamboats laid up during the winter months; usually the city where the owner lived.

"hoop-poles and pumpkins": Slang used by rivermen referring to produce and livestock floated down the river on rafts.

hot engineer: An engineer who was willing to push his engines to (and sometimes beyond) the limits of safety to achieve greater speed, primarily during races.

hulks: The submerged remains of destroyed boats. They presented significant danger to passing vessels.

hurdle: A run through a narrow or difficult channel.

ice box/ice house: Compartments filled with ice to preserve fresh fruits, vegetables and meat, as well as ice for mixed sweet drinks.

insurance: Boats and freight were insured separately. Rates on cargo were determined by the trade (area of operation) and distance traveled. Rates averaged one-half to 1.25 percent of value. Upstream rates on the Mississippi tended to be slightly higher than downstream trips owing to the snag menace. On old boats, insurance rates were several times higher than on boats with a first-class rating. Shippers were responsible for insuring their own cargo (see "common-law doctrine"). Insurance on boats was usually granted only for partial value in order to discourage owners from wrecking their vessels for the insurance money. Throughout the steamboat era, insurance was considered more a luxury than a necessity.

jack-staff: Flagpole.

jack towel: A towel suspended over a wooden roller.

jigsaw and fretwork: Architectural designs on the superstructure of a steamboat.

jingler: The ringing of the bell.

joint-stock company: A formally organized group of individuals with a board of directors and trustees for holding property in a steamboat venture. Overly complicated, this type of association was seldom used in the freewheeling West.

"Jonathan" (or "Brother Jonathan"): A typical American, just as "John Bull" referred to a typical Englishman.

"jumping the boat": Walking off the job (primarily said of deckhands). Such an act was taken to protest poor wages or highlight bad working conditions, usually in reference to a cruel and abusive mate. Synonymous with "mutiny."

keelboat (keel): A raft constructed of heavy planks, sharp at the bow and stern, capable of carrying 20–40 tons of freight. Keelboats received their name because of their heavy, four-foot-square planking that extended from bow to stern to absorb shock from the frequent impact of submerged obstructions. They ranged from 45 to 75 feet long and 7 to 9 feet wide. These craft occasionally carried masts and sails for downriver trips, and were crewed by a steersman and two men at the oars. Because keelboats could not fight the upstream current, they were literally hauled across the water by six to ten polemen, with one crewman acting as a steersman.

ladies' cabin: Situated aft, considered the safer part of the steamboat, where the women were protected from the heat and danger of the boilers, but subjected to far more vibration from the paddlewheels. Married men traveling with their wives stayed in the ladies' cabin.

landing stage: A series of connected wooden planks easily swung into position along the high, slippery riverbanks to facilitate the flow of traffic. Landing stages did not come into use until the late 1870's.

landing whistle: Two long blasts followed by three short ones.

lard: See "resin."

laying by: Staying in the wharf.

lead line: Used in sounding to determine the depth of low water.

levee rats: Local boys living in a wharf town.

light-draft boats: A small class of steamboats built specifically to operate during periods of low water when heavy-draft boats were unable to operate.

lightening: Temporarily removing cargo to allow a boat to get over an obstruction. See also "double-tripping."

light goods: A term of freight classification that included merchandise and manufactured goods, proportionately more valuable than heavy.

lightning pilot: An ace; an expert. Used as a supreme compliment.

little towhead: An infant island.

Livingston, Robert: Robert Fulton's financial backer and a former chancellor to France.

Louisville Pilots Benevolent Association: Founded in 1841, the association dealt with matters such as salary, while stressing professional standards on the river. It had little influence. See also "Pilot's Benevolent Association."

lower between deck: Used for machinery, cargo and deck passengers, and provided the convenience of handling freight as near the water's edge as possible. By 1850, the headroom for a 350-ton boat was 11–12 feet and, on larger boats, 15–20 feet. In contrast, headroom in the hold was reduced to less than five feet.

low-pressure engine: Originally used by Robert Fulton. The engine used on the *Clermont* was built in England by Boulton and Watt. Later, low-pressure engines were locally made. They were widely used on eastern rivers, which were more stable, tended to be straighter and easier to navigate. The low-pressure engines were condensing engines, and generated less power.

Lower Mississippi: Generally considered to extend from the mouth of the Missouri River, past St. Louis and onward to the Gulf of Mexico. It covered 12,000 meandering miles, passing cotton and sugar plantations and many small settlements and squatters' shacks. Past Cairo, the river flowed through a level floodplain from 50 to 100 miles wide. In addition to the Missouri, the main tributaries are the Arkansas and Red rivers.

mail service: The federal government offered contracts to steamboat owners to deliver the U.S. mail as early as 1813, but low payment and indifferent service resulted in loss of contracts, so this never became a major source of revenue for steamboat operators. With the advent of the western railroads, mail contracts to rivermen drastically diminished.

main deck: The lower deck, closest to the water. Used to stow freight and house deck passengers.

Man at the wheel: The pilot.

marks: Points on the river used by pilots to navigate.

mark twain: A false point. An area on the river that is not what it seems. Also used when sounding the depth of the river to indicate twelve feet, or two fathoms.

Marine Hospital Service: In 1798, the federal government established sickness and disability relief for seamen, with dues taken out of their wages for the construction and upkeep of maritime hospitals. The act of 1837 expanded provisions for rivermen, but no hospitals were opened along the Mississippi or Ohio rivers before 1850. Benefits were offered to crews of registered and licensed steamboats, but were not available for keelboatmen who worked on unregistered craft and consequently paid no dues into the system. Nor were slaves working the rivers entitled to benefits, as their owners paid no dues for them.

mate: The mate served directly under the captain in the overall management of the boat. The mate directed the deckhands in regard to cargo and manual labor. He stood alternate watches with the captain and was typically an unsavory character, renowned for his liberal use of profanity and physical prowess.

mickies (also known as "niggers"): Small hoisting engines, introduced in the late 1840's to stow and remove heavier freight from the hold. They were not used to move cargo from shore to the boat or vice versa. Such equipment did not come into use until after the Civil War.

Mississippi: The word means "Great River" in the language of the Ojibway, an Indian tribe of the Algonquin Nation (woodland Indians). In the beginning the name referred only to the headwaters, but it was passed on by French fur traders and missionaries, and eventually came to refer to the entire river.

mosquito fleet: Smaller stern-wheel boats, which were capable of operating throughout the year. During low-water months, passengers were the prime "cargo," traditional freight being too heavy.

moving hotel: See: Floating Palaces."

muffler: See "shroud."

navigators: Guides published for the use of the pilots. The most famous were *The Navigator*, (1801–24), *The Ohio and Mississippi Pilot* (1820), *Western Navigator* (1822 and reissued over a period of 30 years under the title *The Western Pilot*) and the *New River Guide*, 1848.

newspaper puffs: Journalists who extolled their particular city or the excitement of dangerous speeds and steamboat racing, among other evils. An irresponsible reporter. See also "puff."

nightmares: Places of extreme danger. On the Upper Mississippi, these included Cassville, Brownsville, Trempealeau, Rolling-stone, Beef Slough, Prescott, Grey Cloud and Pig's Eye.

"no bottom": A river is deeper than 12 feet, or the length of the sounding pole.

obstructions: Any natural or artificial blockage of the river or riverbed. Typical obstructions were sandbars, rapids, reefs, snags and hulks.

oil: See "resin."

Old Man: A term taken from maritime usage, referring to the captain.

ornamentation: Any of the fancy, superfluous design on the exterior of the steamboat.

oxbow: A U-shaped bend in the river, formed by currents cutting passages through the soft loam and sand of the river plain, generally adding length to the stream. Conversely, when the river cut a channel through the arm, or narrow neck of the wide loop of the oxbow during periods of rising flood waters, it formed a cut-off, drastically shortening the course of the river.

packets: Steamboats that made regular trips along the same trade and kept to a schedule.

paddle-boxes: Wooden frames protecting the paddle-wheels.

paddle-wheel buckets: The slats on the paddle-wheels.

Panic of 1837: Caused primarily by unbridled land speculation and the overextension of credit to buy land. The states were also in debt due to expensive internal improvements. When President Jackson issued the Specie Circular, requiring only gold or silver (specie) be used to pay for public lands, there was a run on the banks. Many banks could not meet the demand and folded. A depression followed.

pantry boy: A member of the cabin crew who washed, wiped and took general care of dishes and silverware.

papers: A pilot's license.

partner: A term used between pilots to signify one-half of a working team. Also, joint ownership between two men on a steamboat, where in each was responsible for the boat and the debts incurred.

partnerships: Approximately two-thirds of all steamboat ownership throughout the era was by four men or fewer. Partners were regarded as agents of the boat.

"perils of the river": An expression typically used by insurance agents and meant to cover all natural accidents that were preventable by prudent action.

pest boat: One thought to be infested with cholera.

pilot: Known as the "king of the rivermen," the pilot served as both navigator and helmsman. A regular officer, he was typically the most skilled crewman aboard the boat and was well paid for his services, receiving a salary only slightly less than the captain.

pilothouse: A small, elevated, often glass-enclosed compartment for the navigator, traditionally off limits to passengers.

Pilots' Benevolent Association: Developed to create standards and preserve wages. Mark Twain described it in his *Life on the Mississippi*, and had a higher opinion of its worth and success than did most authorities.

pine knots: See "resin."

pirogue: A large canoe, often 40–50 feet long and 6–8 feet wide, capable of carrying a family and several tons of household goods.

pole (sounding pole): A 12-foot pole used for sounding in the Upper Mississippi where the water was shallow.

poling: A method requiring manual labor to get a boat over sandbars or across rapids. Men on each side of the boat dug their poles into the river bottom and walked toward the stern, literally pushing the craft forward.

porkopolis: Cincinnati, so called for its abundance of pigs.

posey Country: Indiana. A derisive term for backswoodsmen and families from the interior.

puff: A derogatory term; an irresponsible person; one who puffed up the merits of a particular place or activity.

quarantine grounds: Areas ashore where the sick were treated; places where steamboat passengers suspected of carrying disease were housed until cleared by a physician. Quarantine grounds were often marked by yellow flags.

quarantine stations: Built outside major cities for the inspection of arriving steamboats. If disease were found aboard (particularly cholera and yellow fever), infected persons (most likely deck passengers) were removed from the boat and prohibited from entering the city.

raft: A series of interconnected driftwood, trees and debris that formed an impenetrable river obstruction. Most notable used in conjunction with the Red River.

rank: Twenty cords of wood, standing eight feet high.

rapids: Where a river's current is fast and fraught with obstruction.

reach: A stretch of straight river where the current drove through at a good pace.

reef: A long sandbar, rising abruptly at one side.

resin: An auxiliary fuel used as a temporary measure to increase the power of the engines, often employed during races. Resin created a great deal of smoke, and this type of additive was also used when a steamboat came into the wharf to announce its arrival, or when departing at the start of a voyage to create an aura of grandeur.

ripples (or shoals): A series of sand or gravel bars, over which the river passed with greater speed than that of ordinary current.

riverboat: This word was not used during the steamboat era. Contemporary accounts *always* used the word "steamboat," or "steamer."

robber barons: A derisive term for railroad owners.

rock bars: Chiefly found on the Upper Mississippi and Ohio rivers, formed by debris falling from overhanging ledges and bluffs. These bars caused significant and occasionally fatal damage to unwary steamers.

roustabout: An unskilled laborer who did manual work aboard a steamboat. Also known as a "rouster."

runners: Men and boys employed at the various wharves to help secure passengers, often promising low fares and quick trips. Runners were also used by hotels to promote their accommodations to disembarking passengers.

"running out": Challenging a boat to a race.

"running where there is a heavy dew": Operating the steamboat in very low water.

Sabbatarian Movement: Those involved in prohibiting steamboats from running on Sundays. They had little impact on western commerce.

safety barges: A unique invention of the Upper Mississippi, barges were attached to steamboats and used to carry additional passengers, or for boats lacking proper accommodations. Their introduction was used as a means of assuring passenger safety; in case of a boiler explosion, passengers were protected by being on a separate vessel. They also demonstrated the value of flat-bottomed boats by overcoming the dangers of navigating the rapids.

St. Joe: St. Joseph River, Michigan.

saloon (salon): The central apartment on western steamboats. It was long and narrow; the staterooms were situated on either side. At mealtime, tables were placed back to back in the saloon for dining, and after the meal they were removed. In their place the cabin crew put out sofas, rocking chairs and other accommodations for lounging and entertaining. The bar was situated at the head of the saloon, and during the colder months stoves were placed at bow and stern for warmth. This area of the boat was the most highly decorated, and it was either here or on the promenade that passengers spent most of their time.

sandbars: The most common river obstruction, forming at the head and foot of channel islands, at the mouth of every stream and in areas of the river where the current slowed enough to deposit sand and silt on the river bottom. Sandbars were transient and constantly shifting.

sawyer: A snag positioned with its head downstream, so-called for its swaying motion caused by the current.

scalding: Burns caused by escaping steam or boiler explosions. The engine room crew and deck passengers bore the brunt of this type of injury.

"scale the boilers": Clean the boilers.

scow: A type of raft in use on the rivers before steamboats. Later, they were used as wooding boats.

sea-anchor: Anything dropped overboard to hold the boat in place. When the normal anchor was not substantial enough, other material was lashed together, tied with rope and sent overboard.

set position: Poles at the ready to start a trip; used in keelboating.

setting the boat: Putting the boat in her marks.

shaver: A boy.

shindy: A quarrel.

"shipped a sea": Water spilling over the deck and into the cabins and saloon.

shipping papers: A written contract between the captain and the individual crewmembers, detailing terms of employment. They were seldom, if ever used on western rivers, where crews were hired for indefinite periods and discharged as necessity dictated.

shipping up: The act of shifting the cam rod from the lower pin on the reversing lever to the upper, or vice versa.

shoals (or ripples): A series of sandbars or gravel bars, over which the river passed with greater than ordinary current.

shroud (or muffler): A covering made of heavy canvas, placed around the forward part of the boat in front of the furnaces to block off light when navigating on very dark nights.

sideswipe: A brief collision between two steamboats, usually resulting in little or no damage.

side-wheeler: A steamboat with paddle-wheels on either side; the bow was reserved for freight. Early boats had the engine and smokestack exposed in the center.

slough: A lagoon.

smallpox: A deadly disease carried by emigrants along the waterways. Although epidemic, smallpox never presented a threat similar to that of cholera or yellow fever.

snagboat: Developed by Henry Shreve and put into use in 1829. The purpose of the steam-driven snagboat was to catch a snag on the snag beam and force it up out of the water, where the obstruction was either broken up by the force of the extraction or was sawed by hand. This was the major river tool used for river improvement throughout the steamboat era. Snag boats were nicknamed "Uncle Sam's tooth-pullers."

snag: The root end of a fallen tree that became embedded in the water, weighed down by dirt and gravel. The tree eventually lost its branches and settled in at an angle that often caused it to spear passing boats. See also "sawyer."

soldiered: Lounged on duty.

sparring (sparring off): To get over an obstruction, spars were placed on either side of the vessel and used as crutches to literally lift the boat out of the water. The paddle-wheels were then put in motion to move the boat ahead several feet. The procedure was repeated until the vessel was clear. This was the most common method of freeing a grounded boat, but was physically demanding and often resulted in injury to the deckhands.

speaking tube: A hollow tube running from the pilot house to the engine room used for oral communication.

spreader bars: These joined the chimneys at the chimney tops for stability. In the 1850 and 1860s they were frequently decorated to enhance the looks of the boat.

stage (stage plank): The platform put down when the boat docked for passage across to the shore or wharf.

stateroom rate: The fee for first-class accommodations. Tickets were purchased aboard the boat at the clerk's office, which was usually situated to the right of the bar off the saloon. If no staterooms were available, passengers could opt to sleep on the saloon floor, but were required to pay the same fee as if they had a private cabin.

staterooms: Private quarters for first-class passengers off the saloon to either side. On early boats these quarters were separated by curtains, but as designers made improvements, they were shut off into separate rooms. A typical stateroom in 1850 was 6 feet square. Private quarters were separated into gentlemen's and ladies' areas; married men stayed in the ladies' section with their wife.

Steamboat Acts: Attempts by Congress to remedy some of the ills associated with steamboat travel. The act of 1832 was the first such legislation but it was broadly written and had minimal effect. The next act came in 1852. It attempted to address concerns of the first act as well as regulate other areas of operation, including issues such as escape for deck passengers in case of disaster, limiting the number of deck passengers, and fixing penalties for noncompliance. There were also regulations concerning boiler manufacture and factory inspection, the licensing of pilots and engineers, and specifications limiting steam pressure during racing.

steamboat agent: In the early years of steamboating, private merchants acted as agents for shippers, usually without compensation. As traffic increased and questions of legality and responsibility became an issue, a separate job classification of agent developed. Agents secured freight, forwarded through cargo, or stored it in warehouses until space aboard a boat became available. They also worked for captains, securing cargo, collecting debts, and performing other services such as arranging for fresh provisions to be brought aboard. For these tasks they received small fees.

steamboat bell: The primary means of communicating between two passing boats. The bell was mounted forward of the chimneys on the hurricane deck. It was eventually replaced by the steamboat whistle, which improved messages sent between captains.

steamboat construction: Pittsburg, Cincinnati and Louisville were the principal centers of steamboat construction in the West, joined by St. Louis after 1840.

Steamboat Era: Generally considered to be 1807 through 1861, the beginning of the Civil War.

Steamboat Gothic: A fairly universal catch-all expression of the time, meant to describe the American penchant for decorating Floating Palaces with heavy gilt, gingerbread (tawdry, gaudy or superfluous ornament), an abundance of illumination and red plush. Europeans generally viewed the style as ostentatious; Americans less so.

steamboat insurance: Boats and freight were insured separately and throughout the Era, insurance on a boat was considered a luxury rather than a necessity. (see: insurance).

steamboat lines: Lines consisted of two or more steamboats offering regular service in a given trade: one boat would travel one section and the second and third would provide service to farther connecting sections of the river. With multiple boats, service was arranged to provide daily, thrice-weekly or weekly departures along the better-traveled routes. Boats were independently owned and lines seldom lasted more than a season or two.

steamboat officers: Officers included the captain, pilot, engineer, clerk, mate and, on larger boats, the steward.

Steamboat Scandal: Late in 1817, Secretary of War John C. Calhoun issued three contracts to Colonel James Johnson to transport military supplies, clothing and ordnance from Pittsburg to St. Louis and then deliver them by steamboat up the Missouri River. This was the earliest issued contract issued by the federal government to utilize steamboats, and Calhoun suffered great criticism for attempting to use this new and untried method of transportation. Johnson did not succeed in fulfilling the terms of his contract, but billed the government $256,818.15 for the effort. The invoice was paid, which subsequently brought about a House action to appoint a committee to investigate the scandal.

steamboat's shingle: Typically a blackboard placed on the wharf, announcing departure dates and times, freight fees and passenger fares. During periods of intense competition, these rates could drop drastically within a matter of minutes. Scheduled dates and times were seldom reliable.

steersman: See "cub."

stern: The rear of the boat.

stern-wheeler: A steamboat first designed by Henry Miller Shreve, possibly to get around the Fulton-Livingston patent, which utilized one large paddle-wheel at the stern of the boat.

stevedore: A roustabout; deckhand.

steward: An officer of the first class, responsible for providing meals for passengers. Under the steward were the cooks, waiters and pantrymen. A well-regarded steward might command a salary equal to that of the captain. The steward stood no watch, but was constantly "on duty."

stirrups: Double bolts with nuts that clamped the buckets to the wheel-arms. These were typically made aboard the boat in the blacksmith shop by the engineer.

striker: An apprentice engineer.

stroke: The stroke of the engine was the distance traveled by the cross-head of the piston in making a complete revolution of the wheel.

storm toggery: Waterproof garments.

"stove": Broken up.

sweeps: Oars.

table d'hote: The dining room of a hotel.

"taking in the milk": Bringing aboard a cow to provide fresh milk for first-class passengers.

telegraph: The earliest way to communicate via electricity over long distances. The first telegraph lines reached St. Louis, Keokuk, Burlington and Dubuque in 1848.

terminals: Wharves. Steamboats generally required no more than a gangplank to load and unload freight and passengers. Large cities along well-traveled trades eventually constructed graded landings to improve conditions. The city then charged small fees for the service and upkeep, which became a bone of contention between boat owners and city politicians.

temperance boats: Boats that did not serve alcoholic beverages. While the revenue from the sale of spirits was not substantial, the loss of passengers preferring alcoholic beverages was, and few captains opted to sacrifice profit for principle.

"Texas" (alternately spelled with an upper- or lower-case "T"): An upper deck, narrower than the main deck and shorter in length. It garnered the nickname (according to the Cincinnati *Gazette*, October 11, 1846) because this deck was "annexed" or added on. Also known as the hurricane deck after the mid–1840's. By the 1850's the Texas deck was widened and lengthened, providing accommodations for officers and passengers.

Texas-tender: A servant aboard a steamboat; one of the cabin crew serving passengers.

three-month engineers: A derisive term for inexperienced steamboat engineers. Steamboat owners were often accused of hiring poorly trained engineers in order to save money on wages.

through shipments: Freight going the entire trip without being unloaded at a stop along the way.

ticket, buying "clear through": A ticket enabling the passenger to travel the entire distance advertised by the steamboat "card." Captains frequently decided not to complete the entire trip, however, usually because of a lack of cargo, and passengers often found it difficult to receive a refund for the unused portion.

tiller ropes: Made of hemp in the early to middle years of steamboating, connecting the pilothouse wheel to the rudder. Tiller ropes were often the first nonhuman casualities of a fire, when they burned through, causing an immediate loss of steering control.

toilet facilities: Water closets were constructed into the wheelhouse or over the stern.

tonnage: A boat weighing 200 tons would carry 300–350 tons of weight in cargo. **Torch (smoking jack):** Used for illumination; an iron basket a foot in diameter and 18 inches deep, swung from a forked bar or standard and set in holes on the forward deck, leaning out over the water. Fueled with "light-wood" or "fat-wood." Pulverized resin was also used to flare up, creating a fierce flame.

trades: A field of operation between two or more ports. Local trades might be as short as 20 miles, while others spanned over 1,000 miles, as in the Louisville-to-New Orleans trade. Trades were not fixed and often fluctuated in importance.

tramp steamboat: See "transients."

transfer arrangements: Passengers going beyond the final stop of a steamboat were required to make their own connections. This could not be done in advance, for arrival dates were frequently uncertain.

transients: Usually smaller-tonnage steamboats without a fixed trade that followed whichever trade happened to be the most profitable at any given time. Transients were the most numerous of all steamboats, dominating the rivers until after the Civil War. They generally carried passengers and freight at lower fares than did the gaudier, fancier, faster boats.

trips on end: Pertaining to the crew; two or more voyages, back-to-back. Two "trips on end" were usually as much as a typical deckhand could physically endure. Afterward, the deckhand usually laid up for rest and refreshment.

trumpet: Sounding device used to speak through when issuing orders.

trunk lines: Most significant were the Mississippi and the Ohio. Minor trunk lines were the tributaries of these rivers.

trying-out time: The period or incident when a newly graduated cub proved himself, usually referring to performing a difficult feat alone.

turnover: A term used in reference to the paddle-wheel making one revolution.

turpentine: See "resin."

twisting the wheel: Turning the pilot's wheel one or two spokes at a time.

upper between-deck: An open promenade devoted to the accommodation of cabin passengers. The hurricane deck formed its roof.

Upper Mississippi: The 200 miles between St. Louis and Cairo where the Ohio River comes in. It is also referred to as that portion of the Mississippi above St. Louis and, in later years, the river from St. Louis to St. Paul.

"wad of steam": The steamboat man's expression for reserve power obtained by hard firing and the use of highly combustible substances, such as resin, pine knots, oil, lard, or turpentine.

"walking the boat": See "sparring."

Ward's Island, New York: The initial point of disembarkment for emigrants from Europe, where the thousands affected with diseases were quarantined in separate buildings designed for the purpose.

warp: Rope.

warping (1): Manually working the capstan to aid the paddle-wheels and free a boat from an obstruction.

warping (2): Using a rope to move a keelboat upstream against a swift current. The raftsman would attach the rope to a tree and other land anchor some distance ahead, and walk from bow to stern, pulling as he went. When he reached the stern he would "break off" and run to the bow or another turn.

washrooms: Separated for the use of each gender, these usually adjoined the staterooms. For the most part, they were no more than several washbasins on a bench, filled with river water. Two jack towels habitually served the entire complement, and were seldom clean. Common combs, brushes and more rarely, toothbrushes, were supplied. By the 1850s, most staterooms had their own basins and towels.

watch: The time spent on duty; divided into shifts or rounds of duty and held alternately. A common division of labor was the following:

7:00 A.M.–12:00 noon: captain and chief clerk
12:00 noon–6:00 P.M.: mate and second clerk
6:00 P.M.–12:00 midnight: captain and chief clerk
12:00 midnight–4:00 A.M.: mate and second clerk

The "dog watch" was between the hours of 4:00 A.M. and 7:00 A.M. On some boats the off-duty pilot and/or the engineer took their turn during these hours, so those standing the midnight shift would not get too tired and thus careless.

way-freight; way-passengers: Freight or passengers put ashore along the journey; not going through to the end of the run.

wharf-boat: Every good-sized town had a wharf-boat, usually an old steamer or a barge. It was used to ferry passengers from a steam-boat to shore, or by those awaiting the arrival of an outgoing boat. If they had to wait overnight, sleeping accommodations were provided. Outgoing cargo was also stored on these boats, as was incoming cargo awaiting final disposition.

wharfmaster: The city official in charge of the wharves in the larger cities. His primary duty was to collect fees, for wharf upkeep.

white log: An obstruction in the river that was nearly impossible to see at night (as opposed to a "black log," which was more visible).

wild boats: Transients.

"wildcat money": Bills printed by "wildcat banks" with little or no backing by gold or silver. Passed at face value, their actual worth was nominal. Wildcat money and counterfeiting proliferated in the West.

wind reef: A false reef that looks exactly like a bluff reef, but is caused by wind, and is thus harmless, unless mistaken for a legitimate threat.

wood-boats: Flatboats and scows holding 20–40 cords of wood, which belonged to the woodyard man. These were attached to a steamboat at various points along the river. After unloading, the wood-boats were set free to drift down the river where they were reclaimed by their owners and restocked. Used by steamboats only on upriver trips.

wooding up: Steamboats taking on wood. The typical steamboat wooded up every 30 miles.

wooders: Crew (and passengers who hired themselves out) who either chopped wood along the bank or brought in the stacked wood from woodyards.

woodyards: Places along the river where steamboats stopped in order to buy wood used as fuel. Oak, beech, ash and chestnut were the most favored.

word passers: Crew on the hurricane deck who passed along the leadsman's reading to the pilot.

"wrecked and floundered": A government classification which included boats damaged or destroyed due to storms, grounding and obstructions.

Appendix B: Original Accounts of Steamboat Disasters

Steamboating on the western waterways was first and foremost a commercial venture, but it was the passenger trade that earned its distinction in American lore. The Gothic decoration of a steamboat cabin, the shine of gilt, the excitement of speed and racing all added to the almost mythic regard in which steamboats and their captains were held. Sadly, there was a dark side to the opulence, the profit-loss calculations, the bold explorations and the romance: the many disasters and horrific loss of life which accompanied the trial-and-error period and that expanded well past the Civil War. The following articles, drawn from newspapers of the time, speak for those who never had the chance to write a glowing letter home about the "Grand Tour," live the adventure of a Mississippi pilot or merely make a new life in a new world.

A mere twenty-one years after the *New Orleans* became the first steamboat to travel from Pittsburg to its namesake, thus opening the inland waterways to a new mode of travel, the *Adams Sentinel* of December 2, 1833, published a compilation of disasters, along with a telling editorial:

STEAM BOAT DISASTERS

The history of steam boat navigation affords no parallel to the loss of lives and property on board boats navigating the Western waters, accounts of which have been received within a few days past.—The waste of human life has been horrible;—and how did it happen? For the want of attention; this is the *mildest* term that can be applied to it;—a fine consolation, truly, for orphans deprived of their parents; widows of their husbands; and parents of their children.—There has been an indifference, a lukewarmness, upon this subject, which cannot be longer tolerated; when we mentioned the burning of the Caspian yesterday, and the loss of lives, the reply was—"*Wonder if she was insured!*" There must be some plan adopted by legislation, or otherwise, to put an end to it; it can be borne no longer.—The following list has been received in less than a month, and four of them in the last few days! Columbia—run under and sunk from being too heavily laden. Paul Clifford—run into by the Huron and sunk. Thomas Yeatman—burst a boiler—6 lives lost. Peruvian-struck against a snag, and sunk. New Brunswick—burnt on her passage to St. Louis. St. Martin—burnt, and thirty or forty lives lost. Illinois-burst a boiler,

and scalded between 30 and 40–13 dead. Bonnets of Blue — on her way to Nashville — snagged and sunk.
Doubtful.

The *Republican Compiler,* March 12, 1839, added this to the growing statistics:

The Alton Telegraph publishes a list of the steamboat disasters of importance which occurred on the Western waters during the [year] 1838. By far the greatest number of these mishaps has arisen from *snagging,* by which species of casualty no less than 37 have taken place. "Blown up" and "collapsed" announce the fate of 14 out of the whole number, which is 80. Of total losses there are 34, or nearly one half, to which we may add four that cannot be repaired. Of the 37 snagged, but 4 were on the Ohio, and the whole number 50 were on the Lower Mississippi. No collapse took place on the Ohio, and there was but one blown up. *Balt. Amer.*

The toll for 1838 was updated and printed in the *Adams Sentinel* on December 31, 1838:

The past season of low waters in the Western rivers has been marked by an unusual number of steamboat losses. The following list of disasters is furnished in a Cincinnati paper, but it is believed not to comprise all that occurred. — When to these losses are added those which have arisen to the Western merchants in not receiving their goods bought for the late fall trade, and also those connected with the disappointments of traders who could not send produce to market on account of the protracted drought, succeeded by an early obstruction of navigation by ice — it will be found that the Western States have experienced a season of general and serious disadvantage:

Steamboat Accidents on the Western Waters during the past season.

The Rolla,	sunk
Czar,	"
Clinton,	"
Platte,	"
Logansport,	"
Belle of Missouri,	"
Dart,	"
Ashley,	"
Gov. Dodge,	"
Washington,	"
Gen. Brown,	blown up
Augusta,	"
Motto,	sunk
Chillicothe,	"
Corinthian,	"
Comanche,	"
Renown,	"
Norfolk,	"
Mississippi,	"
Palmyra,	"

It would not have surprised the editor of the *Adams Sentinel* to observe that in the above article, loss to western merchants and the disappointment of traders is lamented, but no mention whatsoever is made of lives lost.

The Wisconsin *Enquirer* of February 26, 1840, reprinted a list from the Cincinnati *Daily Gazette*:

Steam-Boat Accidents

List of Steam-boat accidents on the western waters in the year 1839. It is not claimed that the subjoined list is complete; the writer merely kept a record of such accidents that met his eye in the papers; which, however, were carefully examined, with a view of making it as accurate as possible.

In February.—The Oswego struck a rock on the Ohio, near the mouth of the Kentucky, on her passage up from New Orleans, and sunk, boat and cargo lost. The Victor was destroyed in the Ohio by ice. Another boat, name not given, at about the same time was burnt. The Pawnee, on her passage down the Mississippi, from St. Louis to New Orleans, got aground near the Grand Tower. After taking out her cargo, she swung around on a rock, and sunk. Boat lost.

In March.—The Howard snagged on the Missouri, near St. Charles. Lost. The Livingston struck a snag in Red River, on her passage from New Orleans to Shreveport. Boat and cargo lost. The Alice Maria, from Pearl river, snagged in the Mississippi. Cargo saved-boat lost. The Shylock, on the Mississippi, injured by a rock, and run on shore.

The Othello, bound up the Wabash, was run into on the Ohio, by the Peru, and sunk. Boat and cargo total loss.

The General Brady snagged on the Mississippi, near Laurie Island. Part of her cargo saved-boat lost.

In April.—The Alert, snagged on the Missouri river, on her downward trip.—Total loss. The Pennsylvania was burnt at the landing at Paducah, loaded with cotton and specie.

In May.—The George Collier burst her cylinder on the Mississippi, eighty miles above New Orleans, killed 26, and scalded more or less 45 more.

The _____ (name not recollected) burnt on the Mississippi fifteen miles above the mouth of the Ohio. Passengers and crew rescued by the North Star; 2 drowned.

The Buckeye burst her boilers on the Mississippi, five miles above Randolph; six persons were killed by the explosion and several wounded.

The South Alabaman, from Shreveport to New Orleans, snagged on the Mississippi and sunk in two minutes to her boiler deck. Total loss.

In June.—The Macfarland at Walnut Point on the Mississippi was run into by the Danube, and sunk in five minutes in deep water. Two lives lost.

In July.—The Peru ran aground on Cumberland bar in the Ohio, the falling of the river caused her to break in two.—Cargo saved-boat lost.

The Sylph snagged in the Ohio river twenty miles below Portsmouth; sunk rapidly. Boat lost; no great loss on cargo.

The Casket sunk in the Ohio, near the mouth. No particulars.

The Sultan snagged in the Mississippi at Island No. 8, and sunk in ten feet of water. Total loss, boat and cargo.

In August.—The [name illegible] on her downward passage from St. Louis to New Orleans, struck a snag near the Grand Tower, and was run ashore and sunk, the water coming a few inches above her guards; amount of damage not stated, but very heavy.

The W.L. Robeson snagged near the same place, on her downward trip; sunk in ten feet water. Boat and cargo lost.

The Adventurer burst a boiler near Van Buren, Arkansas; one man killed.

In September.—The Corsican struck a log in the Mississippi, ten miles below Selma, sunk in six feet water; cargo damaged; it was thought the boat could be raised.

In October.—The Josink Nichol struck a rock in the Ohio twenty miles below Shawneetown, and sunk in four feet water. Boat a total loss. The Elk snagged in the Missouri river, eighty miles above the mouth, sunk in seventeen feet water. Loss, boat and cargo.

The Far West was run into neat Bayou Sarah, on the Mississippi, by the Southerner, and sunk to her guards. Cargo saved, boat lost. The Camden, heavily freighted, snagged on her downward passage on the Missouri river, about eighty miles above the mouth, and sunk in eight feet water. Entire loss.

In November.— The Arrow, on her first trip down the Mississippi, struck a snag and sunk after making her landing in deep water, near Bayou Sarah; a valuable freight, and $10,000 in specie, which, with the boat, was lost.

The Wilmington, at Island No. 74, on the Mississippi, burst her middle boiler, throwing the two outside boilers overboard, killing six and drowning and scalding 12 others.

In December.— The North Star, with a valuable cargo, struck the wreck of another boat at the mouth of the Louisville canal [?] and sunk in deep water. An entire loss, boat and cargo.

The Trader, a small packet running between Palquemine and [?] was snagged and sunk in a few minutes in the Mississippi. The Pizarro, burnt upon the Dry Dock at St. Louis.

The Knickerbocker, with a valuable going down to New Orleans, snagged and sank in the Mississippi, near the mouth of the Ohio, in 18 feet water. An entire loss.

The Bridgewater on her downward trip from Vicksburg, snagged and sunk in the Mississippi, in deep water, near Fort Adams. Freighted with cotton; total loss.

The Return, coming down the Yazoo, snagged and sunk in deep water.

The General Gaines snagged near the mouth of the Wabash, in the Ohio, and was afterwards raised. The Danube snagged in the Ohio near the Three Sisters; not much damaged.

The Belle of Missouri, with gunpowder on board, caught fire in the Mississippi near Cape Girardeau, at a wood yard, and blew up; the passengers and crew escaping.

The sum total of losses is 43; of this number, 32 were an entire loss; snagged 21; struck rocks or other obstacles 7; burnt, 5; burst their boilers, 4; run into by other boats, 3 — 49. There were snagged on the Lower Mississippi, 11; on the Missouri, 4;on the Ohio, 4; on the Yazoo, 1; on the Red River, 1. It is remarkable but a majority of the boats were snagged on their downward trips. Lives lost by bursting boilers, 39; by other causes, 6. Total, 45. The amount of property destroyed in boats and their cargoes is certainly not less than a million of dollars. Of this the heaviest burden appears to have fallen on St. Louis. Number of lives lost (though doubtless the above is a very imperfect number) is much below the average of several years past."

The *Settler & Pennon, M'Kean and Potter County Advertiser* (Smethport, M'Kean County, Pennsylvania), of October 7, 1841, reports a current list for its readers:

The Cincinnati Republican gives a list of the accidents to steamboats on the western waters, which have come under its notice, from January 1st to September 1st, by which it appears that twenty-one were snagged, nine sunk or injured by collision, four burnt, two burst their boilers. The estimated loss by these is $1,350,000. The above, it is said, do not include more than two thirds of the accidents which have happened in the months as stated above. There is a marked diminution of accidents occurring by the bursting of boilers, and other disasters connected to the action of steam, attributed to the inspection laws, the introduction of the Evans' Safety Guard, and the general disinclination of the passengers to take passage on boats that are in the habit of racing.

On September 14, 1840, the *Adams Sentinel* (Gettysburg) republished an editorial from the Baltimore *American*. It is reprinted here in full, along with an editorial from the Missouri *Republican,* reprinted in the Alton *Telegraph* of October 16, 1841. Following them is a list of steamboat disasters from 1855, fully fourteen years later. They bear consideration together and serve as a reminder that as much as things change, they stay the same.

Steamboat Disasters—The law of 1838 regarding vessels propelled in whole or in part by steam, was not amended at the late session of Congress, although the subject was under the consideration of the Committee on Commerce, and a report, with an amendatory bill, was submitted to the Senate from the Committee. Like many other things relating to the general interest of the country, this subject was overlooked or crowded out of place by the pressure of political matters which engrossed so much of the attention of both Houses. In reference to the law of 1838, the Committee expresses the opinion that it has contributed in some degree to public security, but add: "We have proof that it falls far short of effectually shielding the public from those disasters which prompted its adoption." It appears that in the course of the past year, about 200 lives have been lost in consequence of steamboat disasters.

In 1839 there were forty-one accidents on the Western waters alone. Of these, 21 were snagged, 7 struck upon rocks, &c. Four only were caused by explosions—six by fire, and three by collisions. Twenty-three were total losses. The value of property destroyed is estimated at about one million of dollars. Forty-six lives were lost, of which thirty-nine were in consequence of explosions.

The principal other steamboat accidents for the year were the burning of the Great Western in Detroit River, and the loss of the Lexington by fire on Long Island Sound. There were other accidents, such as the collapse of flues, &c. by which persons were scalded without causing death.

The whole number of steamboat accidents of a serious nature that have occurred in the United States since the introduction of steam navigation, is stated at 272, of which the following is a summary:

No. of vessels	No. of Lives Lost
103 explosions and collapses of boilers	856
73 striking on snags and sawyers	119
35 shipwrecks, gales and collisions	473
34 fires from various causes	414
27 unascertainable causes	
272	1,921

The returns show about one hundred and fifty wounded. This is truly a fearful array, and exhibits in a startling point of view the recklessness with which human life has been sacrificed on board of steamers in the United States. It was for legislation to interpose—and atone as well as possible for a remissness which should never have been allowed.

The greatest number of lives lost at any time by *explosion* is set down at 138. The occasion was in 1838 when the ill-fated Pulaski perished on the coast of North Carolina. The number destroyed by the explosion of the Moselle opposite Cincinnati is not given. If we recollect rightly, it was greater than that above named. On the Oronoko, in the Mississippi, in 1838, one hundred and thirty lives were lost.

But the most appalling catastrophe in view of human destruction was that of the Monmouth by *collision* on the Mississippi in 1837, *three hundred* are supposed to have perished. On board the Lexington, last winter, one hundred and fifty were destroyed. Another terrible disaster by fire was the burning of the Ben Sherrod on the Mississippi in 1837—one hundred and thirty lives lost. Of the 272 accidents above referred to, 207 occurred on the Western waters—On the Eastern waters and the Lakes, 65. The aggregate loss of property by these disasters is estimated at about six millions of dollars."

* * *

Appendix B

Injustice to Western Interests

On a former occasion, alluding to the neglect of the General Government to provide for the improvement of the rivers of the West, we said that the losses of the St. Louis shippers alone would have been more than sufficient to have rendered the Mississippi and Missouri entirely safe for navigation. We recur again to this subject, because we believe gross and rank injustice has been done the west in this matter. Hundreds of thousands have been expended on the improvement of the Red River and Arkansas, and yet the main stem is totally neglected. Thousands upon thousands have been expended on eastern harbors and inlets, and in the erection of light houses, break-waters, and in dredging, &c. &c. More money has been expended on a single harbor than would be required for the entire improvement of the Mississippi; and yet the whole amount of business done at such harbor will not probably amount to one tenth of the interest constantly at stake on the river. As we have before said, it is not our purpose as to complain of these expenditures. We are pleased to see any and every portion of the country improved and every facility given to its commerce. But what we do contend for is, that *the larger interests should first command the attention of the Government and first receive the necessary appropriations.* The neglect which has been exhibited towards the Mississippi and its principal tributaries may have, in part, originated in ignorance of the necessity for the improvement and the amount of interests at stake. We do not believe that eastern, or even many western, members of Congress know the actual condition of the Mississippi, or how large an amount of property has actually been lost by the snags and other obstructions in the river, nearly all of which could be removed at a comparatively small expense. With a view to a more current understanding of the subject, we submit a statement of the losses of St. Louis shippers since January last. We doubt not that many of our citizens are ignorant of the enormous extent of these losses. The statement we give below has been prepared with great care by several gentlemen engaged in the shipping business, and has been furnished us by the Secretary of one of the Insurance companies. The only error is in fixing the estimates below the real losses.

Since the 1st day of January, 1841, the following boats engaged in the St. Louis trade have been lost.

The Vermont, sunk between St. Louis and the mouth of the Ohio, valued at $5,000

Rienzi	do [= ditto]	do	8,000
Peoria	do	do	5,000
Chester	do	do	20,000
Homer	do	do	6,000
Maid of Orleans	do	do	25,000
Oregon	do	do	20,000
Keokuck	do	do	6,000
Wm. Paris	do	do	12,000
A.M. Phillips	do	do	6,000
Tehula	do	do	15,000
U.S. Mail	do	do	15,000
Brazil on Upper Mississippi			8,000
Caroline, below mouth of Ohio			35,000
Chief Magistrate		do	15,000
Baltic	do	do	12,000
Malta, sunk on the Missouri			15,000
Missouri, burnt at the wharf			50,000
			———
			$200,000

The above we believe is a complete list of all the boats totally lost; and the value of them is set down below the rate they were insured at, and all greatly below their actual cost. The first cost of these vessels would not fall short of $400,000.

Of these vessels, not exceeding fifteen per centum has been saved of the wrecks. To that we believe the actual amount much below this sum. The loss upon hulls alone, within less than nine months, may be safely set down at not less than $257,500.

To the loss of hulls there is to be added the loss of cargoes, which, by a careful estimate made from the best means in the possession of the Insurance offices, may be set down as exceeding the sum of $363,000—making the total losses on hulls and cargoes within the nine months, $620,500.

In this estimate it will be perceived that no allowance is made for incidental losses, such as loss of time, being thrown out of employment, &c. &c; neither have we carried into the estimate any of the partial losses where the boats have been saved and repaired. This last sum would swell the amount to at least $150,000 more.

The Missouri was burnt at the wharf, but no cargo was lost in her. By the Brazil, which was lost on the Upper Mississippi, no cargo — and the loss of cargo on the Malta, on the Missouri, in our estimate is set down at only $10,000. From these facts it appears that the loss on hulls on the Mississippi, between this port and New Orleans, has been $217,000, and on cargoes $353,000. Of these amounts, the whole, except perhaps the loss of the Baltic, is chargeable to snags in the river.

Another fact, presented by the above, is particularly deserving of attention.— Twelve of the above boats were snagged at various points between this port and the mouth of the Ohio river—a distance of two hundred miles, or thereabouts. This distance, with an expenditure of one half the sum actually lost upon it, could be rendered as safe as any other portion of the river. In some places, the expenditures would be comparatively trifling. For example, we are credibly informed, that at Turkey Island, were the Oregon sunk, the removal of a few logs, which, with the aid of a snag boat, would cost not exceeding two or three hundred dollars, the channel might be made safe. The same remark might be applied to many other places.

We submit the above facts to the candor of the community; and ask if the General Government ought not to provide for the removal of theseobstructions? It will be remembered that, whilst the actual losses have in nine months amounted to nearly a million of dollars, there has been, during that period, more than fifty millions of property afloat on the river, and liable to be lost. Are not these interests of sufficient magnitude to demand the attention of the General Government? Are not the number of persons involved sufficient to recommend it to the attention of the nation? If the General Government do not accomplish the work, who can or will effect it? The river is a national high-way; and no State has the right, if she had the inclination, to interfere with the channel.

If the press throughout the country will give the facts here stated, and for all which indisputable testimony can be brought to the nation, we have no fears but that justice will yet be done the West in this matter. (No significant improvements were made to the Mississippi until after the Civil War, by which time the railroads had virtually taken over western commerce.

The Alton *Weekly Courier* of January 10, 1856, provided a detailed glimpse into the steamboat disasters for the year 1855. It makes for sad, but very telling reading.

January, 1855
- 2nd: Westerner sunk in Lower Mississippi; loss $5,000; boat and cargo fully insured
- 4th: Switzerland and J.O. Fremont in collision near Vevay, on the Ohio; Fremont badly injured; a barge sunk, loaded with pork and lard; loss $5,000
- 5th: The hull of the Winfield Scott, in tow by the Switzerland, sunk in the Ohio; the hull a total loss; cargo injured; loss $3,000

18th: Garden City burned near Grand Gulf; loss on boat and cargo $100,000; insured $60,000

20th: Niagara and Obion, collision on the Illinois; Obion sunk; loss $3,000; not insured

21st: Bee, on the Mississippi, struck a snag near Vicksburg; loss $5,000; no insurance

23rd: Eliza sunk on Lower Mississippi; loss on boat and cargo $75,000; one-half covered by insurance

25th: Honduras sunk in Lower Mississippi; boat valued at $30,000; cargo $75,000; partly insured

27th: Several flat boats sunk by wind on the Ohio; loss unknown

29th: Hindou sunk at St. Genevieve; loss $8,000

February, 1855

1st: Toledo sank at St. Louis; loss $3,000. Fanny Farrar sank in Cumberland River

2d: Forest Queen sank in Ohio river neat Lawrenceburgh; total loss

3d: Four pair of coal boats sunk just above Natchez

4th: Jeannie Bealle snagged and sunk on downward trip from Montgomery to Mobile; loss trifling, and raised again

9th: Alhambra, Wenona, Badger State, Alton and Walk-in-the-Water badly damaged at the St. Louis levee, by the breaking loose of the ice. At the same time, a barge at the foot of Morgan street was sunk

7th: Bulletin damaged to the amount of $1,000, by coming in collision with Grand Turk

8th: Magnolia and T.P. Leathers came in collision near Vicksburg; the latter destroyed. Buckeye Belle struck a snag and sank at Hat Island; she was afterward raised

10th: Galena and Rock Island wharf-boat sank at St. Louis wharf, but was raised in a few days

11th: E. Howard, while descending Cumberland river loaded with cotton, took fire and was scuttled and sunk, in several feet water. Boat and cargo both considerably damaged; the boat was subsequently raised

12th: Dresdon, from St. Louis to New Orleans, snagged and sunk near New Madrid; boat and cargo total loss. Boat valued at $20,000; insured for $15,000. Cargo worth $80,000 to $10,000

10th: *D.H. Bacon struck a snag above Cape Giradeau; damaged to amount of $5,000

12th: *Peter Fellon, snagged and sunk at President's Island, lower Mississippi. Boat raised but the cargo was damaged to amount of about $15,000 to $20,000

13th: James Robb struck a snag and sank just above Cape Giradeau. Cargo valuable, and only a small part of it saved. Boat valued at $50,000, and only half insured; total loss

15th to 20th:
During this time from twenty to thirty coal and flat-bottomed boats were sunk in the Ohio and Mississippi, by the breaking up of the ice; value not given

20th: Norma sunk in Choctaw Bend, Lower Mississippi, with a valuable cargo of groceries, totally lost. Probable value $100,000; partially insured

24th: James Trabue, a new boat valued at $30,000, sunk in the Red River, and, with cargo, proved a total loss

25th: Helen Mar sunk in the Ohio River, below Mayville, but was raised at a cost of about $2,000

March, 1855

1st: The Luda, heavily loaded with cotton, snagged and sunk in Red River — total loss

*printed out of order

2d: James Watt sunk near Petersburg, Ky.; boat and cargo of pork and lard, total loss

4th: Mary Clifton struck a snag and sunk near Tait's Shoals, in Alabama River; damage to boat and cargo trifling

5th: Huron struck a snag and sunk in Atchafalya River at Connohan's Landing, and, with cargo of cotton, was totally lost

6th: Altamont, loaded with dry goods, ran aground at Marietta Island, and was badly cut by the ice. Damage to boat $8,000; to cargo $25,000

7th: Louisa sunk in Ouachita River; boat and cargo total loss 10th: Alice sunk at Bilbo's Island, Alabama river, and totally lost. The cargo consisted of 500 bales of cotton, half of which was lost. No insurance

12th: Advance sunk in Ohio river, near Marietta, and a valuable cargo damaged; the boat was raised

13th: Heroine burst her boilers on Bigbee river, killing three persons and wounding several others; and the boat caught fire and burnt to the water's edge

20th: The Georgetown damaged on the Lower Rapids; after being repaired, she went up the Missouri and sunk, a total loss 25th: Bulletin caught fire while under way, sixty miles above Vicksburg, and was totally destroyed, with 8,000 bales of cotton; the boat was worth $65,000, insured for $25,000. Several cabin passengers and sixteen negroes perished in the flames

26th: Huntsville burned at Hamburgh, in Tennessee River; worth $65,000, insured for $35,000

28th: Americus burned at Bath, Illinois River, and, with a valuable cargo, totally destroyed

April, 1855

1st: Keystone State, with a heavy cargo, totally destroyed by fire below Pekin. Two lives lost

6th: Alton packet Reindeer collapsed a flue eleven miles below Alton, killing two persons, and scalding some others

7th: Banner State, worth $10,000, and fully insured, sunk in Missouri River, twelve miles above its mouth, and, with a valuable cargo, was entirely lost

8th: The stern-wheel Evansville, worth $6,000, was sunk and totally lost below Oil Creek, in Ohio River

9th: El Paso struck a snag and sunk below Boonsville, Missouri River. Cargo of hemp and grain badly damaged; boat fully insured

18th: The Conewago struck a rock in the Lower Rapids in four feet of water. She was raised

19th: The Clarion was sunk by the ice between Galena and St. Paul, but was raised again

20th: Fanny Fern had her upper works demolished by a storm while crossing the lower Rapids; loss about $3,000

22nd: The R.H. Lee, valued at $4,000 and uninsured, was cut in two by a collision with the Ocean Wave, and with a cargo of grain totally lost; the accident happened just below the Illinois river. On the same day, the Gazelle and Bridge City came in contact about eleven miles below Alton; the former was badly damaged

23rd: The Wm. Knox, with a large cargo on board, was entirely destroyed by fire; boat worth $35,000, and [cargo] valued at $25,000. Seven lives lost

24th: The Wharf Boat at Aurora sunk

25th: Emerald, a large tow steamer, sunk at St. Claire Falls in the Ohio river, but was afterwards raised

26th: A large wharf boat was sunk at Cincinnati, but was subsequently raised, and valuable merchandise which was on it, recovered in a damaged condition

27th: Falcon was totally destroyed by fire at New Orleans; loss $10,000–half insured
28th: Two coal boats, worth $2,000 each, burned at Uniontown, Ohio river
30th: Mary Cole, a steam ferry boat, on her way to Council Bluffs, sank fifteen miles below Bluff City, and, with a large amount of freight was a total loss

May, 1855

5th: Dan Coners struck a rock and sank on Lower Rapids, but was subsequently raised
7th: Georgetown sunk in Missouri river, near St. Charles. Boat worth $12,000, and valuable cargo of U.S. stores, totally lost
8th: N.W. Sherman snagged and sank in Yazoo river; total loss
9th: Exchange struck a snag and sunk at Hoovey Island, Zazoo river; boat and cargo total loss
10th: Helen, with 800 bales of cotton, entirely destroyed by fire in Mobile Bay; loss $60,000 — no insurance
20th: Glendy Burke, bound for New Orleans, snagged and totally lost near Grand Tower
22nd: Sylvester Weber snagged and sunk five miles above the mouth in Amite river; total loss — no insurance
31st: Midas badly damaged by striking a snag boat near Paducah, but did not sink

June, 1855

8th: Australia, with valuable cargo of government stores, snagged and sunk near Boonville. Cargo lost; boat afterwards raised
10th: Mary C. sunk by striking on rocks of Lower Rapids but was raised again
11th: Fashion ran into the bank and sunk near Evansville, Ohio river, boat was raised and is now running
12th: Young America, with valuable cargo, sunk in Illinois river near Bath, but was raised and run until about a month since, she went down in a storm of wind and was totally lost
14th: Belle Golding snagged and sunk at Devil's Island, in Mississippi river; loss $20,000, three-fourths insured
15th: Express struck a snag and sunk near Spa Island, Missouri river; total loss. Value $6,000 — insured for $3,000

July, 1855

2nd: Lexington, full of passengers, exploded her boilers, about ninety miles below Louisville, while under way, killing thirty or forty of her passengers, and wounding many more. The boat was not insured 8th: Magnolia Banner, with 1,000 bales cotton on board, was totally destroyed by fire at Baton Rouge
10th: Lady Pike ran into a flat boat containing 900 bbls whisky; the flat boat and the whisky were both demolished
12th: Telegraph No. 2 was totally consumed by fire at New Albany, Indiana
20th: Hermann, worth about $3,000, snagged and totally lost near St. Charles, Missouri river
21st: Wade Allen, worth $9,000 and insured for $5,000, destroyed by fire in Mobile Bay; one passenger perished in the flames
23rd: Baltimore ran into the bank at Chester and sunk; she was raised and is now running
24th: Swamp Fox sunk and lost in Red river

August, 1855

5th: Kate Swinney snagged and totally lost above Council Bluffs. She was worth $25,000 — half insured

10th: Helen Mar sunk on Lower Rapids — afterwards raised

12th: Canal boat Prairie State, heavily freighted with wheat and sugar, sunk in Illinois river

16th: Fanny Harris, bound for Galena, sprung a leak below Quincy, and came near sinking, but was finally saved

17th: William Phillips, freighted with wheat, sunk in the Osage river September, 1855

19th: Alliance sunk near Sinn Creak, Osage river, and a valuable cargo totally lost; the boat was raised

25th: Ohio and Golden Gate came in collision on the Ohio river; the latter boat received severe injuries, two lives, and a large amount of freight were lost

November, 1855

10th: The Wenona, a new boat worth $20,000, and fully insured, struck a snag, and sunk ten miles above Jefferson City

11th: Memphis sunk on Louisville Falls, but was afterwards raised

20th: Switzerland sunk by coming in collision with Uncle Sam, but was subsequently raised; boat and cargo both insured

December, 1855

3rd: May Flower, George Collier, and wharf boat Mary Hunt, totally destroyed by fire at Memphis; loss of property near $300,000. The May Flower was the finest boat on the Western waters, and was valued at $130,000. Ten or twelve persons, mostly deck hands, on the Geo. Collier, perished in the flames

7th: Prairie City, Twin City and Parthenia destroyed by fire at St. Louis. The Parthenia was worth $16,000, and had on board $100,000 worth of freight; partially insured. The other two boats were worth about $10,000, and partially insured

8th: Young America was blown ashore about thirteen miles below Alton, and sunk, and, with a cargo, proved a total loss to the amount of $50,000

23rd: The Charleston was burned at Golconds, and 1,000 bales cotton destroyed; loss of cargo $40,000; loss of boat $25,000; half insured

Chapter Notes

Chapter 1

1. H. W. Dickinson, "Robert Fulton, Engineer and Artist," http://www.history.rochester.edu/steam/dickinson, pp. 2–4.
2. Ibid., p. 3.
3. Ibid., pp. 7–8.
4. Ibid., p. 8.
5. Ibid.
6. Ibid., p. 3.
7. Ibid., p. 8.
8. Ibid., p. 6.
9. Ibid., p. 5.
10. Ibid.
11. Ibid., p. 6.
12. Ibid., p. 8.
13. Ibid., p. 13.
14. Alice C. Sutcliffe, *Robert Fulton and the Clermont*, New York: The Century Co., 1909, pp. 220–221; see also Dickinson, p. 6.
15. Dickinson, pp. 9–10.
16. Ibid.
17. Ibid., p. 10
18. *Republican Compiler*, May 3, 1820.
19. *Adams Sentinel* (Gettysburg, Pennsylvania), June 27, 1821.
20. *Republican Compiler*, June 27, 1821

Chapter 2

1. "Flatboats and Keelboats." http://www.encyclopedia.com/doc/1G2-2536600963.
2. Walter Blair and F. J. Meine. *Mike Fink: King of Mississippi Keelboatmen*. 1933. Reprint, Westport, Conn.: Greenwood Press, 1971, p. 56.
3. R. Carlyle Burley, *The Old Northwest: Pioneer Period, 1815–1840*, Indiana Historical Society, 1950.
4. George Byron Merrick, *Old Times on the Upper Mississippi*. Arthur H. Clark Company, 1909. Reprint, Minneapolis: University of Minnesota Press, 2001, p. 29.
5. Dickinson, pp. 5–6.
6. Ibid., p. 6.
7. Ibid., p. 5.

Chapter 3

1. Dickinson, p. 6.
2. Sutcliffe, pp. 174–176.
3. William J. Petersen, *Steamboating on the Upper Mississippi*, New York: Dover Publications, 1968, p. 45.
4. Petersen, p. 59.
5. Hudson River Maritime Museum, www.hrmm.org/steamboats/fulton.
6. *Louisiana Gazette and Advertiser* (New Orleans), February 12, 1812.
7. "Robert Fulton," www.robertfulton.org.
8. Louis C. Hunter, *Steamboats on the Western Rivers: An Economic and Technological History*, New York: Dover Publications, 1977, p. 12.
9. "Henry Miller Shreve" (1), http://freepages.genealogy.rootsweb.ancestry.com/~walker/Henry%20Miller%Shreve.
10. "Henry Miller Shreve" (2), http://xroads.virginia.edu/~hyper/DETOC/transport/shreve.html.
11. Shreve (1).
12. "Flatboats and Keelboats."
13. Shreve (2).
14. Dickinson, p. 7.

Chapter 4

1. Hunter, p. 489.
2. Fanny Trollope, *Domestic Manners of the Americans*, 1832. Reprint, New York: Penguin Books, 1997, p. 9.
3. Frederick Marryat, *A Diary in America*, Vol. II, p. 143.
4. Charles Dickens, *American Notes*, 1883. Reprint, Sandy, Utah: Quiet Vision Publishing, 2003, p. 138.
5. Mark Twain, *Life on the Mississippi*, 1883. Reprint, New York: The Library of America, 1982, p. 417.
6. Ibid., p. 284
7. Ibid.
8. Shreve (1).
9. Ibid.

Chapter 5

1. "Lewis Cass." http://en.wikipedia.org/wiki/Lewis_Cass.
2. Petersen, p. 13.
3. Petersen, pp. 14–16.
4. "Factory (trading post)." http://en.wikipedia.org/wiki/Factory_(trading_post).
5. Dickinson, p. 1.
6. "American State Papers, Military Affairs," Vol. II, 25th Congress, pp. 68, 69, 324, 325; cited in Petersen, pp. 83, 86.
7. Petersen, p. 223.
8. Petersen, p. 100.
9. Petersen, p. 223.

10. Sen. Ex. Doc. 512, 23 Congress, I Session, pp. 687–688.
11. Petersen, pp. 201–203.
12. Ibid., p. 235.
13. Petersen, pp. 207–209.
14. Petersen, p. 235.
15. Merrick, p. 32.
16. Petersen, p. 160.
17. *Madison Express,* March 7, 1840.
18. Petersen, p. 376.

Chapter 6

1. Petersen, p. 130.
2. Hunter, p. 131.
3. Merrick, p. 37.
4. Merrick, p. 102.
5. Hunter, pp. 644, 646.
6. Merrick, p. 59–60.
7. Ibid., p. 34.
8. Hunter, p. 267.
9. Petersen, p. 59.
10. Hunter, p. 35.
11. *Adams Sentinel,* May 11, 1830.
12. *Congressional Globe,* August 28, 1852.
13. Naval Historical Center.
14. Sen. Ex. Doc. 512, 23 Congress, I Session, p. 1.
15. Hunter, p. 250.

Chapter 7

1. Twain, p. 309.
2. Merrick, p. 89.
3. Sen. Ex. Doc. I, 37 Congress, 2 Sess, p. 509.
4. Hunter, p. 226.
5. Merrick, pp. 40–41.
6. Merrick, p. 41.

Chapter 8

1. Hunter, pp. 317–320.
2. Ibid., p. 314.
3. Department of Transportation by Water, Part 1, Office of the Chief of Engineers, "Water Terminal and Transfer Facilities," *House Ex. Doc. 226,* 63rd Congress, 1st Session, pp. 916-919, 939-944, 969-971.

Chapter 9

1. "The Panic of 1819." http://thehistorybox.com/NY_City/panics_article2a, pp.1–3.
2. Ibid., p. 103.
3. "The Panic and Depression of 1832." http://thehistorybox.com/NY_City/panics/panics_article3a.
4. *Adams Sentinel,* October 13, 1834.
5. Edward M. Shepard, *Life of Martin Van Buren,* New York: Houghton Mifflin Co., 1889, pp. 1–6.
6. Stanley Rubenstein, "The Panic of 1837." http://cooperativeindividualism.org/ruberstein_stan_panic_of_1837.
7. Shepard, pp. 1–6.
8. Ibid.
9. "The Panic of 1857." http://thehistorybox.com/NY_City//panics/panics/article7a.
10. Merrick, pp. 175–180.

Chapter 10

1. Margaret Strebel Hartman, "Steamboat Sunflower." http://rootsweb.ancestry.com/~kycampbe/steamboatsunflower.
2. Hunter, p. 362.
3. Ibid., p. 370.
4. Ibid., p. 368.
5. Ibid., pp. 363–369.
6. Ibid., p. 365.
7. Ibid., pp. 363–369.
8. Merrick, p. 167.
9. Ibid.
10. Petersen, pp. 170–171.
11. *The Janesville Gazette* (Wisconsin), July 24, 1847.
12. *The Racine Daily Journal* (Wisconsin), February 10, 1857.
13. Charles Frederick Briggs, *The Story of the Telegraph.* New York: Rudd & Carleton, 1858, pp. 187–188.
14. *St. Paul Daily Minnesotan,* August 19, August 20, 1858.

Chapter 11

1. Hunter, p. 272.
2. *Daily Commercial Register,* August 7, 1852.
3. James T. Lloyd, *Lloyd's Steamboat Directory.* Cincinnati: J. T. Lloyd, 1856, pp. 197–201.
4. Hunter, p. 249.
5. *New Orleans True American,* May 11, 1837.
6. *Natchez Mississippi Free Trader,* May 18, 1837.
7. Hunter, p. 28.
8. Jim Reis, "Death Rides the Waves of the Ohio River." www.rootsweb.ancestry.com/~kycampbe/steamboatdeaths.
9. Twain, pp. 354–359.
10. *Cincinnati Western Spy,* June 14, 1816.
11. Henry Howe, "Historical Recollections of the Ohio." www.rootsweb.ancestry.com/~kycampbe/steamboatmoselle.
12. Lloyd.
13. Ibid.
14. *Cincinnati Republican* (in the Baltimore *Republican),* May 1, 1838.
15. Hunter, p. 287.
16. Merrick, p. 39.
17. Trollope, p. 9.
18. Twain, p. 341.

Chapter 12

1. Dickinson, p. 4.
2. *Cincinnati Western Spy,* May 30, 1817.
3. *Adams Sentinel,* June 11, 1817.
4. Data from Hunter.
5. *Northwestern Gazette and Galena Advertiser,* March 28, April 24, May 30, 1845.
6. Petersen, pp. 421–422.
7. Ibid., pp. 266–270.
8. Merrick, p. 148.
9. Carl Schurz, *The Reminiscences of Carl Schurz,* New York: McClure, 1907.
10. Merrick, p. 39.
11. Reis.

Chapter 13

1. Hunter, pp. 442–443.
2. Petersen, p. 445.
3. Merrick, p. 163.
4. Hunter, p. 399.
5. Petersen, p. 403.
6. Merrick, p. 72.
7. Hunter, p. 411.
8. Ibid.
9. Twain, p. 314.
10. Ibid., p. 319.
11. Ibid., p. 267.
12. Ibid., p. 269.
13. Hunter, p. 245.
14. Pittsburg *Gazette,* June 3, 1841.
15. Hunter, p. 262.
16. Louisville *Gazette,* September 21, 1859.

Chapter 14

1. Hunter, 448.
2. Ibid., p. 448–450.
3. Ibid., p. 450.

4. Merrick, p. 64.
5. Hunter, p. 282.
6. Trollope, pp. 45–46.
7. Ibid., p. 49.
8. Richard Biddle, "A Review of Captain Basil Hall's Travels in America, by an American," London, 1830, p. 66.
9. Dickens, p. 185.
10. Nashville *Union*, June 10, 1841.
11. William Edward Henry, "The Great Western Land Pirate, Again." http://51Illinois.org/Murrell.
12. Ibid.
13. John Habermehl, *Life on the Western Rivers*, Pittsburgh: McNary & Simpson, 1901, pp. 72–73.
14. Merrick, p. 129.
15. Cincinnati *Gazette*, May 20, 1856.
16. Ibid., October 17, 1872.
17. Merrick, pp. 69–70.
18. Twain, p. 494.
19. Ibid., p. 293.
20. Lafayette Bunnell, *History of Winona County*, Chicago: H.H. Hill and Company, Publishers, 1883.
21. Cincinnati *Gazette*, May 17, 1820.
22. Merrick, p. 301.

Chapter 15

1. Trollope, p. 18.
2. Dickens, pp. 1–2.
3. Theodore Rodolf, *Pioneering in the Western Lead Region*, Wisconsin Historical Collection, Vol. 15, pp. 339–340.
4. Ole Munch Raeder, *America in the Forties*, pp. 121–122.
5. Cincinnati *Times*.
6. Dickens, p. 126.
7. Hunter, p. 396.
8. Twain, p. 267.
9. Trollope, p. 18.

Chapter 16

1. Twain, p. 267.
2. "George Catlin." http://en.wikipedia.org/wiki/George_Catlin.
3. George Catlin, *The North American Indians*, New York: Penguin Books, 1989 (reprint); Father Louis Hennepin, *A New Discovery of a Vast Country in America*, Chicago: A.C. McClurg & Co., 1903.
4. *Northwestern Gazette and Galena Advertiser*, April 22, May 13, 1837.
5. Ibid., June 29, 1839.
6. From a manuscript in the Missouri Historical Society, Jefferson Memorial Library.
7. Friedrich Arends, *Schilderung des Mississippithales, oder des Westen der Vereinigten Staaten von Nordamerika*, Emden, 1838.
8. Cincinnati *Gazette*, January 13, 1848.
9. Dickens, p. 127.
10. Merrick, p. 129.
11. Dickens, p. 107.
12. Trollope, p. 20.
13. Merrick, p. 126.
14. Petersen, p. 376.
15. Ibid., p. 381.
16. Cincinnati *Gazette*, April 21, August 23, 1859.
17. Steven Soifer, "The Evolution of the Bathroom and the Implications for Paruresis." http://paruresis.org/evolution.htm.
18. Trollope, p. 34.
19. Soifer.
20. "Thomas Crapper." http://en.wikipedia.org/wiki/Thomas_Crapper.
21. "History of the Flush Toilet." http://toiletology.com.
22. John H. Lienhard, "Engines of Our Ingenuity No. 157: Thomas Crapper." http://www.uh.edu/engines/epi157.htm.
23. Dickens, p. 104.

Chapter 17

1. "History of Gambling in the United States." http://www.library.ca.gov/CRB/97/03/Chapt2.html.
2. *Adams Sentinel*, February 20, 1828.
3. *Hagerstown Mail*, December 25, 1835.
4. "The History of Poker in the Old West." http://legendsofamerica.com/WE-Poker2.
5. "History of Gambling in the United States."
6. Habermehl, p. 45.
7. Merrick, p. 138.
8. Twain, p. 450.
9. Merrick, p. 140.
10. Merrick, pp. 140–141.
11. "Old West Legends." http://legendsofamerica.com.
12. "Edmund Hoyle." http://en.wikipedia.org/wiki/Edmund_Hoyle.
13. Twain, p. 267.
14. Merrick, p. 155.
15. Ibid., pp. 152–156.
16. Ibid., p. 157.

Chapter 18

1. James Simmons, *Star-Spangled Eden*, New York: Carroll & Graf Publishers, 2002, p. 11.
2. Edward K. Spann, *New Metropolis: New York City 1840–1857*. New York: Columbia University Press, 1981, p. 122.
3. David Armstrong and Elizabeth Metzger Armstrong, *The Great American Medicine Show*, New York: Prentice Hall, 1991, p. 41.
4. Ibid., pp. 115–117.
5. Sharon Peregrine Johnson and Byron A. Johnson, *Authentic Guide to Drinks of the Civil War*, Gettysburg, Pa.: Thomas Publications, 1992, pp. 12–17 (all recipes cited are from this work; original sources noted).
6. Johnson, p. 92.
7. Jerry Thomas, *How to Mix Drinks, or the Bon-Vivant's Companion*. New York: Dick & Fitzgerald, 1862.
8. Johnson, p. 92
9. Dickens, p. 48.
10. Daniel Poole, *What Jane Austen Ate and Charles Dickens Knew*, New York: Simon and Schuster, 1993, pp. 211–212.
11. Thomas.
12. Ibid.
13. *Cooling Cups and Dainty Drinks*, New York, 1869.
14. Thomas.
15. Merrick, p. 132.
16. Ibid., p. 132.
17. Thomas.
18. Twain, p. 440.
19. Merrick, p. 135.
20. Johnson, p. 39.
21. Ibid., p. 40.
22. Armstrong, p. 40.
23. Trollope, p. 187.
24. Thomas.
25. Ibid.
26. Dickens, p. 131.

Chapter 19

1. Hunter, pp. 421–422.
2. Trollope, p. 19.
3. Dickens, p. 179.
4. Ibid., pp. 179–180.
5. "Immigration and the Commissioners of Emigration of the State of New York," New York, pp. 62–64.

6. Petersen, p. 321.
7. Hunter, p. 420.
8. Habermehl, p. 56.
9. Marryat, p. 143.
10. Merrick, p. 285.

Chapter 20

1. Trollope, p. 275.
2. "Basil Hall." http://en.wikipedia.org/wiki/Basil_Hall.
3. Onondaga *Standard* (Syracuse, New York), September 30, 1829.
4. Trollope, p. 277.
5. Pamela Neville-Sington, Introduction to Fanny Trollope, *Domestic Manners of the Americans*, New York: Penguin Books, 1997, p. xii.
6. Basil Hall, *Travels in North America in the Years 1827 and 1828*. Philadelphia: Carey, Lee & Carey, 1829.
7. Trollope, p. 30.
8. Trollope.
9. Twain, p. 501.
10. Hall.
11. Trollope, p. 279.
12. Dickens, p. 90.
13. Ibid., p. 200.
14. Neville-Sington, p. xxxii.

Chapter 21

1. Armstrong, p. 12.
2. Ibid., p. 3.
3. Trollope, p. 34.
4. Dickens, p. 135.
5. Armstrong, p. 3.
6. Trollope, p. 20.
7. Armstrong, p. 162.
8. Ibid., p. 175.
9. Ibid., p. 9.
10. Clayton L. Thomas, editor, *Taber's Cyclopedic Medical Dictionary*, 13th ed., Philadelphia: F. A. Davis, 1977, p. C61-62.
11. Robert Hooper, *Lexicon Medicum; or Medical Dictionary*, New York: Harper & Brothers, 1842, p. 220.
12. Galena *Daily Advertiser*, April 2, April 3, 1849; St. Louis *Republican*, September 20, 1824; *Northwestern Gazette and Galena Advertiser*, August 24, 1847.
13. New York *Enquirer* (reprinted in the *Republican Compiler*) November 27, 1832.
14. *The People's Press* (Gettysburg, Pennsylvania), July 24, 1835.
15. Advertisement in the *Hagerstown Mail*, September 11, 1840.
16. Hunter, p. 431.
17. Wyeth, John B., *Oregon, or a Short History of a Long Journey*, http://www.xmission.com/~drudy/mtman/html/jwyeth.html.
18. Hooper, pp. 353–355.
19. Twain, pp. 410–411.
20. Thomas, p. Y1.
21. Thomas, p. M5–6.
22. Hooper, p. M40.
23. Thomas, p. S58.
24. "Henry Schoolcraft." http://en.wikipedia.org/wiki/Henry_Schoolcraft.
25. Madison (Wisconsin) *Express*, March 7, 1840.
26. Thomas, p. D43–45.

Chapter 22

1. "New Orleans." http://wikipedia.org/wiki/New_Orleans%2C_Louisiana.
2. "New Orleans French Quarter History." http://intours.com.
3. Gaspar Cusachs, "Lafitte, Louisiana Pirate and Patriot." http://penelope.uchicago.edu/Thayer/E/Gazetteer/Places/America/United_States/Louisiana.
4. "Jean Lafitte." http://en.wikipedia.org/wiki/Jean_Lafitte.
5. Coin World, "State Treasures Waiting to be Found." http://www.coinworld.com/NewCollector/StateTreasures2.asp.
6. "New Orleans."
7. William F. Switzler, "Report on the Internal Commerce of the United States," *House Ex. Doc 6*, 50th Congress, 1st Session, Part II, pp. 184, 191, 199, 215.
8. Hunter, p. 644.
9. "New Orleans."
10. Ole Rynning, *True Account of America for the Information and Help of Peasant and Commoner* (English translation), Christiana, 1838, p. 99.
11. "Marie Laveau." http://en.wikipedia.org/wiki/Marie_Lavea; David Arbury, "Voodoo Dreams," http://ame2.asu.edu/sites/voodoodreams.
12. "Haunted New Orleans." http://haunteddoghouse.com.
13. "The Pirate House and Jean Lafitte." Hancock County Historical Society. http://hancockcountyhistoricalsociety.com/history/lafittepirate.
14. *Adams Sentinel*, May 5, 1834.
15. "Natchez Trace." http://en.wikipedia.org/wiki/Natchez_Trace.
16. Ibid.
17. Trollope, p. 21.
18. Hunter, pp. 16, 35.
19. Twain, p. 342.
20. Petersen, p. 211.
21. "Henry Miller Shreve."
22. Merrick, p. 233.
23. Letter in Risvold Collection, as cited in Peterson, p. 80.
24. Petersen, pp. 80a & b.
25. Merrick, p. 250.

Chapter 23

1. Pittsburg *Gazette*, May 18, 1853; August 2, 1854.
2. Hunter, p. 485.
3. George Rogers Taylor, *The Transportation Revolution 1815–1860; Volume IV, The Economic History of the United States*, New York: M.E. Sharpe, 1951, p. 85.
4. Taylor, pp. 95–96.
5. Hunter, p. 486.
6. Petersen, pp. 271–281.
7. Robert C. Black III, *The Railroads of the Confederacy*, Chapel Hill: The University of North Carolina Press, 1952, pp. 9–10.
8. Hunter, p. 489.
9. Ibid., p. 490.
10. Dickens, p. 50.
11. Petersen.
12. Twain, p. 360.

Bibliography

"American State Papers, Military Affairs," Vol. II. 25th Congress. 7 volumes.

Arbury, David. "Voodoo Dreams." http://ame2.asu.edu/sites/voodoodreams.

Arends, Friedrich. *Schilderung des Mississippithales, oder des Westen der Vereinigten Staaten von Nordamerika*. Emden, 1838.

Armstrong, David, and Elizabeth Metzger Armstrong. *The Great American Medicine Show*. New York: Prentice Hall, 1991.

"Basil Hall." http://en.wikipedia.org/wiki/Basil_Hall.

"Basil Hall's Copy-Right Fees." Onondaga *Standard*, September 30, 1829.

"The Beaver and Ontario." Madison *Express*, March 7, 1840.

Biddle, Richard. "A Review of Captain Basil Hall's Travels in North America, by an American." London, 1830.

Black, Robert C., III. *The Railroads of the Confederacy*. Chapel Hill: University of North Carolina Press, 1952.

Blair, Walter, and F. J. Meine. *Mike Fink: King of Mississippi Keelboatmen*. 1933. Reprint, Westport, Conn.: Greenwood Press, 1971.

Briggs, Charles Frederick. *The Story of the Telegraph*. New York: Rudd & Carleton, 1858.

Bunnell, Lafayette. *History of Winona County*. Chicago: H.H. Hill and Company, Publishers, 1883.

Burley, R. Carlyle. "The Old Northwest: Pioneer Period, 1815–1840." Indiana Historical Society, 1950.

Catlin, George. *The North American Indians*. New York: Penguin Books, 1989 (reprint).

"Cholera Outbreak." *People's Press*, July 24, 1835.

Congressional Globe, August 28, 1852.

Cooling Cups and Dainty Drinks. New York, 1869.

Cusachs, Gaspar. "Lafitte, Louisiana Pirate and Patriot." http://penelope.uchicago.edu/Thayer/E/Gazetteer/Places/America/United_States/Louisiana.

"Daniel Smith Harris." *Northwestern Gazette and Galena Advertiser*, April 22, May 13, 1837; March 28, April 24, May 30, 1845.

"Delaware and Maryland Lotteries." Hagerstown *Mail*. December 25, 1835.

Department of Transportation by Water, Part 1. Office of the Chief of Engineers. "Water Terminal and Transfer Facilities." *House Ex. Doc. 226*, 63rd Congress, 1st Session.

Dickens, Charles. *American Notes*. 1883. Reprint, Sandy, Utah: Quiet Vision Publishing, 2003.

Dickinson, H. W. "Robert Fulton, Engineer and Artist." http://www.history.rochester.edu/steam/dickinson.

"Die Vernon." St. Paul *Daily Minnesotan*. August 19, 20, 1858.

"Drowning of a Free Negro." St. Louis *Republican*, September 20, 1824.

"Drowning Others by Waves." Natchez, Mississippi *Free Trader*, May 18, 1837.

"Edmund Hoyle." http://en.wikipedia.org/wiki/Edmund_Hoyle.

"Extorting from Steamboats." Pittsburg *Gazette*, June 3, 1841.

"Factory (trading post)." http://en.wikipedia.org/wiki/Factory_(trading_post).

"Flatboats and Keelboats." http://www.encyclopedia.com/doc/1G2-2536600963.

"George Catlin." http://en.wikipedia.org/wiki/George_Catlin.

Habermehl, John. *Life on the Western Rivers*. Pittsburgh: McNary & Simpson, 1901.

Hall, Basil. *Travels in North America in the Years*

1827 and 1828. Philadelphia: Carey, Lee & Carey, 1829.
Hartman, Margaret Strebel, "Steamboat Sunflower." http://rootsweb.ancestry.com/~kycampbe/steamboatsunflower.
"Haunted New Orleans." http://haunteddoghouse.com.
Hennepin, Father Louis. *A New Discovery of a Vast Country in America*. Chicago: A.C. McClurg & Co., 1903.
Henry, William Edward. "The Great Western Land Pirate, Again." http://51Illinois.org/Murrell.htm.
"Henry Miller Shreve." http://freepages.genealogy.rootsweb.ancestry.com/~walker/Henry%20Miller%Shreve.
"Henry Miller Shreve." http://xroads.virginia.edu/~hyper/DETOC/transport/shreve.html.
"Henry Schoolcraft." http://en.wikipedia.org/wiki/Henry_Schoolcraft.
"History of Gambling in the United States." http://www.library.ca.gov/CRB/97/03/Chapt2.html.
"The History of Poker in the Old West." http://legendsofamerica.com/WE-Poker2.
"History of the Flush Toilet." http://toiletology.com.
Hooper, Robert. *Lexicon Medicum; or Medical Dictionary*. New York: Harper & Brothers, 1842.
Howe, Henry. "Historical Recollections of the Ohio." www.rootsweb.ancestry.com/~kycampbe/steamboatmoselle.
Hudson River Maritime Museum. www.hrmm.org/steamboats/fulton.
Hunter, Louis C. *Steamboats on the Western Rivers: An Economic and Technical History*. New York: Dover Publications, 1977.
"Jean Lafitte." http://en.Wikipedia.org/wiki/Jean_Lafitte.
Johnson, Sharon Peregrine, and Byron A. Johnson. *Authentic Guide to Drinks of the Civil War*. Gettysburg, Pa.: Thomas Publications, 1992.
Kapp, Friedrich. *Immigration and the Commissioners of Emigration of the State of New York*. New York: The Nation Press, 1870.
"Lewis Cass." http://en.wikipedia.org/wiki/Lewis_Cass.
Lienhard, John H. "Engines of Our Ingenuity No. 157: Thomas Crapper." http://www.uh.edu/engines/epi157.htm.
Lloyd, James T. *Lloyd's Steamboat Directory*. Cincinnati: J. T. Lloyd, 1856.
"Long Lines of Caskets." Baltimore *Republican*, May 1, 1838.
"Marie Laveau." http://en.wikipedia.org/wiki/Marie_Laveau.
Marryat, Frederick. *A Diary in America*. Vol. II. 1830.

Merrick, George Byron. *Old Times on the Upper Mississippi*. Arthur H. Clark Company, 1909. Reprint, Minneapolis: University of Minnesota Press, 2001. Missouri Historical Society, Jefferson Memorial Library. Manuscript.
"Most Respectable Passengers." Cincinnati *Western Spy*, May 30, 1817.
"Natchez Trace." http://en.wikipedia.org/wiki/Natchez_Trace.
Neville-Sington, Pamela. "Introduction," in Trollope, Fanny, *Domestic Manners of the Americans*. New York: Penguin Books, 1997.
"New Orleans." http://wikipedia.org/wiki/New_Orleans%2C_Louisiana.
"The New Orleans." Louisiana *Gazette and Advertiser*, February 12, 1812.
"New Orleans French Quarter History." http://inetours.com.
"New York Cholera Epidemics." New York *Enquirer*, November 12, 1832.
"A Noble Experiment." *Adams Sentinel*, August 14, 1816.
"Old West Legends." http://legendsofamerica.com.
"The Panic of 1819." http://thehistorybox.com/NY_City/panics/panics_article3a.
"The Panic of 1857." http://thehistorybox.com/NY_City/panics/panics/article7a.
"Pecuniary Loss." *Daily Commercial Register*, August 7, 1852.
Petersen, William J. *Steamboating on the Upper Mississippi*. New York: Dover Publications, 1968.
"The Pirate House and Jean Lafitte." Hancock County Historical Society. http://hancockcountyhistoricalsociety.com/history/lafittepirate.
Poole, Daniel. *What Jane Austen Ate and Charles Dickens Knew*. New York: Simon and Schuster, 1993.
"Practices for the Profession." Louisville *Gazette*, September 21, 1859.
Raeder, Ole Munch. *America in the Forties*. Minneapolis: University of Minnesota Press, 1929.
Reis, Jim. "Death Rides the Waves of the Ohio River." www.rootsweb.ancestry.com/~kycampbe/steamboats.
"Robert Fulton." www.robertfulton.org.
Rodolf, Theodore. *Pioneering in the Western Lead Region*. Wisconsin Historical Collection, Vol. 15.
Rubenstein, Stanley. "The Panic of 1837." http://cooperativeindividualism.org/rubenstein_stan_panic_of_1837.
Rynning, Ole. *True Account of America for the Information and Help of Peasant and Commoner* (English translation). Christiana, 1838.
Schurz, Carl. *The Reminiscences of Carl Schurz*. New York: McClure, 1907.

"The Screams of Men." New Orleans *True American*, May 11, 1837.

"Senator Robert F. Stockton." *Congressional Globe*, August 28, 1852.

Sen. Ex. Doc. 512. 23rd Congress, 1st Session (1833–35).

Sen. Ex. Doc. 37th Congress, 2nd Session (1861–63).

Shepard, Edward. *Life of Martin Van Buren*. New York: Houghton Mifflin Co., 1889.

"Ships Built at Cincinnati." Racine *Daily Journal*, February 10, 1857.

Simmons, James. *Star Spangled Eden*. New York: Carroll & Graf, 2002.

"Slaves Hired on Steamboats." Nashville *Union*, June 10, 1841.

Soifer, Steven. "The Evolution of the Bathroom and the Implications for Paruresis." http://www.paruresis.org/evolution.htm.

Spann, Edward K. *New Metropolis: New York City 1840–1857*. New York: Columbia University Press, 1981.

"State Treasures Waiting to Be Found." Coin World. http://www.coinworld.com/NewCollector/StateTreasure2.asp.

"Steam Ship Robert Fulton." *Republican Compiler*, May 3, 1820.

"Supposed to Be a Deck Hand." Cincinnati *Gazette*, May 20, 1856.

Sutcliffe, Alice C. *Robert Fulton and the Clermont*. New York: The Century Co., 1909.

Switzler, William F. "Report on the Internal Commerce of the United States." *House Ex. Doc. 6*, 50th Congress, 1st Session, Part II.

Taylor, George Rogers. *The Transportation Revolution 1815–1860. Volume IV: The Economic History of the United States*. New York. M.E. Sharpe, 1951.

Thomas, Clayton L., editor. *Taber's Cyclopedic Medical Dictionary*, 13th ed. Philadelphia: F.A. Davis, 1977.

Thomas, Jerry. *How to Mix Drinks, or the Bon-Vivant's Companion*. New York: Dick & Fitzgerald, 1862.

"Thomas Crapper." http://en.wikipedia.org/wiki/Thomas_Crapper.

Trollope, Fanny. *Domestic Manners of the Americans*. 1832. Reprint, New York: Penguin Books. 1997.

Twain, Mark. *Life on the Mississippi*.1883. Reprint, New York: The Library of America, 1982.

"The Washington Union." Janesville *Gazette*, July 24, 1847.

"Wright's Indian Vegetable Pills." Galena *Daily Advertiser*, April 2, 3, 1849.

Wyeth, John B. "Oregon, or a Short History of a Long Journey." http://www.xmission.com/~drudy/mtman/html/jwyeth.html.

Index

Adams, John Quincy 12, 39
alcohol 169, 206–10, 213–16, 230, 234–35, 253, 274
Algonquin 121
Allegheny River 45, 62
Allen, John 5
Alton 123
Astor, John Jacob 42
Augusta 125

Baker, Nicholas 23
Baldwin, Ruth 8
Bank of the United States 88, 90–1, 99, 223, 249
barges 16–7, 21, 52, 65, 145, 244
Barlow, John 8
Baton Rouge 24, 84
Bayou Sara ("the Coast") 21
Belle of the West 123–24, 127
Ben Sherrod 122–23
Biddle, Nicholas 90
Bierce, Ambrose 53
Big Sandy River 21
Black Hawk 130, 190, 261
Black Hawk (Indian) 47–8
Black River 21
Bonaparte, Napoleon 20
Boulton and Watt 8–9, 11, 58
brag boats 112, 127, 138, 140, 146, 161, 220
Bramah, Joseph 5
Brazil 188, 282
Browne, Charles 9
Brownsville 24–6, 42, 58, 73
Buchanan, James 97–8, 114

Calhoun, John C. 44, 112
Car of Neptune 12
Caspian 130, 227
Cass, Lewis 41, 46
Catlin, George 55, 186–9
Chancellor Livingston 12
Charlotte Dundas 7

Chief Justice Marshall 117, 161
cholera 148, 207, 231, 235–40
Chouteau, Auguste 24, 102
Cincinnati 18, 21, 23, 30, 89, 96, 100, 127–8, 132, 159, 162–3, 184, 191, 195, 216–17, 232–3, 238–40, 244–5, 259–60, 262; commerce 66, 78, 80, 85, 87, 107, 109–11, 119, 146; distance/costs 62, 139, 143–4, 193, 263; shipbuilding 112, 139, 180, 188, 202
Claiborne, William C. C. 22, 248
Clark, William 186
Clermont 8–13, 20, 43, 66, 89
Clipper 126
Columbus 110, 123
Comet 24, 25, 43, 58
Constitution 43, 120, 137, 138, 239
Cooper, James Fenimore 227
cotton 66, 85, 92, 94, 108, 122, 171, 197, 217, 245, 246
cotton gin 5
counterfeiting 99, 172, 173

Davis, Jefferson 48
Demologos (Fulton I) 43, 44
Depression of 1837 95
Despatch 58
Dickens, Charles 30, 80, 171, 178, 180–91, 196, 209–10, 216, 218–19, 227–29, 233, 263
Die Vernon 115, 141–2, 169
diphtheria 243–4
Dr. Franklin 49, 140
Dodd, Daniel 27
Dubuque, Julien 50

Edward Bates 128, 255–7
Eliza 118
Enterprise 25, 26, 47, 58, 89, 124, 150, 206, 248–9

Etna 24, 26, 127
Evans' Safety Guard 132
Express 119

Factor System 42
Falls of Ohio 33, 42
Falls of St. Anthony 29, 55, 89, 140, 187, 189, 202, 261
"Fashionable Tour" 55, 140, 186–9, 203
Fever River 24, 50, 141, 189, 255, 261
Fink, Mike 18–19
Firefly 12
Fitch, John 5–6
flatboats 5, 16–17, 21, 23–4, 64, 68, 111, 116, 145, 152, 167, 244, 246
"Floating Palaces" 11, 23, 116, 146, 158, 169, 183, 185, 258, 262
Ford, Rufus 141
French, Daniel 24–5
Fulton, Mary 7
Fulton, Robert 5, 7, 11–16, 19–20, 24, 28, 43, 58, 65, 89, 100, 137, 145, 186, 246, 253
Fulton, Robert, Sr. 7
Fulton-Livingston: concern 20, 23, 26; line 11, 50, 137, 177; monopoly 25, 27, 36, 42, 58, 66, 89, 145, 259; patent 22, 24, 60

Galena 17, 48, 50–2, 54–7, 62, 89, 108, 115, 139–41, 167, 179, 187, 189, 192, 221, 224, 245, 256, 261, 264
gambling 97, 148, 175, 198–201, 203, 218, 253
Gibbons, Thomas 28
"Grand Tour" 151
Grey Eagle 114–15, 142

Hall, Basil 171, 193, 223–5, 227–8
Harris, Daniel Smith 114–15, 139–41, 150, 188, 193
Harris, Robert Scribe 188
Heliopolis 36
Hempstead, William 48, 179
Hornet 119–20
Hudson Steamboat Company 12

Illinois 50, 126
immigration 88, 215, 230
Ione 50

Jack, Andrew 23
Jackson, Andrew 26, 88, 90, 94, 207, 223, 232, 248
Jessup, Thomas S. 44
Johnson, James (Col.) 44

Kanawha River 121
keelboatmen 18, 21, 43, 53, 100, 145, 152, 167, 174, 218
keelboats 16–8, 21, 23–24, 26, 41–42, 50, 52, 65, 107, 112, 182, 218, 246, 252
Keokuk 45, 55, 167, 264
Keokuk Packet Company 140–1, 152
Key City 142

Lafitte, Jean 89, 125, 247–9, 251
lead (mines) 46, 50, 89, 151, 245
Lewis, Merriwether 186, 252
Lincoln, Abraham 17, 173, 243
Liquest, Pierre Laclede 41
Livingston, Robert 22, 28
"local trade" 77, 179
Lodwich, M. W. 140
Louisville 17–8, 25–6, 33, 35, 87, 96, 120, 122, 126, 128, 157, 162–3, 170, 238, 244–5; commerce 66, 78, 81, 100, 113–4, 203, 206, 217, 253, 259, 260; distance/costs 23, 43, 60–2, 107, 143–4, 193, 219, 263

malaria 231, 235, 242
Malta 179
Marine Hospital Service 148, 244
Marryat, Frederick 29, 211, 220
Marshall, John 28, 117, 161
Memphis 29, 49, 69, 114, 125–7, 135, 170, 221, 238, 241, 252, 254, 259–60
Merrick, George Byron 72, 76, 99, 129, 155, 170, 174–76, 191–92, 199–201, 203, 213, 220, 258
Minnesota Packet Company 140, 142, 152, 191–2, 202
Mississippi River 5, 12, 16, 22, 26, 36, 65, 110, 193; early history 5, 22, 40–1, 246, 252
Missouri 60, 119

Missouri River 29, 41, 59, 78, 102, 110; exploration, 43–5, 254–5
Monmouth 120–1, 281
Monongahela River 23
Monroe, James 90
Moselle 103, 127–8, 133
Murdoch, William 7
Murel, John A. (Murrell, Murrel) 125, 172–3, 252

Natchez 16, 18, 24–5, 29, 31, 37, 66, 77, 123, 130, 143, 173, 177, 244–6, 252–3; Natchez-Over-the-Hill 251; Natchez-Under-the-Hill 18, 252
Native Americans 16, 41–2, 46–50, 86, 88, 92, 120–1, 186–7, 189, 213, 226, 243, 245, 252, 253–54; lead mining 50, 150; relations with government 41–2, 47–8, 50, 57, 187, 243, 255; relations with whites 46, 55, 116, 188–9; removal 49, 120–1
Nautilus 8
navigation difficulties rapids 18, 33, 40, 49–50, 52, 54, 57, 73, 115, 120, 137, 150; shoals 32–3, 73–5, 120, 157
New Orleans 5, 12–3, 16–9, 21, 23–6, 29, 31, 40, 46–7, 64, 85, 110, 156, 173, 175, 182, 197, 199, 208, 244–54, 256, 258–59; association with Fulton 13, 23, 25, 77, 177; Battle of New Orleans 36, 40, 58; commerce 66, 81, 87, 89, 107, 111, 113–14; prior to steamboats 5, 17
New Orleans (I) 23–4, 26, 65, 77, 100, 253
New Orleans (II) 155, 158
New Orleans–Natchez trade 16, 143–4, 170
New Orleans trade 24, 31, 46, 55–7, 78, 146, 152, 170, 179, 193, 219, 262–3

Ogden, Aaron (Col.) 27–8
Ohio River 21, 25–6, 42, 53, 58, 65, 73, 102, 110–11, 167, 253; commerce 78, 179, 206, 259; disease 237–8; obstructions 32–3
Ohio Steamboat Navigation Company 23–4
Ouachita River 21, 285

paddlewheel 5, 7, 34, 43, 58, 60–1, 66–8, 73, 101, 105, 117, 148, 178, 182, 189, 197, 200, 203
Panic of 1832 107
Panic of 1857 97, 264
Paragon 12

Pennsylvania 125–6, 189
Perrot, Nicholas 50
Phoenix 86, 134, 149, 151
Pike, Zebulon 41
pilothouse 60, 66–7, 74–5, 115, 122, 130, 140, 142, 157–9, 161
pirates 18–20, 89, 172, 174, 246–52
Pittsburg(h) 17–18, 21, 23–4, 30, 54, 57, 65, 89, 119, 131–2, 150, 159, 163, 190, 238, 240, 244–5; commerce 50, 78, 80, 111, 146, 180, 217, 246, 258–60, 262; distance/costs 31, 56, 62, 73, 107, 144, 193, 219, 263; shipbuilding 26, 42, 45–6, 48, 100, 179, 256
Pizarro 188–9, 280
Polander 119–20
political satire 96, 227
Potosi 220–1
Pulaski 127, 163–5, 281

rafts (obstruction) 30, 37–9
rafts/raftsmen 5, 16–8, 21, 36, 53, 64, 66, 75, 110, 128, 146, 150, 175, 244
railroads 27, 57, 61, 65, 81, 91, 97, 106, 114, 190, 259–64
Red River 31, 35–40, 45, 78, 84, 130, 258
Red Stone 144
Red Wing 179–80, 257
Richmond 12
river obstructions driftwood 34, 37, 40, 61, 64, 74, 246; gravel bars 32, 33; hulks 33, 34, 133; sandbars 32–3, 54, 61, 117, 183; snags 18, 24, 33, 35–40, 42, 54, 69, 102–03, 113, 116–19, 129, 131, 147, 157, 183
Rob Roy 117
Robert Fulton 13
Roosevelt, Nicholas J. 23
Rumsey, James 5

safety barge 48–9
St. Croix 18, 21, 41, 139, 142
St. Louis 17, 21, 24, 34–5, 41–3, 45–8, 52, 89, 117, 119, 123, 126, 133, 147, 150–2, 156, 163, 167, 176, 186–7, 189–90, 213, 220–1, 239–40, 244–5, 253–4; commerce 50–1, 54–5, 77–8, 85, 87, 100, 111, 113, 132, 141–2, 146, 170, 238, 255; distances/costs 29, 53, 56–7, 62, 73, 84, 107, 127, 139, 143–4, 193–4, 219, 259–60, 262–64; Great Fire 105, 257–8
salvage/business 19, 24, 34, 103–05, 133–35
sanitation 220, 232, 244
Schoolcraft, Henry 41, 243

Scott, Winfield 48
Shallcross, S. (Capt.) 49, 150
Shreve, Henry Miller 24–6, 35–7, 39–40, 50, 58, 60, 89, 126–27, 137, 177, 248, 255
sidewheeler 23, 48, 60–1, 68, 188, 202
slavery 5, 18, 70–1, 88, 92, 96, 170–73, 197, 201, 230–31, 240, 244, 246–48, 251, 253
small pox 231, 243
Smelter 139, 188
Smith, Orren 176
Species Circular 93, 99, 136
speculation 5, 88, 97, 100, 106, 231; land 90–3, 230; land sales 92, 94–5
steamboat accidents 14, 19, 35, 40, 44, 57, 61, 102–3, 116–27, 129–35, 138, 143, 148, 161–63, 169, 184, 215, 220–21, 264
steamboat accommodations 9, 23, 26, 30, 43, 46, 48–9, 55, 58–9, 66–7, 86, 118, 122, 130, 144, 155, 166, 177–85, 188, 196, 201–03, 218
steamboat boilers 9, 11, 14, 23–4, 40, 43, 46, 48, 59–60, 65, 67, 69, 71, 116, 177–79, 182, 184, 189, 197; accidents 6, 42, 68, 70, 86, 103, 117, 123–33, 138, 143–44, 150, 162–64, 168–69, 216, 219–21; cleaning 59, 132, 161, 184
steamboat captain 48, 51, 55, 60, 68, 77–9, 83, 87, 89, 113, 120, 122–23, 138–40, 144, 146, 155, 157, 163, 170, 177, 180, 188, 192, 203, 219, 223, 229, 232, 239–40; earnings 49–53, 66, 80, 147, 205, 213, 253; racing 59, 128, 133, 142; rank 145, 160; reputation 54, 106, 125, 135, 143, 149–52
steamboat cargo 5–6, 16–8, 24, 26, 52–7, 61, 63, 66–7, 75, 78, 80–90, 92, 96, 101, 103, 105–08, 114, 117, 119, 122, 130, 133–40, 143, 145, 148, 151, 155–56, 165, 168, 174, 176–77, 180–81, 197, 205, 214, 218, 221, 248, 257–58, 261–64
steamboat clerk 83, 147 48, 156, 165, 175, 182; rank 128, 145–46; responsibilities 64, 99, 135, 197, 205; risks 126, 144, 150
steamboat crew steward 46, 126, 128, 146, 166, 169; stoker 141, 146, 168
steamboat deck passengers 63–9, 108, 125, 156, 174, 177–78, 180, 204, 217, 263; danger 118, 120, 122, 169, 184, 219, 220–21; disease 237; fares 56, 138

steamboat deckhands 18, 114, 146, 167, 170, 174, 219, 221; rank 145
steamboat discipline 174–75
steamboat doctor 68, 143
steamboat engineer 7, 10, 23, 65–6, 69, 70–1, 129, 147, 155–58; blame 125, 131, 154, 162–63; cub 159, 199; danger 67, 87, 120, 126, 128, 130, 132, 169–70, 219; racing 59, 68, 143, 162; rank 145–46, 160–65
steamboat engines high-pressure 59, 66, 68, 116, 127, 179; low-pressure 11, 43, 58–60, 66, 116, 127
steamboat entertainment 178, 197, 203, 211
steamboat fares 12, 55–7, 71, 83, 95–6, 112, 138, 140–42, 148, 166, 184, 186, 192–93, 204–05, 217–21
steamboat food 56, 63, 169, 174, 179, 181, 190–192, 199, 204, 206, 218–20, 223, 236
steamboat fuel: coal 13, 21, 38, 53–4, 65, 143, 198, 262; wood 59–60, 63–5, 68, 70, 101, 108, 121, 141–43, 148, 163
steamboat Gothic 65, 177, 181–82, 203
steamboat humor 221, 228
steamboat insurance 61, 82, 103–06, 111, 118–19, 124, 133, 257
steamboat intraship communication 145, 161
steamboat maintenance 59, 65, 86, 102, 184
steamboat mates 19, 125, 147, 163, 197; rank 145–46, 165, 169
steamboat overhead 101–02, 107, 182–83, 193, 234, 264
steamboat ownership 77, 82–3, 101, 104, 137
steamboat pilot 31, 40, 66, 72, 80, 104, 106, 113, 117, 119, 125, 142–43, 145–48, 152–53, 155–59, 163, 268; cub pilot 74, 125, 160–61, 182, 186; danger 87, 117, 123, 125, 128, 130; licensing 71, 160; navigation 32–3, 35, 60, 64, 71, 73–5, 78, 115, 122, 129–30, 140, 205
steamboat profits/losses 12, 24–5, 49–57, 65–7, 77, 80, 82–3, 90, 94, 101, 108–9, 112–13, 138, 140, 142, 147–48, 155, 180, 192, 202, 217, 239, 253–54, 264
steamboat saloon 55, 165, 169, 177–82, 198, 213

steamboat steam gauges 69, 71, 116, 131, 143, 162
steamboat stock 5, 81, 83, 100
steamboat tonnage 61, 67, 91, 101, 107, 110–11, 146, 165, 253, 258; *see also* individual boats
steamboat value 23, 44, 50, 60, 66, 77, 104, 111, 141, 180, 188, 203
sternwheeler 25, 60–1
Stockton, Robert F. 71
Symington, William 7

Taylor, Zachary 48
telegraph 48, 97, 114, 158, 257
temperance 215–16, 230, 239
terminal facilities 84; *see also* wharves
texas (design) 67, 119, 145
Throckmorton, Joseph 48, 150–51, 179
Trollope, Frances 29, 133, 171, 178, 180, 184, 191, 195, 215, 218, 223–9, 232–3, 250, 252
Twain, Mark 30–1, 72, 133, 143, 156–60, 172, 175, 182, 186, 199, 201, 213, 226, 239, 241, 246, 250, 252, 264

Uncle Sam's toothpuller 35–6

Van Buren, Martin 40, 94, 204
Vanderbilt, Cornelius 28
Vesuvius 24, 26, 177
Vicksburg, MS 29, 87, 121, 238, 244, 259, 262
Voodoo 250–1

War Eagle (I, II) 139, 180, 190, 261, 264
War of 1812 23–4, 42, 44, 47, 71, 89–90, 187, 248, 251
Warren 19, 120–1
Warrior 19, 48–9, 189
Washington 58–60, 126, 137–38, 177
West Newton 141, 169
Western Engineer 43, 45
wharf boat 86
wharfage fees 86–7
wharves 6, 83–6, 108, 132, 147, 156, 165, 175, 178, 245, 255, 259
White Cloud 255–57
Whitney, Eli 5
wildcat banks 88, 98

yellow fever 6, 231, 234, 238, 240 42

Zebulon M. Pike 42–3